BUSINESS MATHEMATICS

BUSINESS MATHEMATICS

By

Dr. R.K. Rajput

D P H

DISCOVERY PUBLISHING HOUSE PVT. LTD.
NEW DELHI-110 002

First Published – 2008

Reprinted – 2025

ISBN: 978-81-8356-338-3

Business Mathematics

Published by:

DISCOVERY PUBLISHING HOUSE PVT. LTD.
4383/4B, Ansari Road, Darya Ganj
New Delhi-110 002 (India)
Phone: +91-11-23279245; 23253475; 43596065
Mobile: +91 9811179893 / +91 9871656464
E-mail: discoverybooksindia@gmail.com
orderdphbooks@gmail.com
namitwasan9@gmail.com
web: www.discoverypublishinggroup.com

Printed at:
Infinity Imaging Systems
Delhi (INDIA)

Preface

This book "Business Mathematics" has been written to provide a comprehensive and updated foundation course in Mathematics and its varied applications in management, commerce and economics.

This book has been written from studies point of view, so that they can easily understand various mathematical concepts, techniques and tool needed for their course. Each new concept and technique is properly supported by suitable solved examples. In solving the questions cave has been taken to explain each step so that students can follow the subject matter themself without even consulting others.

I believe that the book should serve specific need of the students appearing in B.Com. (Hons.), B.A. (Eco. Hons.), M.Com., (M.A.) (Eco.) B.B.A., B.C.A., M.B.A. and other competitive examinations.

Suggestions and comments for the improvement of the book from readers will be thankfully received and duly incorporated in the subsequent editions.

Author

CONTENTS

1

Limits, Continuity, Differentiability and Function of Real Variable

INTRODUCTION OF LIMIT

A function $f(x)$ is said to tend to the limit A as x tends to a, if corresponding to any arbitrary chosen positive number ε, however small (but not zero), there exists a positive number δ, such that

$$|f(x) - A| < \varepsilon,$$

for all values of x for which $0 < |x - a| < \delta$.

We write it as $\lim_{x \to a} f(x) = A$.

Right Hand Limit

A function f(x) is said to tend to the limit A as $x \to a$ from the right, if corresponding to any arbitrary chosen +ive number ε, however small (but not zero), there exists a +ive number δ, such that

$$|f(x) - A| < \varepsilon,$$

for all values of x for which $0 < x - a < \delta$.

The limit of f(x) as $x \to a$ from the right is called the *right hand limit* of f(x) and is denoted by $\lim_{x \to a+0} f(x)$ or by $f(a + 0)$. We calculate $f(a + 0)$ by evaluating $\lim_{h \to 0} f(a + h)$, where h is +ive and sufficiently small.

Left Hand Limit

A function f(x) is said to tend to the limit A as $x \to a$ from the left, if corresponding to any arbitrary chosen +ive number ε, however small (but not zero), there exists a +ive number δ, such that

$$|f(x) - A| < \varepsilon,$$

for ail values of x for which $0 < a - x < \delta$.

The limit of f(x) as $x \to a$ from the left is called the *left hand limit* of f(x) and is denoted by $\lim_{x \to a-0} f(x)$ or by f(a + 0).

We calculate f(a – 0) by evaluating $\lim_{h \to 0} f(a - h)$, where h is +ive and sufficiently small.

From the definition it is clear that the limit of f(x) as $x \to a$ has no necessary connection with the value of the function at a except that when the function is continuous at x = a. The two are equals; even in the case of such functions the limit is distinct in idea from the value.

Note 1: $\lim_{x \to a} f_1(x)$ exists only if, limit on the left = limit on the right.

Note 2: If $f(x) \to A$ and $f_2(x) \to B$, as $x \to a$, where A and B are both finite, we have

(i) $\lim_{x \to a} \{f_1(x) \pm f_2(x)\} = A \pm B$,

(ii) $\lim_{x \to a} \{f_1(x) \,.\, f_2(x)\} = A \,.\, B$,

(iii) $\lim_{x \to a} \left\{\frac{f_1(x)}{f_2(x)}\right\} = \frac{A}{B}$, provided $B \neq 0$,

(iv) $\lim_{x \to a} cf(x) = cA$ where c is constant.

LIMIT OF INEQUALITIES

It $\phi(x) + A$ and $\psi(x) \to B$ as $x \to a$ and if

$$\phi(x) < \psi(x),\ x \neq u$$

Then $A \leq B$

Since $\phi(x) < \psi(x)$ for all values of a an which the limit depend one would naturally expect that A < B. But it is also possible that A may be equal to B. Let $\phi(x) = 1 + x^2$. and $\psi(x) = 1 + 2u^2$. We have $\lim_{x \to a} \phi(x) = 1$, $\lim_{x \to a} \psi(x) = 1$. Also if $x \neq 0$ we have $\phi(x) < \psi(x)$ for all values of x on which the limit depend.

ALGEBRA OF LIMITS

Example:

The limit of a sum is equal to the sum of the limits.

Solution:

Let $\lim_{x \to a} f_1(x) = A$ and $\lim_{x \to a} f_2(x) = B$.

Let ε be any arbitrary chosen positive number.

We can choose δ_1, δ_2 such that

$$|f_1(x) - A| < \frac{1}{2}\varepsilon, \text{ when } 0 < |x - a| < \delta_1,$$

and $$|f_2(x) - B| < \frac{1}{2}\varepsilon, \text{ when } 0 < |x - a| < \delta_2,$$

Let δ be any positive number smaller than both δ_1 and δ_2. Then

$$|f_1(x) - A| < 1/2\ \varepsilon \text{ and } |f_2(x) - B| < 1/2\varepsilon, \text{ for } 0 < |x - a| < \delta$$

Now from algebra, we have

$$|\{f_1(x) + f_2(x)\} - (A + B)| = |\{f_1(x) - A\} + f_2(x)\} - B)\}|$$

$$\leq |\{f_1(x) - A| + |f_2(x)\} - B)\}| < \frac{\varepsilon}{2} + \frac{\varepsilon}{2} \; i.e., < \varepsilon$$

$$\therefore \lim_{x \to a} \{f_1(x) + f_2(x)\} = A + B.$$

VARIABLES

It is a quantity which during any set of mathematical operation, does not retain the same value but is capable of assuming different value. The variable are generally denoted by the last letters of the alphabet *i.e.*, u, v, w, x, y, z etc., the variable are of following kinds:

(a) Continuous and Discrete Variables

If a variable can take all values from one given numbers to another, then it is called a continuous variable, but if it cannot take all values from one given number to another, then it is called a discontinuous or discrete variable.

(b) Independent Variable

The variable which may take up any arbitrary value that may be assigned to it is called an independent variables.

(c) Dependent Variable

This is an variable which assumes its value in consequence of some second variable or a system of variables taking up any set of arbitrary value that may be assigned to them.

CONSTANT

It is a quantity which during any set of mathematical operations, retains the same value. The constant are generally denoted by earlier letter a, b, c, . . . , or by numbers 1, 2, 3, ...

FUNCTION

If x is a variable then a symbol which has one definite value for every value which x can have, is called 0 function of x.

Notation: A function of x is generally denoted by a single letter y. More generally it is denoted by symbols like f(x), F(x), $\phi(x)$, $\psi(x)$, $f_1(x_2)$, $f_2(x)$ etc. If a is any particular value of x, the value of the function f(x) obtained by putting x = a in it, is called the functional value of f(x) and is denoted by f(a).

A function of x is generally written as

$$y = f(x)$$

Here x and y are called the related variable because the variation of y depends upon the variation of x.

To be more clear about variation, consider the function.

$$y = x^2 + 2x + 6 \qquad ...(1)$$

For different values of x, y (or the expression $x^2 + 2x + 6$) assumes different values as shown below:

x	1	2	3	4	5
y	9	14	21	20	41

Here y is a dependent variable and x is the independent variable.

If y = f(x) = c *i.e.* the value of function always remain 'c', whatever by the x, then such a function is called 0 constant function.

TYPES OF FUNCTIONS

(i) **Algebraic Functions:** A function consisting a finites number of term involving powers and roots of the variable x and the four basic mathematical operations (addition, subtraction, multiplication and division) is called an algebraic function.

For example:

$2x^3 + 5x^2 + 6x + 1$, $(x + 8)^7$, $\dfrac{x}{(x^2 - 1)}$ etc. are all algebraic functions.

(ii) **Transcendental Functions:** The functions which are not algebraic are called transcendental functions. These functions include:

(a) Trigonometric function such as

$\sin x$, $\cos x$, $\tan x$, etc.

(b) Inverse trigonometric functions such as

$\sin^{-1} x$, $\cos^{-1} x$, $\cot^{-1} x$, etc.

(c) Logarithmic functions such as

$\log_e x \ \log_a x \ \log\left(\sqrt{x} + \frac{1}{\sqrt{x}}\right)$ etc.

(d) Exponential functions such as

e^x, a^x, $e^{\tan x}$, $x^{\cos x}$, x^x etc.

(e) In commensurable power functions of x as

x^6, $a^{\sqrt{7}}$, $x^{5/2}$, etc.

(iii) Explicit Functions: A function which is directly expressed in terms of the independent variable, is called an explicit function.

For example in $y = x^2$, or $y = e^x + 2\cos x$, y is an explicit function of x.

(iv) Implicit Functions: A function in which the variable x and y both occur together in an equation, but y cannot be directly expressed in terms of x, then y is called and implicit functions of x.

For example $2 + x^4 + 6x^2y^2 + 4xy + y^2 = 0$ is a implicit function of x. The implicit functions are usually denoted by $f(x, y) = 0$.

(v) Even or Odd Functions: If a functions does not change its sign when the sign of its independent variable is changed, then it is said, to be on even functions such as x^6, $\cos x$, etc., are even functions of x because if we put $(-x)$ in place of x, their signs are unchanged. But if a function changed its sign when the sign of its variable is changed, then it is called an odd function such as x^7, $\sin x$, $\cot x$ etc., are odd function because their signs are changed if we put $(-x)$ in place of x in these.

(vi) Periodic Functions: If $f(x + T) = f(x)$, where T is a constant then f(x) is called a periodic function of x with period T. For example $\sin(2\pi + x) = \sin x$, and hence $\sin x$ is a periodic functions with period 2π.

(vii) Single-Valued and Many-Valued Functions: The function is a single-valued function when to every value of the independent variable there corresponds only one value of the function, otherwise it is a many-valued function. An implicit function is often a many-valued function.

Example 1:

The limit of a product is equal to the product of the limits.

Solution:

Let $\lim_{x \to a} f_1(x) = A$ and $\lim_{x \to a} f_2(x) = B$.

Let $A \neq 0$ and $B \neq 0$

Again let, $f_1(x)f_2(x) - AB$

$$= [f_1(x) - A]\,[f_2(x)\} - B] + A\,[f_2(x) - B] + B[f_1(x) - A].$$

$\therefore$ by algebra, we have $|f_1(x) - f\,|\,f_2(x) - AB|$

$$\leq |f_1(x) - A\,|\,.\,|\,f_2(x) - B)\,| + |A|\,.\,|f_2(x) - B\,| + |\,B\,|\,f_1(x) - A\,|$$

Now, $\lim_{x \to a} f_1(x) = A$ and $\lim_{x \to a} f_2(x) = B$. Therefore, for a given ε we can take δ such that

$$|f_1(x) - A\,| < \frac{\varepsilon}{3|B|} \text{ and } |\,f_2(x) - B\,| < \frac{\varepsilon}{3|A|},$$

when $0 < |x - a| < \delta$. Then

$$|f_1(x)f_2(x) - AB\,| < \frac{\varepsilon}{3}.\frac{\varepsilon}{3|A|\,|B|} + \frac{\varepsilon}{3} + \frac{\varepsilon}{3}.$$

Since only small values of ε need by considered, we can take ε such that $\frac{\varepsilon}{3|A|\,|B|} < 1$. Then

$$|f_1(x)f_2(x) - AB\,| < \frac{\varepsilon}{3} + \frac{\varepsilon}{3} + \frac{\varepsilon}{3} \text{ i.e, } < \varepsilon, \text{ when } 0 < |x - a| < \delta.$$

Hence $\lim_{x \to a} \{f_1(x)\,.\,f_2(x)\} = AB$.

CONTINUITY

Definition: *A function f(x) defined for x = a is said to be continuous at x = a if*

(i) f(a) i.e., the value of f(x) at x = a is a definite number.

(ii) the limit of the function f(x) as x → a exists and is equal to the value of f(x) at x = a.

Otherwise the function is **discontinuous** at x = a.

A function is said to be continuous in an interval (a, b) if it is continuous at every point of that interval.

Arithmetical Definition of Continuity: *A function f(x) is said to be continuous at x = a, if for any arbitrarily chosen positive number ε, however small (but not zero), we can find a corresponding number δ such that*

$$|f(x) - f(a)| < \varepsilon$$

for all values of x for which $|x - a| < \delta$.

This definition, as given by Cauchy, is known as Cauchy's definition of continuity.

Remark:

One comparing the definitions of limit and continuity, we find that a function f(x) is continuous at x = a if

$$\lim_{x \to a} f(x) = f(a).$$

Thus f(x) is continuous at x = a if we have L.H.L. = R.H.L. = f(a); otherwise it is discontinuous at x = a.

DIFFERENTIABILITY

Definition: *A function f(x) is said to be differentiable at x = a if both*

$$\lim_{h \to 0} \frac{f(a+h) - f(a)}{h}, h > 0$$

and

$$\lim_{h \to 0} \frac{f(a-h) - f(a)}{-h}, \text{ (L.H.L. = R.H.L.)}$$

exist and have a common value (finite or infinite). This common value is called the *derivative* of f(x) at the point x = a and is denoted by f'(a). Another name for derivative is known as differential co-efficient.

Progressive and Regressive Derivatives: Progressive derivative or *the right hand differential coefficient of* f(x) at x = a is given by

$$\lim_{h \to 0} \frac{f(a+h) - f(a)}{h}, h > 0.$$

It is denoted by Rf'(a).

Regressive derivative or *the left hand differential coefficient* of f(x) at x = a is given by

$$\lim_{h \to 0} \frac{f(a-h) - f(a)}{-h}, h > 0.$$

It is denoted by Lf'(a).

A function f(x) is said to be differentiable at x = a if both Rf'(a) and Lf '(a) exist and are equal; otherwise it is said to be not differential.

Theorem:

Continuity is necessary but not a sufficient condition for the existence of a finite derivative.

Proof:

Let f(x) have a finite derivative f'(a) at x = a. Then

$$\lim_{h\to 0}\frac{f(a+h)-f(a)}{h}=f'(a). \qquad ...(1)$$

Now $f(a + h) - f(a) = \dfrac{f(a-h)-f(a)}{h} \times h,\ h \neq 0.$

Taking limit of both sides when h → 0, we get

$$\lim_{h\to 0}\left[f(a+h)-f(a)\right]=\lim_{h\to 0}\frac{f(a+h)-f(a)}{h}\times\lim_{h\to 0}h$$

$= f'(a) \times 0,$ [from (1)]

$= 0$, since f'(a) is finite.

Now, $\lim_{h\to 0}\ [f(a + h) - f(a)] = 0$

$\Rightarrow \lim_{h\to 0}\ f(a + h) = f(a).$ ...(2)

From the condition (2) we see that f(x) is necessarily continuous at x = a. Thus continuity is a necessary condition for the existence of a finite derivative.

In order to show that continuity is not a sufficient condition for differentiability, we should give an example of a function which is continuous at a point but is not differentiable at that point.

CONTINUITY OF ELEMENTARY FUNCTION

The student as supposed to be familiar with the elementary functions and to know that:

(i) x^n is continuous for all values of x when n is positive.

(ii) x^n is continuous for all values of x except x = 0 when n is negative. For when n is negative and equal to –m say, where m iis positive we can write x^n as $\dfrac{1}{x^m}$ and as $x \to 0$ $\dfrac{1}{x^m}$ either does not tends to a limit or $\to \infty$.

(iii) sin x and cos x are continuous for all values of x.

(iv) tan x is continuous for all values of x except $\pm\frac{\pi}{2}$, $\pm\frac{3\pi}{2}$, $\pm\frac{5\pi}{2}$ etc.

as $x \to \frac{\pi}{2}$ from the right tan $x \to -\infty$ as $x \to \frac{\pi}{2}$ from the left tan $x \to +\infty$. Thus, $\lim_{x\to\pi/2}$ tan x does not exist.

(v) Similarly, statement can be made for cot x, sec x and cosec x.

(vi) $\sin^{-1}x$ and $\cos^{-1}x$ are not real for value of x out side the domain [– 1, 1] similarly, $\sec^{-1} x$ and $\text{cosec}^{-1} x$ are not real for values of x between + and.

(vii) All the inverse circular function are continuous at every point with in the domain in which they are real.

(viii) Log x is continuous for all values of $x > 0$. For values of $x < 0$ the function is not real and we shall not consider these values as x tends to zero from the right log x tends to $-\infty$.

CONTINUITY OF SUM, PRODUCIS etc.

(i) The sum of two continuous function is continuous.

(ii) The product of two continuous function is continuous.

(iii) The quotient of two continuous function is a continuous function except for value of x which make the denominator zero.

LAGRANGE'S MEAN VALUE THEOREM OR FIRST MEAN VALUE THEOREM

If a function f(x) is

(i) continuous in the closed interval $a \le x \le b$, and

(ii) differentiable in the open interval (a, b) i.e., $a < x < b$, then there exists at least one value 'c' of x lying in the open interval $a < x < b$ such that

$$\frac{\mathbf{f(b) - f(a)}}{\mathbf{b - a}} = \mathbf{f'(c)}.$$

Proof:

Consider the function $\phi(x)$ defined by

$$\phi(x) = f(x) + Ax, \quad ...(1)$$

where A is a constant to be determined such that $\phi(a) = \phi(b)$ *i.e.*,

$$f(a) + Aa = f(b) + Ab$$

or $$-A = \frac{f(b) - f(a)}{b - a} \qquad ...(2)$$

Now f(x) is given to be continuous in $a \le x \le b$ and differential in $a < x < b$.

Again, A being a constant, Ax is also continuous in $a \le x \le b$ and differentiable in $a \le x \le b$.

$\therefore$ $\phi(x) = f(x) + Ax$ is continuous in $a \le x \le b$ and differentiable in $a < x < b$. Also by our choice of A, we have $\phi(a) = \phi(b)$. Thus, $\phi(x)$ satisfies all the conditions of Rolle's Theorem in the interval [a, b]. Hence there exists at least one point, say x = c, of the open interval $a < x < b$, such that $\phi'(c) = 0$.

But $\phi'(x) = \phi'(x) + A$, from (1).

$\therefore$ $\phi'(c) = 0$ gives $f'(c) + A = 0$

or $$f'(c) = -A = \frac{f(b) - f(a)}{b - a}, \text{ from (2).}$$

This proves the theorem.

Another Form of Lagrange's Mean Value Theorem

If a function f(x) is

(i) continuous in the closed interval [a, a + h], and

(ii) differentiable in the open interval (a, a + h), then there exists at least one number θ lying between 0 and 1 such that

$$f(a + h) = f(a) + hf'(a + \theta h).$$

Proof:

Let a + h = b. Then b – a = h = the length of the interval.

Now give the complete proof of Lagrange's mean value theorem.

Since c lies between a and a + h, therefore it is greater than a by a fraction of h and may be written as $c = a + \theta h$, where $0 < \theta < 1$.

Hence the result of Lagrange's mean value theorem can be written as

$$f(a + h) - f(a) = hf'(a + \theta h), \qquad [0 < \theta < 1].$$

Geometrical Interpretation of the Mean Value Theorem

In the figure let ACB be the graph of f(x) in (a, b) and let the chord AB make an angle α with the x-axis so that

$$\tan\alpha = \frac{f(b) - f(a)}{b - a}$$

$= f'(c)$, by the Mean Value Theorem where $a < c < b$.

Thus, there is some point c within (a, b) such that the tangent to the curve at the point [c, f(c)] is parallel to the chord AB.

SOME IMPORTANT DEDUCTIONS FROM MEAN VALUE THEOREM

Theorem 1:

If f(x) is

(i) *continuous in the closed interval [a, b],*

(ii) *differentiable in the open interval (a, b), and*

(iii) *$f'(x)$ is –ive in $a < x < b$, then f(x) is a monotonically decreasing function in the closed interval [a, b].*

Proof:

Let x_1, x_2 be any two points belonging to the closed interval [a, b] such that $x_2 > x_1$.

Applying Lagrange's mean value theorem to f(x) in the interval $[x_1, x_2]$, we have

$$f(x_2) - f(x_1) = (x_2 - x_1)\, f'(c),$$

where $x_1 < c < x_2$. ...(1)

Now $x_2 - x_1 > 0$. Since by hypothesis $f'(x)$ is negative for every x in (a, b), therefore $f'(c) < 0$. Hence from (1), we have

$$f(x_2) - f(x_1) < 0, \text{ i.e., } f(x_2) < f(x_1).$$

Thus, f(x) is a decreasing function of x in [a, b].

Similarly, we can prove that *a function having a positive derivative for every value of x in an interval is a monotonically increasing function in that interval.*

Corollary: *The function f(x) is strictly decreasing or increasing in [a, b] if $f'(x) < 0$ or ($f'(x) > 0$) for every x in (a, b) except for a finite number of points where the derivative is zero.*

Theorem 2:

If a function f(x) be such that $f'(x)$ is zero throughout the interval (a, b), then f(x) must be constant throughout the interval.

Proof:

Let x_1, x_2 be any two points in the interval (a, b), such that $x_2 > x_1$. Since $f'(x)$ exists through out the interval (a, b), therefore f(x) satisfies the conditions of Lagrange's mean value theorem in the interval $[x_1, x_2]$. So applying this theorem for f(x) in the interval $[x_1, x_2]$, we get

$$\frac{f(x_2) - f(x_1)}{x_2 - x_1} = f'(c), \text{ where } x_1 < c < x_2.$$

But by hypothesis $f'(x) = 0$ throughout the interval (a, b).

$\therefore \quad f'(c) = 0$ or $f(x_2) - f(x_1) = 0$

or $\quad f(x_2) = f(x_1)$.

Thus, the values of f(x) at every two points of (a, b) are equal.

Hence f(x) must be constant throughout (a, b)

Theorem 3:

If f(x) and $\phi(x)$ be two functions such that $f'(x) = \phi'(x)$ throughout the interval (a, b), then f(x) and $\phi(x)$ differ only by a constant.

Proof:

Consider the function $F(x) = f(x) - \phi(x)$.

Throughout the interval (a, b), we have

$$F'(x) = f'(x) - f'\phi(x) = 0, \qquad [\because f'(x) = \phi'(x)].$$

Therefore, from theorem 1, we have

$F(x) = $ const, or $f(x) - \phi(x) = $ const.

CAUCHY'S MEAN VALUE THEOREM OR SECOND MEAN VALUE THEOREM

If two function f(x) and g(x) are

(i) continuous in the closed interval [a, b]

(ii) differentiable in the open interval (a, b), and

(vii) $g'(x) \neq 0$ for any point of the open interval (a, b), then there exists at least one value c of x in the open interval (a, b), such that

$$\frac{f(b) - f(a)}{g(b) - g(a)} = \frac{f'(c)}{g'(c)}, \ a < c < b.$$

Proof:

First we note that $g(b) - g(a) \neq 0$. For if g(b) – g(a0 = 0 *i.e.*, g(b) = g(a), then the function g(x) satisfies the conditions of Rolle's Theorem and so its derivative g '(x) should vanish for at least one value of x lying in the open interval (a, b). But this is contrary to our hypothesis.

Now consider the function F(x) defined by

$$F(x) = f(x) + Ag(x), \qquad ...(1)$$

where A is a constant to be determined such that F(a) = F(b) *i.e.*,

$$f(a) + Ag(a) = f(b) + Ag(b)$$

or $$-A = \frac{f(b) - f(a)}{g(b) - g(a)}. \qquad ...(2)$$

Since g(b) – g(a) ≠ 0, therefore A is a definite real number.

Now, the function F(x) obviously satisfies the conditions of Rolle's Theorem in the interval [a, b]. Therefore, there exists, at least one value, say c, of x in the open interval (a, b) such that F '(c) = 0.

But $F'(x) = f'(x) + Ag'(x)$, from (1).

∴ $F'(c) = 0$ gives $f'(c) + Ag'(c) = 0$

or $$-A = \frac{f'(c)}{g'(c)}. \qquad ...(3)$$

From (2) and (3), we get

$$\frac{f(b) - f(a)}{g(b) - g(a)} = \frac{f'(c)}{g'(c)}.$$

Another Form: Let b = a + h. Then a + θh = a when θ = 0 and a + θh = b when θ = 1. Therefore a + θh, where 0 < θ < 1, means some value between a and b.

So putting b = a + h and c = a + θh, the result of the above theorem can be written as

$$\mathbf{\frac{f(a+h) - f(a)}{g(a+h) - g(a)} = \frac{f'(a+\theta h)}{g'(a+\theta h)}, \ 0 < \theta < 1.}$$

Note: Lagrange's mean value theorem is a particular case of Cauchy's mean value theorem.

Let us set g(x) = x in Cauchy's mean value theorem which is justified because g(x) = x means g(b) = b, g(a) = a, g'(x) = 1 and so g'(c) = 1. Putting these values in Cauchy's mean value theorem, we get which is nothing but the result of Lagrange's mean value theorem.

$$\frac{f(b) - f(a)}{b - a} = f'(c), \ (a < c < b)$$

TAYLOR'S THEOREM WITH LAGRANGE'S FORM OF REMAINDER AFTER n TERMS

If f(x) is a single valued function of x such that

(i) all the derivatives of f(x) upto $(n - 1)^{th}$ are continuous in $a \leq x \leq a + h$.

(ii) $f^{(n)}(x)$ exists in $a < x < a + h$, then

$$f(a + h) = f(a) + hf'(a) + \frac{h^2}{2!}f''(a) + \ldots$$

$$+ \frac{h^{n-1}}{(n-1)!}f^{(n-1)}(a) + \frac{h^n}{n!}(a + \theta h), \text{ where } 0 < \theta < 1.$$

Proof:

Consider the function $\phi(x)$ defined by

$$\phi(x) = f(x) + (a + h - x) f'(x) + \frac{(a+h-x)^2}{2!}f''(x) + \ldots$$

$$+ \frac{(a+h-x)^{n-1}}{(n-1)!}f^{(n-1)}(x) + \frac{A}{n!}(a+h=x)^n,$$

where A is a constant to be determined such that

$$\phi(a) = \phi(a + h).$$

Now $$\phi(a) = \phi(a) + hf'(a) + \frac{h^2}{2!}f''(a) + \ldots$$

$$+ \frac{h^{n-1}}{(n-1)!}f^{(n-1)}(a) + \frac{A}{n!}h^n,$$

and $$\phi(a + h) = f(a + h).$$

Therefore A is given by

$$f(a + h) = f(a) + hf'(a) + \frac{h^2}{2!}f''(a) + \ldots$$

$$+ \frac{h^{n-1}}{(n-1)!}f^{(n-1)}(a) + \frac{h^n}{n!}A \qquad \ldots(1)$$

Now, by hypothesis, all the functions

$$f(x), f'(x), f''(x), \ldots, f^{n-1)}(x)$$

are continuous in the closed interval [a, a + h]

and differentiable n the open interval (a, a + h).

Also $(a + h - x)$, $(a + h - x)2/2!$, ..., $(a + h - x)^n/n!$, all being polynomial, are continuous in the closed interval [a, a + h] and differentiable in the open interval (a, a + h). Further, A is a constant.

$\therefore$ $\phi(x)$ is continuous in the closed interval [a, a + h] and differentiable in the open interval (a, a + h). Also by our choice of A, $\phi(a) = \phi(a + h)$. Thus $\phi(x)$ satisfies all the conditions of Rolle's Theorem.

$\therefore$ $\phi'(a + \theta h) = 0$, where $0 < \theta < 1$.

Now $\phi'(x) = f'(x) - f'(x) + (a + h - x) f''(x) - (a + h - x) f''(x) + \ldots$

$$+ \frac{(a+h-x)^{n-1}}{(n-1)!} f^{(n)}(x) - \frac{A}{(n-1)!}(a+h=x)^{n-1}$$

$$= \frac{(a+h-x)^{n-1}}{(n-1)!} [f^{(n)}(x) - A]$$, since other terms cancel in pairs.

$\therefore$ $\phi'(a + \theta h) = 0$ gives

$$\frac{[a+h-(a+\theta h)]^{n-1}}{(n-1)!} [f^{(n)}(a + \theta h) - A] = 0$$

or $$\frac{h^{n-1}(1+\theta)]^{n-1}}{(n-1)!} [f^{(n)}(a + \theta h) - A] = 0.$$

Now $h \neq 0$. Also $(1 - \theta) \neq 0$ because $0 < \theta < 1$.

$\therefore$ $f^{(n)}(a + \theta h) - A + 0$ or $A = f^{(n)}(a + \theta h)$.

Substituting this value of A in (a), we get

$$\mathbf{f(a + h) = f(a) + hf'(a) + \frac{h^2}{2!} f''(a) \ldots}$$

$$\mathbf{+ \frac{h^{n-1}}{(n-1)!} f^{(n-1)}(a) + \frac{h^n}{n!} f^{(n)}(a + \theta h).}$$

This is **Taylor's Development** of f(a + h) in ascending integral powers of h. The $(n + 1)^{th}$ term $\mathbf{\frac{h^2}{n!} f^{(n)}(a + \theta h)}$ is called **Lagrange's Form of Remainder** after n terms in Taylor's expansion of f(a + h).

Note: If we take n = 1, we observe that Lagrange's mean value theorem is a particular case of Taylor's theorem.

Corollary. *(Maclaurin's Development):* Instead of considering the interval [a, a + h], let us take the interval [0, x]. Then changing a to 0 and h to x in Taylor's theorem, we get

$$\mathbf{f(x) = f(0) + xf'(0) + \frac{x^2}{2!} f''(0) + \ldots + \frac{x^{n-1}}{(n-1)!} f^{(n-1)}(0) + \frac{x^n}{2!} f^{(n)}(\theta x).}$$

which is known as *Maclaurin's Theorem* or *Maclaurin's Development* of f(x) in the interval [0, x] with *Lagrange's Form of Remainder* $\mathbf{\frac{x^n}{n!} f^{(n)}(\theta x)}$.

TAYLOR'S THEOREM WITH CAUCHY'S FORM OF REMAINDER

If f(x) is a single valued function of x such that

(i) all the derivatives of f(x) upto $(n-1)^{th}$ are continuous in $a \le x \le a+h$, and

(ii) $f^{(n)}(x)$ exists in $a < x < a+h$, then

$$f(a+h) = f(a) + hf'(a) + \frac{h^2}{2!}f''(a) + \dots + \frac{h^{n-1}}{(n-1)!}f^{(n-1)}(a) + \frac{h^n}{(n-1)!}(1-\theta)^{n-1}(a+\theta h),$$

where $0 < \theta < 1$.

Proof:

Consider the function $\phi(x)$ defined by

$$\phi(x) = f(x) + (a+h-x)f'(x) + \frac{(a+h-x)^2}{2!}f''(x) + \dots + \frac{(a+h-x)^{n-1}}{(n-1)!}f^{(n-1)}(x) + (a+h-x)A,$$

where A is a constant to be determined such that

$$\phi(a) = \phi(a+h).$$

Now $\quad \phi(a) = \phi(a) + hf'(a) + \frac{h^2}{2!}f''(a) + \dots + \frac{h^{n-1}}{(n-1)!}f^{(n-1)}(a) + \frac{A}{n!}hA,$

and $\quad \phi(a+h) = f(a+h).$

Therefore A is given by

$$f(a+h) = f(a) + hf'(a) + \frac{h^2}{2!}f''(a) + \dots + \frac{h^{n-1}}{(n-1)!}f^{(n-1)}(a) + hA. \qquad \dots(1)$$

As shown in (2.5), we can easily show that $\phi(x)$ satisfies all the conditions of Rolle's Theorem.

$\therefore \quad \phi'(a+\theta h) = 0$, where $0 < \theta < 1$.

Now

$\phi'(x) = \frac{(a+h-x)^{n-1}}{(n-1)!}f^{(n)}(x) - A$, since other terms cancel in pairs.

$\therefore \quad \phi'(a+\theta h) = 0$ gives

$$\frac{[a+h-(a+\theta h)]^{n-1}}{(n-1)!} f^{(n)}(a + \theta h) - A = 0$$

or $$A = \frac{h^{n-1}(1+\theta)]^{n-1}}{(n-1)!}(1 - \theta)^{n-1} f^{(n)}(a + \theta h).$$

Substituting this value of A in (a), we get

$$\mathbf{f(a + h) = f(a) + hf'(a) + \frac{h^2}{2!} f''(a) ... + \frac{h^{n-1}}{(n-1)!} f^{(n-1)}(a)}$$
$$\mathbf{+ \frac{h^n}{(n-1)!}(1-\theta)^{n-1} f^{(n)}(a + \theta h).}$$

The $(n + 1)^{th}$ term $\mathbf{\frac{h^n}{(n-1)!}(1-\theta)^{n-1} f^{(n)}(a + \theta h)}$ is called *Cauchy's Form of Remainder* after n terms in Taylor's expansion of f(a + h) in ascending integral powers of h.

Corollary. *(Maclaurin's Development with Cauchy's Form of Remainder):* Changing a to 0 and h to x in above result, we get

$$\mathbf{f(x) = f(0) + xf'(0) + \frac{x^2}{2!} f''(0) + ... + \frac{x^{n-1}}{(n-1)!} f^{(n-1)}(0)}$$
$$\mathbf{+ \frac{x^{n-1}}{(n-1)!}(1 - \theta)^{n-1} f^{(n)}(\theta x),}$$

which is known as *Maclaurin's Development* of f(x) in the interval [0, x] with *Cauchy's Form of Remainder* after n terms.

ROLLE'S THEOREMS

If a function f(x) which is continuous in the closed interval [a, b] and differentiable in the open interval (a, b), be zero at x = a and also at x = b. Then f(x) is zero for at least one value of x between a and b.

Or

If function f(x) is such that

(i) f(x) is continuous in the closed interval $a \leq x \leq b$,

(ii) f '(x) exists for every point in the open interval $a < x < b$,

(iii) f '(a) = f(b), then there exits at least one value of x, say c, where $a < c < b$, *such that f '(c) = 0.*

Proof:

Since f(a) = f(b), unless the function f(x) is a constant in which case the theorem is at once established, f(x) should either increase or decrease when x takes values greater than a. Suppose it increases; then since it again takes a value f(b) = f(a), it must cease to increase and begin to decrease at some point c, such that $a < c < b$.

At this point c the function f(x) has a maximum value and so $f(c + h) - f(c)$ and $f(c - h) - f(c)$ are both negative, h being small and positive.

$$\therefore \quad \frac{f(c+h)-f(c)}{h} < 0 \text{ and } \frac{f(c-h)-f(c)}{-h} > 0.$$

Obviously as $h \to 0$, the above expressions tend to being –ive and +ive respectively unless each of them has the limit zero.

If they have different limits, then $Rf'(c) \neq Lf'(c)$ and therefore $f'(c)$ does not exist, contradicting the hypothesis.

Hence each of the above limits must be zero,

i.e., $f'(c) = 0$

where $a < c < b$.

Note 1 : There may be more than one point like c at which f '(x) vansihes.

Note 2 : Rolle's theorem will not hold good

(i) if f(x) is discontinuous at some point in the interval

$a \leq x \leq b$,

or (ii) if f '(x) does not exist at some point in the interval

$a \leq x \leq b$,

or (iii) if $f(a) \neq f(b)$.

Geometrical Interpretation of Rolle's Theorem

Suppose the function f(x) satisfies the conditions of Rolle's theorem in the interval [a, b]. Then its geometrical interpretation is that on the curve y = f(x) there is at least one point lying in the open interval (a, b) the tangent at which is parallel to the axis of x.

SOLVED EXAMPLES

Example 1: *Define the limit of a function. Discuss the existence of the limit of the function f(x) = 1, if x < 1; f(x) = 2 – x, if 1 < x < 2; f(x) = 2, if x ≥ 2 at x = 1 and x = 2.*

Solution: For definition of limit (see definition)

R.H.L. At x = 1,

i.e., $f(1+0) = \lim_{h\to 0} f(1+h)$,

where h is +ive and sufficiently small

$$= \lim_{h\to 0} [2-(1+h) = \lim_{h\to 0}(1-h) = 1;$$

L.H.L. At x = 1,

i.e., $f(1-0) = \lim_{h\to 0} f(1-h) = \lim_{h\to 0}(1) = 1$

since R.H.L. = L.H.L. = 1,

therefore = $\lim_{x\to 1} f(x)$ exists and is equal to 1.

R.H.L At x = 2,

i.e., $f(2+0) = \lim_{h\to 0} f(2+h)$

$$= \lim_{h\to 0}(2) = 2;$$

L.H.L. At x = 2,

i.e., $f(2-0) = \lim_{h\to 0}(2-h)$

$$= \lim_{h\to 0}[2-(2-h)]$$

$$= \lim_{h\to 0} h = 0.$$

$\because$ left hand limit ≠ right hand limit,

$\therefore$ $f = \lim_{x\to 2} f(x)$ does not exist.

Example 2: *Prove that* $\lim_{h\to 0} (1+x)^{1/x} = e$.

Solution: Let $f(x) = (1+x)^{1/x}$.

We have R.H.L.= $f(0+0) = \lim_{h\to 0} f(0+h)$

$$= \lim_{h\to 0} f(h) = \lim_{h\to 0}(1+h)^{1/h}$$

$$= \lim_{h\to 0}\left[1+\frac{1}{h}.h+\frac{\frac{1}{h}\left(\frac{1}{h}-1\right)}{1.2}h^2+\frac{\frac{1}{h}\left(\frac{1}{h}-1\right)\left(\frac{1}{h}-2\right)}{1.2.3}h^3+\ldots\right]$$

$$= \lim_{h\to 0}\left[1+\frac{1}{1!}+\frac{1.(1-h)}{2!}+\frac{1.(1-h)(1-2h)}{3!}+\ldots\right]$$

$$= 1+\frac{1}{1!}+\frac{1}{2!}+\frac{1}{3!}+\ldots\infty = e.$$

We have, R.H.L. $f(0 - 0) = \lim_{h\to 0} f(0 - h) = \lim_{h\to 0} f(-h)$

$$= \lim_{h\to 0} f(1 - h)^{-1/h} = e.$$

Since both f(0 + 0) and f(0 – 0) exist and are equal to e, therefore

$\lim_{x\to 0}(1 + x)^{1/x} = e.$

Example 3: *Evaluate the following limits if they exist:* $\lim_{x\to 2}\frac{x^2-4}{x-2}$.

Solution: Let $f(x) = \frac{x^2-4}{x-2} = \frac{(x-2)(x+2)}{x-2}$

R.H.L. $x = 2$, $f(2 + 0) = \lim_{h\to 0} f(2 + h)$

$$= \lim_{h\to 0}\frac{(2+h-2)(2+h+2)}{2+h-2} = \lim_{h\to 0}\frac{h(h+4)}{h}$$

$$= \lim_{h\to 0}(h + 4) = 4.$$

L.H.L. $x = 2$, *i.e.*, $f(2 - 0) = \lim_{h\to 0} f(2 - h)$

$$= \lim_{h\to 0}\frac{(2-h-2)(2-h+2)}{2-h-2} = \lim_{h\to 0}\frac{-h(-h+4)}{-h}$$

$$= \lim_{h\to 0}(-h + 4) = 4.$$

Since both f(2 + 0) f = f(2 – 0), therefore $\lim_{x\to 2} f(x)$ exists and is equal to 4.

Example 4: *Evaluate the limits:* $\lim_{x\to 0}\frac{1+3x^2}{x}$.

Solution: We have $\lim_{x\to 0}\frac{1+3x^2}{x}$

We get, R.H.L. *i.e.*, $f(0 + 0) = \lim_{h\to 0} f(0 + h) = \lim_{h\to 0} f(h)$

$$= \lim_{h\to 0}\frac{1+3h^2}{h} = \lim_{h\to 0}\left(\frac{1}{h}+3h\right) = \infty$$

Again, L.H.L. *i.e.*, $f(0-0)=\lim_{h\to 0} f(0-h)=\lim_{h\to 0} f(-h)$

$$=\lim_{h\to 0}\frac{1+3h^2}{-h}=\lim_{h\to 0}\left(-\frac{1}{h}-3h\right)=-\infty$$

R.H.L. ≠ L.H.L

Since $f(0+0)\neq f(0-0)$, therefore $\lim_{x\to 0} f(x)$ does not exist.

Example 5: *Evaluate:* $\lim_{x\to 1}\frac{x^3-1}{x^2-1}$.

Solution: *We have* $\lim_{x\to 1}\frac{x^3-1}{x^2-1}=\lim_{x\to 1}\frac{(x-1)(x^2+x+1)}{(x-1)(x+1)}$

$$\lim_{x\to 1}\frac{x^2+x+1}{x+1}=\frac{3}{2}\text{ Ans.}$$

Example 6: *Evaluate:* $\lim_{x\to 2}\frac{x^2+3x+2}{x-2}$

Solution: We have, R.H.L. *i.e.*, $f(2+0)=\lim_{h\to 0} f(2+h)$

$$=\lim_{h\to 0}\frac{(2+h)^2+3(2+h)+2}{2+h-2}=\lim_{h\to 0}\frac{12+7h+h^2}{h}$$

$$=\lim_{h\to 0}\left(\frac{12}{h}+7+h\right)=\infty$$

Also, L.H.L. *i.e.*, $f(2-0)=\lim_{h\to 0} f(2-h)$

$$=\lim_{h\to 0}\frac{(2-h)^2+3(2-h)+2}{2-h-2}=\lim_{h\to 0}\frac{12-7h+h^2}{-h}$$

$$=\lim_{h\to 0}\left(-\frac{12}{h}+7-h\right)=-\infty$$

Since $f(2+0)\neq f(2-0)$, therefore $\lim_{x\to 2} f(x)$ does not exist.

Example 7: *Evaluate:* $\lim_{x\to 0}\frac{(1+x)^{1/3}-(1+x)^{1/3}}{x}$

Solution: We have $\lim_{x\to 0}\frac{(1+x)^{1/3}-(1-x)^{1/3}}{x}=\lim_{h\to 0}\frac{(1+h)^{1/3}-(1-h)^{1/3}}{h}$

$$= \lim_{h\to 0} \frac{1+\frac{1}{3}h + \ldots - (1-\frac{1}{3}h+\ldots)}{h} = \lim_{h\to 0} \frac{\frac{2}{3}h + \ldots}{h} = \frac{2}{3}.$$

Example 8: *Evaluate:* $\lim_{x\to 0} e^{-1/x}$.

Solution: *We have* $\lim_{x\to 0} e^{-1/x}$ *let* $f(x) = e^{-1/x}$.

$$\text{R.H.L. } f(0+0) = \lim_{h\to 0} f(0+h)$$

$$= \lim_{h\to 0} f(h) = \lim_{h\to 0} e^{-1/h} - \infty$$

$$= \lim_{h\to 0} \frac{1}{e^{1/h}} = 0.$$

Again, L.H.L. $f(0-0) = \lim_{h\to 0} f(-h) = \lim_{h\to 0} e^{1/h} - \infty$

Since $f(0+0) \neq f(0-0)$ therefore $\lim_{x\to 0} e^{-1/x}$ does not exist.

Example 9: *Evaluate:* $\lim_{x\to 0} \frac{1}{1-e^{1/x}}$.

Solution: *We have* $\lim_{x\to 0} \frac{1}{1-e^{1/x}}$

Let $f(x) = \frac{1}{1-e^2x}$

We have $f(0+0) = \lim_{h\to 0} f(h) = \lim_{h\to 0} \frac{1}{1-e^{1/h}} = 0$

Again $f(0-0) = \lim_{h\to 0} f(-h) = \lim_{h\to 0} \frac{1}{1-e^{-1/h}}$

$$= \lim_{h\to 0} \frac{1}{1-(1/e^{1/h})} = 1$$

L.H.L $\neq$ R.H.L

$\because f(0+0) \neq f(0-0) \therefore \lim_{x\to 0} \frac{1}{1-e^{1/x}}$ does not exist.

Example 10: *Evaluate:* $\lim_{x\to 0} \frac{\sin x}{x}$

Solution: *Let* $f(x) \frac{\sin x}{x}$

We have R.H.L. $= f(0+0) = \lim_{h\to 0} f(0+h) = \lim_{h\to 0} f(h) = \lim_{h\to 0} \frac{\sin h}{h}$

$$= \lim_{h\to 0} \frac{h - \frac{h^3}{3!} + \frac{h^5}{5!} - \dots}{h} = \lim_{h\to 0}\left(1 - \frac{h^2}{3!} + \frac{h^4}{5!} - \dots\right) = 1$$

L.H.L., $= f(0 - 0) = \lim_{h\to 0} f(0 - h) = \lim_{h\to 0} f(-h)$

$$= \lim_{h\to 0} \frac{\sin(-h)}{h} = \lim_{h\to 0} \frac{\sin h}{h} = 1.$$

L.H.L. = R.H.L.

Since $f(0 + 0) \neq f(0 - 0) = 1$ therefore $\lim_{x\to 0} \frac{\sin x}{x} = 1.$

Example 11: *Evaluate:* $\lim_{x\to\infty} \frac{\sin x}{x}$.

Solution: *Put* $x = \frac{1}{y}$. *Then* $= \lim_{h\to\infty} = \frac{\sin x}{x} = \lim_{y\to 0} y \sin\left(\frac{1}{y}\right)$

Now let $f(y) = y \sin\left(\frac{1}{y}\right)$.

We have R.H.L. $f(0 + 0) = \lim_{h\to 0} f(0 + h) = \lim_{h\to 0} f(h) = \lim_{h\to 0} h \sin\left(\frac{1}{h}\right) = 0$

Since $\sin\left(\frac{1}{h}\right)$ lies between –1 and 1.

Also L.H.L. $f(0 - 0) = \lim_{h\to 0} f(0 - h) = \lim_{h\to 0} f(-h) = \lim_{h\to 0} (-h)$

$$= \lim_{h\to 0} f \sin\left(-\frac{1}{h}\right) = \lim_{h\to 0} h \sin\left(\frac{1}{h}\right) = 0$$

Since both the limits $f(0 + 0)$ and $f(0 - 0)$ exist and are equal to zero, therefore $\lim_{y\to 0} y \sin\left(\frac{1}{y}\right) = 0,$

Hence $\lim_{x\to\infty} \frac{\sin x}{x} = 0.$

Example 12: *Evaluate:* $\lim_{x\to 0} = \sin\frac{1}{x}$.

Solution: *Let* $f(x) = \sin\frac{1}{x}$

We have $f(0 + 0) = \lim_{h\to 0} f(0 + h) = \lim_{h\to 0} f(h) = \lim_{h\to 0} \sin\left(\frac{1}{h}\right)$

As $h \to 0$, the value of $\sin\left(\frac{1}{h}\right)$ oscillates between + 1 and –1, passing through zero and intermediate values an infinite number of times. Thus, there is no definite number A to which $\sin\left(\frac{1}{h}\right)$ tends as h hends to zero. Therefore the right hand limit f(0 + 0) does not exist.

Similarly, the left hand limit f(0 – 0) also does not exist. Hence $\lim_{x\to 0} \sin\frac{1}{x}$ does not exist.

Example 13: *Evaluate:* $\lim_{x\to a} = \frac{x^m - a^m}{x - a}$.

Solution: $\lim_{x\to a} \frac{x^m - a^m}{x - a}$

Let $f(x) = \frac{x^m - a^m}{x - a}$

We have R.H.L. $f(a + 0) = \lim_{h\to 0} f(a + h) = \lim_{h\to 0} = \frac{(a-h)^m - a^m}{a+h-a}$

$$= \lim_{h\to 0} \frac{a^m\left[\left(1+\frac{h}{a}\right)^m - 1\right]}{h}$$

$$= \lim_{h\to 0} \frac{a^m}{h}\left[1 + m.\frac{h}{a} + \frac{m(m-1)}{2!}\frac{h^2}{a^2} + ... - 1\right]$$

L.H.L. $= f(a - 0) = \lim_{h\to 0} f(a - h) = \lim_{h\to 0} \frac{(a-h)^m}{a-h-a} = ma^{m-1}$.

Left hand limit = Right hand limit

Thus, $f(a + 0) = f(a - 0) = ma^{m-1}$

Therefore $\lim_{x\to a} f(x) = ma^{m-1}$.

Example 14: *Evaluate:* $\lim_{x\to 0} \frac{1}{x}e^{1/x}$.

Solution: *Let* $f(x) = \frac{1}{x}e^{1/x}$

R.H.L. $= f(0 + 0) = \lim_{h\to 0} f(0 + h) = \lim_{h\to 0} f(h) = \lim_{h\to 0} \frac{1}{h}e^{1/h} = \infty$

[Since both and $e^{1/h}$ tend to ∞ as $h\to 0$]

Again we have

$$\text{L.H.L.} = f(0-0) = \lim_{h\to 0} f(0-h) = \lim_{h\to 0} f(-h) = \lim_{h\to 0} -\frac{1}{h}e^{-1/h}$$

$$= \lim_{h\to 0} \frac{-1}{he^{1/h}} = \lim_{h\to 0} \frac{-1}{h\left(1+\frac{1}{h}+\frac{1}{2!}\frac{1}{h^2}+\ldots\right)}$$

$$= \lim_{h\to 0} \frac{-1}{h+1+(1/2h)+\ldots} = 0.$$

L.H.L. ≠ R.H.L. Since both the limits f(0 + 0) and f(0 – 0) are unequal, therefore $\lim_{x\to 0} \frac{1}{x} e^{1/x}$ does not exit.

Example 15: *Here the right hand limit i.e., f(0 + 0) = h → 0 f(0 – h), where h is +ive and sufficiently small*

Evaluate: $\lim_{x\to 0} \frac{e^x - 1}{x}$.

$= \lim_{h\to 0} f(h) = \lim_{h\to 0} h$, *[∵ h > 0 and f(x) = x when x > 0]*

$= 0.$

Again the left hand limit i.e., f(0 – 0) $= \lim_{h\to 0} f(0-h)$,

where h is +ive and sufficiently small

$= \lim_{h\to 0} f(-h) = \lim_{h\to 0} -(-h)$, *[∵ – h < 0 and f(x) = – x when x < 0]*

$= \lim_{h\to 0} h = 0$.

Since both the limits f(0 + 0) and f(0 – 0) exist and are equal to zero, therefore $\lim_{x\to 0} f(x)$ *exists and is equal to zero.*

Solution:

Let $$f(x) = \frac{e^x - 1}{x} = \frac{1+x+\frac{x^2}{2!}+\ldots-1}{x} = \frac{x\left(1+\frac{x}{2!}+\ldots\right)}{x}.$$

$$\text{R.H.L. } f(0+0) = \lim_{h\to 0} f(0+h) = \lim_{h\to 0} f(h)$$

$$= \lim_{h\to 0} \frac{h\left[1+(h/2!)+\ldots\right]}{h} = \lim_{h\to 0}\left(1+\frac{h}{2!}+\ldots\right) = 1.$$

L.H.L. $f(0-0) = \lim_{h\to 0} f(0-h) = \lim_{h\to 0} f(-h)$

$$= \lim_{h\to 0} \frac{-h[1 = (h/2!)+...]}{-h} = 1.$$

Example 16: *Thus, $f(0 + 0) = f(0 - 0) = 1$.*

Therefore $\lim_{x\to 0} (e^x - 1)/x = 1$.

Evaluate: $\lim_{x\to 0} \frac{a^x - 1}{x}$.

Solution: $\lim_{x\to 0} \frac{a^x - 1}{x}$. Let $f(x) = \frac{a^x - 1}{x}$

$$= \frac{1 + x \log a + (x^2/2!)(\log a)^2 + ... - 1}{x}$$

Now, $= \frac{x\left[\log a + \frac{1}{2} x (\log a)^2 + ...\right]}{x}$

$$\text{R.H.L.} = f(0+0) = \lim_{h\to 0} f(h) = \lim_{h\to 0} \frac{h\left[\log a + \frac{h}{2}(\log a)^2 + ...\right]}{h} = \log a.$$

L.H.L. $= f(0-0) = \lim_{h\to 0} f(0-h) = \lim_{h\to 0} f(-h) = \log a$.

Thus, $f(0 + 0) = f(0 - 0) = \log a$. Therefore

$$\lim_{x\to 0} \frac{(a^x - 1)}{x} = \log a.$$

Example 17: *Find the right hand and the left hand limits in the following cases and discuss the existence of the limit in each case:*

(i) $\lim_{x\to 2} \frac{2x^2 - 8}{x - 2}$;

(ii) $\lim_{x\to 0} \frac{e^{1/x} - 1}{e^{1/x} + 1}$;

(iii) $\lim_{x\to 0} f(x)$, *where $f(x)$ is defined as*

$f(x) = x, \quad$ *when* $x > 0$

$f(x) = 0,\quad$ *when* $x = 0$

$f(x) = -x,\quad$ *when* $x < 0$

Solution: Let $f(x) = \dfrac{2x^2 - 8}{x - 2}$.

We have L.H.L. $= f(2+0) = \lim_{h\to 0} f(2+h) = \lim_{h\to 0} \dfrac{2(2+h)^2 - 8}{2+h-2}$

$$= \lim_{h\to 0} \frac{2(4+4h+h^2) - 8}{h} = \lim_{h\to 0} \frac{8h+2h^2}{h} = \lim_{h\to 0} \frac{h(8+2h)}{h}$$

$$= \lim_{h\to 0} (8 + 2h) = 8.$$

Again L.H.L. *i.e.* $= f(2-0) = \lim_{h\to 0} f(2-h) = \lim_{h\to 0} \dfrac{2(2-h)^2 - 8}{2-h-2}$

$$= \lim_{h\to 0} \frac{2(4-4h+h^2)}{-h} = \lim_{h\to 0} \frac{-8h+2h^2}{-h}$$

$$= \lim_{h\to 0} \frac{-h(8-2h)}{-h} = \lim_{h\to 0} (8-2h) = 8.$$

Since $f(2 + 0) = f(2 - 0) = 8$, therefore $\lim_{x\to 0} \dfrac{2x^2 - 8}{x - 2}$ exists and is equal to 8.

(ii) Let f(x) $f(x) = \dfrac{e^{1/x} - 1}{e^{1/x} + 1}$

We have R.H.L.

$$f(0+0) = \lim_{h\to 0} f(0+h) = \lim_{h\to 0} f(h) = \lim_{h\to 0} \frac{e^{1/h} - 1}{e^{1/h} + 1}$$

$$= \lim_{h\to 0} \frac{e^{1/h}\left[1 - \left(1/e^{1/h}\right)\right]}{e^{1/h}\left[1 + \left(1/e^{1/h}\right)\right]} = \lim_{h\to 0} \frac{1 - \left(1/e^{1/h}\right)}{1 + \left(1/e^{1/h}\right)} = 1.$$

Again L.H.L. $= f(0-0) = \lim_{h\to 0} f(0-h) = \lim_{h\to 0} f(-h) = \lim_{h\to 0} \dfrac{e^{-1/h} - 1}{e^{-1/h} + 1}$

$$= \lim_{h\to 0} \frac{\left(1/e^{1/h}\right) - 1}{\left(1/e^{1/h}\right) + 1} = \frac{0-1}{0+1} = -1.$$

L.H.L. $\neq$ R.H.L.

Since f(0 + 0) $\neq$ f(0 – 0), therefore $\lim\limits_{x\to 0} \frac{e^{1/x}-1}{e^{1/x}+1}$ does not exist.

Example 18: *Evaluate:* $\lim\limits_{x\to 0} \frac{(1+x)^n - 1}{x}$.

Solution: $\lim\limits_{x\to 0} \frac{(1+x)^n-1}{x}$. Let $f(x) = \frac{(1+x)^n-1}{x}$

R.H.L. $f(0+0) = \lim\limits_{x\to a} f(h) = \lim\limits_{x\to a} \frac{(1+h)^n - 1}{h}$

$$= \lim_{h\to 0} \frac{1 + nh + \frac{n(n-1)}{2!}h^2 + \ldots - 1}{h}$$

$$= \lim_{h\to 0} \frac{h\left[n + \frac{n(n-1)}{2!}h + \ldots\right]}{h}$$

$$= \lim_{h\to 0}\left[n + \frac{n(n-1)}{2!}h + \ldots\right] = n.$$

$$\text{L.H.L.} = f(0-0) = \lim_{h\to 0} f(-h) = \lim_{h\to 0} \frac{(1-h)^n - 1}{-h}$$

$$= \lim_{h\to 0} \frac{1 + n(-h) + \frac{n(n-1)}{2!}(-h)^2 + \ldots - 1}{-h} = n.$$

R.H.L. = L.H.L.

Thus, f(0 + 0) = f(0 – 0) = n. Therefore, $\lim\limits_{x\to 0} f(x) = n$.

Example 19: *If f(x) - x tan$^{-1}\left(\frac{1}{x}\right)$ when x $\neq$ 0 and f(0) = 0 show that f is continuous but not deriable for x = 0.*

Solution: $\lim\limits_{x\to 0} f(x) = \lim\limits_{x\to 0} x \tan^{-1}\left(\frac{1}{x}\right) = 0\left(-\frac{\pi}{2}\right) = 0$

$$\lim_{x\to 0} f(x) = \lim_{x\to 0} + x \tan^{-1}\left(\frac{1}{x}\right) = 0\left(\frac{1}{2}\pi\right) = 0$$

Also f(0) = 0 Hence f is continuous at x = 0

Now $Lf'(0) = \lim_{x\to 0-} \frac{x\tan^{-1}\left(\frac{1}{x}\right) - 0}{x - 0} = \lim_{x\to 0-} \tan^{-1}\left(\frac{1}{x}\right) = \frac{-\pi}{2}$

$Rf'(0) = \lim_{x\to 0+} \tan^{-1}\left(\frac{1}{x}\right) = \frac{\pi}{2}$

$\therefore Lf'(0) \neq Rf'(0)$

Hence f is not deriable at x = 0.

Example 20: *Show that $f(x) = (x^2 - 1)/(x - 1)$ is continuous for all values of x except x = 1.*

How may this function be defined to make it continuous at x = 1?

Solution: At first we have to test continuity of f(x) at x = a where a ± 1 $f(a) = \frac{a^2-1}{a-1}$ = a definite real number because $a - 1 \neq 0$;

We have R.H.L. = $f(a+0) = \lim_{h\to 0} f(a+h) = \lim_{h\to 0} \frac{(a+h)^2-1}{a+h-1} = \frac{a^2-1}{a-1}$;

L.H.L. $f(a-0) = \lim_{h\to 0} f(a-h) = \lim_{h\to 0} \frac{(a-h)^2-1}{a-h-1} = \frac{a^2-1}{a-1}$.

L.H.L. = R.H.L.

Thus f(a + 0) = f(a – 0) = f(a). Therefore, f(x) is continuous at x = a where $a \neq 1$.

From the formula definiting f(x), we have $f(1) = \frac{1-1}{1-1} = \frac{0}{0}$ which is meaningless. Thus f(x) has no definite value at x = 1. Therefore, f(x) is not continuous at x = 1. To make f(x) continuous at x = 1, we shall find $\lim_{x\to 1} f(x)$. We have

$$f(1+0) = \lim_{h\to 0} f(1+h) = \lim_{h\to 0} \frac{(1+h)^2-1}{1+h-1} = \lim_{h\to 0} \frac{2h+h^2}{h}$$

$$= \lim_{h\to 0} \frac{h(2+h)}{h} = \lim_{h\to 0}(2+h) = 2;$$

and $$f(1-0) = \lim_{h\to 0} f(1-h) = \lim_{h\to 0} \frac{(1-h)^2-1}{1-h-1}$$

$$= \lim_{h\to 0} \frac{-2h+h^2}{-h}$$

$$= \lim_{h \to 0} (2 - h) = 2.$$

Thus, $\lim_{x \to 1} f(x) = 2$. So if we take $f(x) = \frac{(x^2 - 1)}{(x - 1)}$ when $x \neq 1$ and $f(x) = 2$ when $x = 1$, it will become continuous at $x = 1$.

Example 21: *Define continuity of function at a point and in an interval. Discuss the continuity of $f(x) = 2, x \leq 0$; $f(x) = 3x + 2, 0 < x \leq 1$; $f(x) = x/(x - 1), 1 < x$ at $x = 0$ and at $x = 1$.*

Solution: See definition of continuity.

(i) To test f(x) for continuity at x = 0. We have

$$f(0) = 2; \text{ R.H.L.} = f(0 + 0) = \lim_{h \to 0} f(0 + h)$$

$$= \lim_{h \to 0} f(h) = \lim_{h \to 0} (3h + 2) = 2$$

$$\text{L.H.L.} = f(0 - 0) = \lim_{h \to 0} f(0 - h) = \lim_{h \to 0} f(-h)$$

$$= \lim_{h \to 0} (2) = 2.$$

Since L.H.L. = R.H.L. = f(0) = 2, the function is continuous at x = 0.

(ii) To test f(x) for continuity at x = 1. We have

$$f(1) = 3.1 + 2 = 5.$$

$$\text{R.H.L.} = f(1 + 0) = \lim_{h \to 0} f(1 + h)$$

$$= \lim_{h \to 0} \frac{1 + h}{(1 + h) - 1} = \lim_{h \to 0} \frac{1 + h}{h} = \infty,$$

$$\text{L.H.L.} = f(1 - 0) = \lim_{h \to 0} f(1 - h)$$

$$= \lim_{h \to 0} [3(1 - h) + 2] = \lim_{h \to 0} [5 - 3h] = 5.$$

Since R.H.L. ≠ L.H.L. the function is not continuous at x = 1.

Example 22: *Show that the function $f(x) = |x|$ is continuous at $x = 0$, but not differentiable at $x = 0$, where $|x|$ means the numerical value of x.*

Solution: To test the continuity of f(x) at x = 0.

We have $f(0) = |x| = 0$;

$$\text{R.H.L. } f(0 + h) = \lim_{h \to 0} f(0 + h) = \lim_{h \to 0} f(h) = \lim_{h \to 0} |h|$$

L.H.L. $f(0-0) = \lim_{h \to 0} f(0-h)$

$= \lim_{h \to 0} f(-h) = \lim_{h \to 0} |-h|$

$= \lim_{h \to 0} h = 0$

L.H.L. = R.H.L.

Since $f(0+0) = f(0-0) = f(0)$, therefore $f(x)$ is continuous at $x = 0$.

To test the differentiability of f(x) at x = 0.

We have $Rf'(0) = \lim_{h \to 0} \frac{f(0+h)-f(0)}{h}$, where h is +ive and sufficiently small

$$= \lim_{h \to 0} \frac{f(h)-f(0)}{h} = \lim_{h \to 0} \frac{|h|-|0|}{h} = \lim_{h \to 0} \frac{h}{h}$$

$$= \lim_{h \to 0} 1 = 1;$$

and $Lf'(0) = \lim_{h \to 0} \frac{f(0-h)-f(0)}{-h}$. h being +ive and sufficiently small

$$= \lim_{h \to 0} \frac{f(-h)-f(0)}{-h} = \lim_{h \to 0} \frac{|-h|-|0|}{-h} = \lim_{h \to 0} \frac{h-0}{-h}$$

$$= \lim_{h \to 0} (-1)$$

$$= -1.$$

Since $Rf'(0) \neq Lf'(0)$, therefore the function $f(x)$ is not differentiable at $x = 0$.

Example 23: *Test the following function for continuity*

$f(x) = x \cos(1/x)$ for $x \neq 0$; $f(0) = 0$ at $x = 0$.

Solution: We have

$$f(0+0) = \lim_{h \to 0} f(0+h) = \lim_{h \to 0} f(h) = \lim_{h \to 0} h \cos \frac{1}{h} = 0.$$

Again $f(0-0) = \lim_{h \to 0} f(0-h)$

$$= \lim_{h \to 0} f(-h) = \lim_{h \to 0} -h \cos\left(\frac{-1}{h}\right) = 0.$$

Also $f(0) = 0$. Thus $f(0+0) = f(0-0) = f(0)$. Therefore $f(x)$ is continuous at $x = 0$.

Example 24: *If a function f is continuous at x = a and if y is a function continuous at f(a) show that the function got is continuous at x = a.*

Solution: If f: R = R and g: R → R then the product or composite of t and g. Written as fog: R → R is defined as

$$(fog)\ (x) = f[g(x)] \text{ for all } x \ \forall\ R.$$

Similarly (gof) (x) = g[f(x)] for all x ∀ R. Since g is continuous at f(a), so for any $\varepsilon > 0$.

There exist some $\delta_1 > 0$, such that

$|g[f(x)] - g(fog)] < \varepsilon$ when

$|f(x) - f(a)| < \delta_1$ ...(1)

Since f is continuous at a fog $\delta_1 > 0$ there exist some $\delta > 0$ such that

$|f(x) - f(a)| < \delta$ when $|x - a| < \delta$ ...(2)

from (1) and (2) we obtain

$|g(f(x)) - g(f(a))| < \varepsilon$, when $|x - a| < \delta$

$|(g\ of)\ (x) - (gof)\ (a)| < \varepsilon$ when $|x - a| < \delta$

If follows that gof is continuous at x = a.

Let y = [x] or E(x) denotes the integral part of x. Prove that this function is discontinuous where x has an integral value. Also draw the graph.

Example 25: *Give* $\in - \delta$ *definition of continuity of a function f at a point use this definition to show that if f is continuous at c, then so |B| |f| is the converse true 3 justify.*

Solution: A function f is said to be continuous at a point c_1 if for any $\in > 0$ three exist some $\delta > 0$ such that

$|f(x) - f(c)| < \in$ when $|x - c| < \delta$ ...(1)

Now we show that if f is continuous at c then |f| is continuous at c.

We know

$||x| - |f|| \leq |x - y|$ for all x, y ∈ R

$||f(x)| - |f(c)|| \leq |f(x) - f(c)|$ ...(2)

from equation (1) and (2) we get

$\therefore |f(x) - |f(c)| < \in$ when $|x - c| < \delta$

Hence |f| is continuous at c

However, the converse is not true *i.e.* If |t| is continuous at a point.

Then f may not be continuous at that point

Let $f(x) = \begin{cases} \dfrac{|x|}{x} & \text{if } x \neq 0 \\ 1 & \text{if } x = 0 \end{cases}$

We have $\lim\limits_{x \to 0} f(x) = \lim\limits_{x \to 0} \dfrac{-x}{x} = -1$

$\lim\limits_{x \to 0} f(x) = \lim\limits_{x \to 0} \dfrac{x}{n} = 1$

$f(0) = 1$

Hence f is discontinuous at x = 0

We have [| f | (x) = 1 ∀ x ∈ n]

i.e., | f | is a constant function and so | f | is continuous at ever point of R including x = 0.

Example 26: *Discuss the continuity of the function f(x) at* $x = \frac{1}{2}$ *where* $f(x) = x,\ 0 \leq x < \frac{1}{2};\ f(x) = 1,\ x = \frac{1}{2};\ f(x) = 1 - x,\ \frac{1}{2} < x \leq 1.$

Solution: We have f(x) = 1, when $x = \dfrac{1}{2}$.

Now R.H.L. = $\lim\limits_{h \to 0} f\left(\dfrac{1}{2} + h\right)$, where h is +ive and sufficiently small

$$= \lim_{h \to 0}\left[1 - \left(\frac{1}{2} + h\right)\right], \qquad \left[\because f(x) = 1 - x \text{ when } \frac{1}{2} < x \leq 1\right]$$

$$= \lim_{h \to 0}\left(\frac{1}{2} - h\right) = \frac{1}{2},$$

$$\text{L.H.L.} = \lim_{h \to 0} f\left(\frac{1}{2} - h\right), \qquad \left[\because f(x) = x \text{ when } 0 \leq x < \frac{1}{2}\right]$$

Since R.H.L. = L.H.L. ≠ the value of f(x) at $x = \dfrac{1}{2}$, therefore the function is discontinuous at $x = \dfrac{1}{2}$.

Example 27: *A function is defined by*

$f(x) = -x^2$ *if* $x \leq 0$

$= 5x - 4$ *if* $0 < x < 1$

$= 4x^2 - 3x$ *if* $1 < x < 2$

$= 3x + 4$ *if* $x \geq 2$.

Examine f for continuity at x = 0, 1, 2. Also discuss the kind of discontinuity if axis.

Solution: At x = 0

$$\lim_{x \to 0} - f(x) = \lim_{x \to 0} \left(-x^2\right) = 0$$

$$\lim_{x \to 0} + f(x) = \lim_{x \to 0} (5x - 4) = -4$$

Also f(0) = 0

Thus, $\lim_{x \to 0} + f(x) \pm f(0)$ and so f has a does continuity of the first and from the right at x = 0

At x = 1

$$\lim_{x \to 1} - f(x) = \lim (5x - 4) = 1$$

$$\lim_{x \to 1} + f(x) = \lim_{x \to 1} \left(4x^2 - 3x\right) = 1$$

Also f(1) = 5 × 1 – 4 = 1

Thus, $1 \lim_{x \to 1-} f(x) = \lim_{x \to 1+} f(x)$ and so f(x) is continuous let x ≤ 1

At x = 2

$$\lim_{x \to 2-} f(x) = \lim_{x \to 2} (4x^2 - 3x) = 4 \times 4 - 3 \times 2 = 10$$

$$\lim_{x \to 2+} f(x) = \lim_{x \to 2} (34 + 4) = 10$$

Thus, $\lim_{x \to 2-} f(x) = \lim_{x \to 2+} f(x) = f(2)$

and so f(x) is continuous at x = 2.

Example 28: *Let* $f(x) = x \dfrac{e^{1/x} - e^{-1/x}}{e^{1/x} + e^{-1/x}}$,

$x \neq 0$; $f(0) = 0$.

Show that f(x) is continuous but not derivable at x = 0.

Solution: We have f(0) = 0;

$$\text{R.H.L.} = f(0+0) = \lim_{h \to 0} f(0+h) = \lim_{h \to 0} f(h)$$

$$= \lim_{h\to 0} h\frac{e^{1/h}-e^{-1/h}}{e^{1/h}+e^{-1/h}} = \lim_{h\to 0} h\frac{1-e^{-2h}}{1+e^{-2h}}, \quad \text{dividing Nr. and Dr. by } e^{1/h}$$

$$= 0\times\frac{1-0}{1+0} = 0\times 1 = 0;$$

and L.H.L. $f(0-0) = \lim_{h\to 0} f(0-h) = \lim_{h\to 0} f(-h)$

$$= \lim_{h\to 0} -h.\frac{e^{-1/h}-e^{1/h}}{e^{-1/h}+e^{1/h}}$$

$$= \lim_{h\to 0} -h\frac{e^{-2/h}-1}{e^{-2/h}+1} = 0\times\frac{0-1}{0+1} = 0.$$

R.H.L. = L.H.L.

$\Rightarrow$ f(0 + 0) = f(0 – 0) = f(0), the function is continuous at x = 0

Now to test f(x) for differentiability at x = 0, e have

$$Rf'(0) = \lim_{h\to 0}\frac{f(0+h)-f(0)}{h} = \lim_{h\to 0}\frac{f(h)-f(0)}{h}$$

$$= \lim_{h\to 0}\left[h\frac{e^{1/h}-e^{-1/h}}{e^{1/h}-e^{-1/h}}-0\right]/h = \lim_{h\to 0}\frac{1-e^{-2/h}}{1+e^{-2/h}} = \frac{1-0}{1+0} = 1,$$

and $Lf'(0) = \lim_{h\to 0}\frac{f(0-h)-f(0)}{-h} = \lim_{h\to 0}\frac{f(-h)-f(0)}{-h}$

$$= \lim_{h\to 0}\left[(-h).\frac{e^{-1/h}-e^{1/h}}{e^{-1/h}+e^{1/h}}-0\right]/(-h)$$

$$= \lim_{h\to 0}\frac{e^{-2/h}-1}{e^{-2/h}+1} = \frac{0-1}{0+1} = -1.$$

Since Rf '(0) $\neq$ Lf '(0), the function is not differentiable at x = 0.

Example 29: *Consider the function f(x) = x – [x], where x is a positive variable, and [x] denotes the integral part of x and show that it is discontinuous for all integral values of x, and continuous for all others.*

Solution: From the definition of the function f(x), we have

f(x) = x – (n –1)	for n – 1 < x < n,
f(x) = 0	for x = n,
f(x) = x – n	for n < x < n + 1

and so on, where n is an integer.

We shall test the function f(x) for continuity at x = n. We have f(n) = 0;

$$f(n + 0) = \lim_{h\to 0} f(n+h) = \lim_{h\to 0} [(n+h)-n] = \lim_{h\to 0} h = 0;$$

$$[\because n < n + h < n + 1]$$

and L.H.L. $= f(n - 0) = \lim_{h\to 0} f(n-h) = \lim_{h\to 0} [(n-h)-(n-1)],$

$$[\because n - 1 < n - h < n]$$

$$= \lim_{h\to 0} (1-h) = 1.$$

Since f(n + 0) ≠ f(n – 0), the function f(x) is discontinuous at x = n. Thus, f(x) is discontinuous for all integral values of x. It is obviously continuous for all other values of x.

Since x is a positive variable, putting n = 1, 2, 3, 4, 5, ..., we observe that the graph of f(x) consists of the following:

$y = x$	when $0 < x < 1$,	$y = 0$ when $x = 1$
$y = x - 1$	when $1 < x < 2$,	$y = 0$ when $x = 2$
$y = x - 2$	when $2 < x < 3$,	$y = 0$ when $x = 3$
$y = x - 3$	when $3 < x < 4$,	$y = 0$ when $x = 4$

and so on.

The graph of the function thus obtained is shown by thick lines from x = 0 to x = 4.

Example 30: *If* $f(x) = \dfrac{xe^{1/x}}{1+e^{1/x}}$ *for* $x \neq 0$ *and* $f(0) = 0$, *show that f(x) is continuous at x = 0,*

but f '(0) does not exist.

Solution: Here f(0) = 0. Also right hand limit *i.e.*,

R.H.L. $f(0 + 0) = \lim_{h\to 0} f(0+h) = \lim_{h\to 0} f(h)$

$$= \lim_{h\to 0} \frac{he^{1/h}}{1+e^{1/h}} = \lim_{h\to 0} \frac{h}{e^{-1/h}+1} \text{ (Note)}$$

$$= \frac{0}{0+1} = 0,$$

and L.H.L. $f(0 - 0) = \lim_{h\to 0} f(0-h) = \lim_{h\to 0} f(-h)$

$$= \lim_{h\to 0}\left[\frac{-he^{-1/h}}{1+e^{-1/h}}\right] = \frac{0}{1+0} = 0.$$

Since f(0 + 0) = f(0 – 0) = f(0), the function is continuous at x = 0.

Now we have to the derivative of f(x) at x = 0.

We have,

$$Rf'(0) = \lim_{h\to 0}\frac{f(0+h)-f(0)}{h} = \lim_{h\to 0}\frac{f(h)-f(0)}{h}$$

$$= \lim_{h\to 0}\frac{\left(\frac{he^{1/h}}{1+e^{1/h}}\right)}{h} = \lim_{h\to 0}\frac{e^{1/h}}{1+e^{1/h}} = \lim_{h\to 0}\frac{1}{e^{1/h}+1}$$

$$= \frac{1}{0+1} = 0$$

and $Lf'(0) = \lim_{h\to 0} = \frac{f(0-h)-f(0)}{-h} = \lim_{h\to 0}\frac{f(-h)-f(0)}{-h}$

$$= \lim_{h\to 0}\frac{\frac{-he^{-1/h}}{1+e^{-1/h}}}{-h} = \lim_{h\to 0}\frac{e^{-1/h}}{1+e^{-1/h}} = \frac{0}{1+0} = 0\cdot$$

Since Rf '(0) ≠ Lf ' (0), the derivative of f(x) at x = 0 does not exist.

Example 31: *Let*

$$f(x) = \frac{x^2}{1+x^2} + \frac{x^2}{(1+x^2)^2} + \frac{x^2}{(1+x^2)^3} + \frac{x^2}{(1+x^2)^4} = \ldots \infty.$$

Is f(x) continuous at the origin? Give reasons for your answer.

Solution: Here the function f(x) forms an infinite G.P whose first terms a is $\frac{x^2}{1+x^2}$ and common ratio r is $\frac{1}{1+x^2}$ which is less than one when x ≠ 0.

$$\therefore\ f(x) = \frac{a}{1-r} = \frac{x^2/(1+x^2)}{1-1/(1+x^2)} = \frac{x^2/(1+x^2)}{x^2/(1+x^2)} = 1, \qquad \text{provided } x \neq 0.$$

Thus, f(x) = 1 for x ≠ 0 and f(x) = 0 for x = 0.

We have

$$f(0+0) = \lim_{h\to 0} f(0+h) = \lim_{h\to 0} f(h) = \lim_{h\to 0} 1 = 1.$$

Thus f(0 + 0) ≠ f(0) and so f(x) is not continuous at x = 0.

Example 32: *A function f(x) is defined as follows:*

$f(x) = 0$ *for* $x = 0$

$\phi(x) = \frac{1}{2} - x$ *for* $0 < x < \frac{1}{2}$

$\phi(x) = \frac{1}{2}$ *for* $x = \frac{1}{2}$

$\phi(x) = \frac{3}{2} - x$ *for* $\frac{1}{2} < x < 1$

$\phi(x) = 1$ *for* $x = 1$.

Find the points of discontinuity. Also draw the graph of the function.

Solution: We shall test the function for continuity at x = 0, 1/2 and 1.

(i) For x = 0, we have $\phi(0) = 0$, $\phi(0 + 0) = \lim_{h\to 0} \phi(0 + h)$

$= \lim_{h\to 0} \phi(h) = \lim_{h\to 0}\left(\frac{1}{2} - h\right) = \frac{1}{2}$. Since $\phi(0) \neq \phi(0 + 0)$, the function $\phi(0)$ is discontinuous at x = 0.

(ii) For $x = \frac{1}{2}$, we have $\phi\left(\frac{1}{2}\right) = \frac{1}{2}$, $\phi\left(\frac{1}{2} - 0\right) = \lim_{h\to 0} \phi\left(\frac{1}{2} - h\right)$

$= \lim_{h\to 0}\left[\frac{1}{2} - \left(\frac{1}{2} - h\right)\right]$, $\left(\text{Note that } 0 < \frac{1}{2} - h < \frac{1}{2}\right)$

$= \lim_{h\to 0} h = 0$

Since $\phi\left(\frac{1}{2} - 0\right) \neq \phi\left(\frac{1}{2}\right)$, the function $\phi(x)$ is discontinuous at $x = \frac{1}{2}$.

(iii) For x = 1, we have $\phi(1) = 1$, $\phi(1 - 0) = \lim_{h\to 0} \phi(1 - h)$

$= \lim_{h\to 0} [3/2 - (1 - h)]$, $\left[\text{Note that } \frac{1}{2} < 1 - h < 1\right]$

$= \lim_{h\to 0}\left(\frac{1}{2} + h\right) = \frac{1}{2}$.

Since $\phi(1) \neq \phi(1 - 0)$, the function $\phi(x)$ is discontinuous at x = 1.

Hence the function $\phi(x)$ has three points of discontinuity at x = 0, $\frac{1}{2}$ and 1.

The function consists of the point (0, 0); the segment of the line $y = \frac{1}{2}, 0 < x < \frac{1}{2}$;

the point $\left(\frac{1}{2}, \frac{1}{2}\right)$; the segment of the line

$$y = (3/2) - x, \frac{1}{2} < x < 1; \text{ and the point } (1, 1).$$

We observe that the function is discontinuous at x = 0, 1/2 and 1.

Example 33(a): *If f(x) = – x for x ≤ 0 and f(x) = + x for x ≥ 0, prove that f(x) is continuous but not differentiable at x = 0.*

Solution:

(i) *To test f(x) for continuity at x = 0.* We have f(0) = 0; f(0 + 0)

$$= \lim_{h\to 0} f(0+h) = \lim_{h\to 0} f(h) = \lim_{h\to 0} -(-h)$$

$$= \lim_{h\to 0} h = 0.$$

Since f(0 + 0) = f(0 – 0) = f(0), the function f(x) is continuous at x = 0.

(ii) *To test f(x) for differentiability at x = 0.*

We have $Rf'(0) = \lim_{h\to 0} \frac{f(0+h)-f(0)}{h} = \lim_{h\to 0} \frac{f(h)-f(0)}{h}$

$$= \lim_{h\to 0} \frac{h-0}{h} = \lim_{h\to 0} 1 = 1.$$

Again $Lf'(0) = \lim_{h\to 0} \frac{f(0-h)-f(0)}{-h} = \lim_{h\to 0} \frac{f(-h)-f(0)}{-h}$

$$= \lim_{h\to 0} \frac{-(-h)-0}{-h} = \lim_{h\to 0} \frac{h}{-h} = \lim_{h\to 0} -1 = 1.$$

Since Lf '(0) ≠ Lf '(0) the function f(x) is not differentiable at x = 0.

Example 33(b): *Discuss the continuity of the function f(x) defined as follows:*

f(x) = x^2 for x < – 2, f(x) = 4 for – 2 ≤ x ≤ 2, f(x) = x^2 for x > 2.

Solution: We shall test f(x) for continuity only at the points x = – 2 and 2. It is obviously continuous at all other points.

(i) *Continuity of f(x) at x = – 2.* We have f(– 2) = 4;

$$f(-2+0) = \lim_{h\to 0} f(-2+h) = \lim_{h\to 0} 4 = 4;$$

$$f(-2-0) = \lim_{h\to 0} f(-2-h) = \lim_{h\to 0} (-2-h)^2 \qquad [\because -2-h < -2]$$

= 4.

Thus f(– 2 + 0) = f(– 2 – 0) = f(– 2). Therefore f(x) is continuous at x = – 2.

(ii) *Continuity of f(x) at x = 2.* We have f(2) = 4;

$$f(2+0) = \lim_{h\to 0} f(2+h) = \lim_{h\to 0} (2+h)^2 = 4.$$

and $f(2-0) = \lim_{h\to 0} f(2-h) = \lim_{h\to 0} (2-h)^2 = 4 = 4.$

Thus, f(x) is also continuous at x = 2.

Example 33(c): *Discuss the continuity and differentiability of the function f(x) defined as follows:*

f(x) = 1 for – ∞ < x 0,

f(x) = 1 + sin x for 0 ≤ x < π/2,

$$f(x) = 2 + \left(x - \frac{\pi}{2}\right)^2 \text{ for } \frac{\pi}{2} \le x < \infty.$$

Solution: We shall test f(x) for continuity and differentiability at x = 0 and π/2. It is obviously continuous as well as differentiable at all other points.

(i) *Continuity and differentiability of f(x) at x = 0.* We have

$$f(0) = 1 + \sin 0 = 1;\ f(0+0) = \lim_{h\to 0} f(0+h)$$

$$= \lim_{h\to 0} f(h) = \lim_{h\to 0} (1+\sin h) = 1;\ \text{and}$$

$$f(0-0) = \lim_{h\to 0} f(0-h)$$

$$= \lim_{h\to 0} f(-h) = \lim_{h\to 0} 1 = 1.$$

Thus, f(0) = f(0 + 0) = f(0 – 0) and therefore f(x) is continuous at x = 0.

$$\text{Now } Rf'(0) = \lim_{h\to 0} \frac{f(0+h)-f(0)}{h} = \lim_{h\to 0} \frac{f(h)-f(0)}{h}$$

$$= \lim_{h\to 0} \frac{(1+\sin h)-(1+\sin 0)}{h} = \lim_{h\to 0} \frac{\sin h}{h} = 1;$$

$$\text{and } Lf'(0) = \lim_{h\to 0} \frac{f(0-h)-f(0)}{-h} = \lim_{h\to 0} \frac{f(-h)-f(0)}{-h}$$

$$= \lim_{h \to 0} \frac{1-(1+\sin 0)}{-h} = \lim_{h \to 0} \frac{0}{-h} = \lim_{h \to 0} 0 = 0.$$

Since $Rf'(0) \neq Lf'(0)$, $f(x)$ is not differentiable at $x = 0$.

(ii) *Continuity and differentiability of $f(x)$ at $x = \pi/2$.* We have

$$f\left(\frac{\pi}{2}\right) = 2 + \left(\frac{\pi}{2} - \frac{\pi}{2}\right)^2 = 2;$$

$$f\left(\frac{\pi}{2}+0\right) = \lim_{h \to 0} f\left(\frac{\pi}{2}+h\right) = \lim_{h \to 0}\left[2 + \frac{\pi}{2}\left\{\left(\frac{\pi}{2}+h\right) - \frac{\pi}{2}\right\}^2\right]$$

$$= \lim_{h \to 0} \left(2 + h^2\right) = 2;$$

and $f\left(\frac{\pi}{2}-0\right) = \lim_{h \to 0} f\left(\frac{\pi}{2}-h\right) = \lim_{h \to 0}\left[1 + \sin\left(\frac{\pi}{2}-h\right)\right]$

$$= \lim_{h \to 0} (1 + \cos h) = 1 + 1 = 2.$$

Thus, $f\left(\frac{\pi}{2}+0\right) = f\left(\frac{\pi}{2}-0\right) = f\left(\frac{\pi}{2}\right)$

and therefore $f(x)$ is continuous at $x = \frac{\pi}{2}$.

Now $Rf'\left(\frac{\pi}{2}\right) = \lim_{h \to 0} \frac{f(\pi/2+h) - f(\phi/2)}{h}$

$$= \lim_{h \to 0} \frac{\left[2 + \{\pi/2 + h - \pi/2\}\right] - \left[2 + (\pi/2 - \pi/2)^2\right]}{h}$$

$$= \lim_{h \to 0} \frac{2 + h^2 - 2}{h} = \lim_{h \to 0} h = 0;$$

and $Lf'\left(\frac{\pi}{2}\right) = \lim_{h \to 0} \frac{f(\pi/2-h) - f(\pi/2)}{-h}$

$$= \lim_{h \to 0} \frac{1 + \sin(\pi/2 - h) - 2}{-h}$$

$$= \lim_{h \to 0} \frac{-1 + \cos h}{-h} = \lim_{h \to 0} \frac{1 - \cos h}{h} = \lim_{h \to 0} \frac{2\sin^2(h/2)}{h}$$

$$= \lim_{h \to 0} \left[\frac{\sin h/2}{h/2} \sin h/2\right] = 1 \times 0 = 0.$$

Thus, Rf '(0) = Lf '(0) and so f(x) is differentiable at x = π/2.

Example 34: *A function f(x) is defined as follows:*

$f(x) = 1 + x$ if $x \le 2$ and $f(x) = 5 - x$ if $x \ge 2$.

Is the function continuous at x = 2?

Solution: We have f(2) = 1 + 2 or 5 – 2 = 3;

$f(2 + 0) = \lim_{h\to 0} f(2+h)$, where h is +ive and sufficiently small

$= \lim_{h\to 0} [5-(2+h)]$, $\quad$ [∵ 2 + h > 2 and f(x) = 5 – x if x > 2]

$= \lim_{h\to 0} (3-h) = 3$

and $f(2 - 0) = \lim_{h\to 0} f(2-h)$, where h is +ive and sufficiently small

$= \lim_{h\to 0} [1+(2-h)]$, $\quad$ [∵ 2 – h < 2 and f(x) = 1 + x if x < 2]

$= \lim_{h\to 0} (2-h) = 3$.

Since f(2 + 0) = f(2 – 0) = f(2), the function f(x) is continuous at x = 2.

Example 35: *If $f(x) = x^2 \sin(1/x)$ for x ' 0 and $f(0) = 0$, show that f(x) is continuous and differentiable at x = 0.*

Solution:

(i) To test f(x) for continuity at x = 0. We have f(0) = 0;

$$f(0+0) = \lim_{h\to 0} f(0+h) = \lim_{h\to 0} f(h) = \lim_{h\to 0} h^2 \sin\frac{1}{h} = 0;$$

and $$f(0-0) = \lim_{h\to 0} f(0-h) = \lim_{h\to 0} f(-h)$$

$$= \lim_{h\to 0} (-h)^2 \sin(-1/h)$$

$$= \lim_{h\to 0} h^2 \sin\frac{1}{h} = 0.$$

Now f(0 + 0) = f(0 – 0) = f(0) implies that f(x) is continuous at x = 0.

(ii) To test f(x) for differentiability at x = 0. We have

$$Rf'(0) = \lim_{h\to 0} \frac{f(0+h)-f(0)}{h} = \lim_{h\to 0} \frac{f(-h)-f(0)}{-h}$$

$$= \lim_{h\to 0} \frac{h^2 \sin(1/h) - 0}{h} = \lim_{h\to 0} h \sin\frac{1}{h} = 0.$$

$$\text{Again } Lf'(0) = \lim_{h\to 0}\frac{f(0-h)-f(0)}{-h} = \lim_{h\to 0}\frac{f(-h)-f(0)}{-h}$$

$$= \lim_{h\to 0}\frac{(-h)^2\sin(-1/h)-0}{-h} = \lim_{h\to 0}\frac{-h^2\sin(1/h)}{-h}$$

$$= \lim_{h\to 0} h\sin\frac{1}{h} = 0.$$

Now, Rf (0) = Lf '(0) implies that f(x) is differentiable at x = 0. The derivative of f(x) at x = 0 has the value zero.

Example 36: *Examine the following curve for continuity and differentiability:*

$y = x^2$ for $x \le 0$, $y = 1$ for $0 < x < 1$, $y = 1/x$ for $x > 1$.

Solution: Let y = f(x). We need to test f(x) for continuity and differentiability at the points x = 0 and 1. It is obviously continuous and differentiable at all other points.

At x = 0. We have $f(0) = 0^2 = 0$; $f(0 + 0) = \lim_{h\to 0} f(0+h)$

$= \lim_{h\to 0} f(h) = \lim_{h\to 0} 1 = 1$. Since $f(0 + 0) \neq f(0)$, the function f(x) is not continuous at x = 0. Consequently it is also not differentiable at x = 0.

At x = 1. The function f(x) is not defined at x = 1. Therefore it is discontinuous as well as non-differentiable at x = 1.

Example 37(a): *Give an example to show that a continuous function need not be differentiable.*

Solution: Here we shall give one more such example. Consider the function f(x) defined as follows:

$$f(x) = x\sin(1/x) \text{ for } x \neq 0,\ f(0) = 0.$$

As already proved in part (b) this function f(x) is continuous at x = 0. Now we shall show that the function f(x) is not differentiable at x = 0. We have

$$Rf'(0) = \lim_{h\to 0}\frac{f(0+h)-f(0)}{h} = \lim_{h\to 0}\frac{f(h)-f(0)}{h}$$

$$= \lim_{h\to 0}\frac{h\sin(1/h)-0}{h}$$

$$= \lim_{h\to 0}\sin\frac{1}{h}, \text{ which does not exist.}$$

Similarly, Lf (0) does not exist Hence f(x) is not differentiable at x = 0.

Example 37(b): *Examine the following curve for continuity and differentiability at the points x = ± 1:*

$$y = x - 1 \text{ for } x \geq 1,\ y = \frac{3}{2} + x \text{ for } x \leq -1,$$

$$y = 1 + x + x^2 + x^3 + \ldots \infty \text{ for } -1 < x < 1.$$

Solution: Obviously $y = \frac{1}{1-x}$ for $-1 < x < 1$. Let $y = f(x)$.

(i) *At x = 1.* We have $f(1) = 1 - 1 = 0$;

$$f(1+0) = \lim_{h\to 0} f(1+h) = \lim_{h\to 0}(1+h-1) = 0;$$

and $f(1-0) = \lim_{h\to 0} f(1-h) = \lim_{h\to 0}\frac{1}{1-(1-h)} = \lim_{h\to 0}\frac{1}{h} = \infty.$

Since f(1 + x) ¹ f(1 − 0), the function f(x) is not continuous at x = 1. Consequently f(x) is also not differentiable at x = 1.

(ii) *At x = − 1.* We have $f(-1) = \frac{3}{2} - 1 = \frac{1}{2}$;

$$f(-1+0) = \lim_{h\to 0} f(-1+h) = \lim_{h\to 0}\frac{1}{1-(-1+h)}$$

$$= \lim_{h\to 0}\frac{1}{2-h} = \frac{1}{2};$$

and $f(-1-0) = \lim_{h\to 0} f(-1-h) = \lim_{h\to 0}\left[\frac{3}{2}+(-1-h)\right]$

$$= \lim_{h\to 0}\left(\frac{1}{2}-h\right) = \frac{1}{2}.$$

Since f(− 1 + 0) = f(− 1 − 0) = f(− 1), the function f(x) is continuous at x = − 1.

Now $Rf'(-) = \lim_{h\to 0}\frac{f(-1+h)-f(-1)}{h}$

$$= \lim_{h\to 0}\frac{\frac{1}{1-(-1+h)}-\left(\frac{3}{2}-1\right)}{h} = \lim_{h\to 0}\frac{\frac{1}{2-h}-\frac{1}{2}}{h}$$

$$= \lim_{h\to 0}\frac{2-2+h}{2h(2-h)} = \lim_{h\to 0}\frac{1}{2(2-h)} = \frac{1}{4}.$$

Again $Lf'(-1) = \lim_{h\to 0}\frac{f(-1-h)-f(-1)}{-h}$

$$= \lim_{h\to 0} \frac{\left[\frac{3}{2}+(-1-h)\right]-\left(\frac{3}{2}-1\right)}{-h}$$

$$= \lim_{h\to 0} \frac{\frac{1}{2}-h-\frac{1}{2}}{-h} = \lim_{h\to 0} 1 = 1\cdot$$

Since Rf' (– 1) ≠ Lf' (– 1), the function f(x) is not differentiable at x = – 1.

Example 37(c): *If* $f(x) = \frac{xe^{1/x}}{1+e^{1/x}}$ *for* $x \neq 0$ *and* $f(0) = 0$, *show that* $f(x)$ *is continuous at* $x = 0$, *but* $f'(0)$ *does not exist.*

Solution: Here f(0) = 0. Also right hand limit *i.e.*,

$$\text{R.H.L.} = f(0+0) = \lim_{h\to 0} f(0+h) = \lim_{h\to 0} f(h)$$

$$= \lim_{h\to 0} \frac{he^{1/e}}{1+e^{1/h}} = \lim_{h\to 0} \frac{h}{e^{-1/h}+1} \text{ (Note)} = \frac{0}{0+1} = 0,$$

$$\text{and L.H.L.} = f(0-0) = \lim_{h\to 0} f(0-h) = \lim_{h\to 0} f(-h)$$

$$= \lim_{h\to 0}\left[\frac{-he^{1/h}}{1+e^{-1/h}}\right] = \frac{0}{1+0} = 0$$

Since f(0 + 0) = f(0 – 0) = f(0), the function is continuous at x = 0.

Now we have to the derivatives of f(x) at x = 0.

We have,

$$Rf'(0) = \lim_{h\to 0} \frac{f(0+h)-f(0)}{h} = \lim_{h\to 0} \frac{f(h)-f(0)}{h}$$

$$= \lim_{h\to 0} \frac{\left(\frac{he^{1/h}}{1+e^{1/h}}\right)}{h} = \lim_{h\to 0} \frac{e^{1/h}}{1+e^{1/h}} = \lim_{h\to 0} \frac{1}{e^{1/h}+1} = \frac{1}{0+1} = 0$$

$$\text{and } Lf'(0) = \lim_{h\to 0} \frac{f(0-h)-f(0)}{-h} = \lim_{h\to 0} \frac{f(-h)-f(0)}{-h}$$

$$= \lim_{h\to 0} \frac{\frac{-he^{-1/h}}{1+e^{-1/h}}-0}{-h} = \lim_{h\to 0} \frac{e^{-1/h}}{1+e^{-1/h}} = \frac{0}{1+0} = 0\cdot$$

Since Rf '(0) ≠ Lf '(0), the derivative of f(x) at x = 0 does not exist.

Example 38: *The function for continuity*

$f(x) = 2^{1/x}$, *when* $x \neq 0$; $f(0) = 0$ *at* $x = 0$.

Solution: We have

$$f(0 + 0) = \lim_{h \to 0} f(0 + h) = \lim_{h \to 0} f(h) = \lim_{h \to 0} 2^{1/h} = \infty.$$

Again $f(0 - 0) = \lim_{h \to 0} f(0 - h) = \lim_{h \to 0} f(-h) = \lim_{h \to 0} 2^{1/h} = 2^{-\infty} = 0.$

Since $f(0 + 0) \neq f(0 - 0)$, the function f(x) is discontinuous at x = 0.

Note: A function f(x) is said to have an *infinite discontinuity* at a point x = a, if one or both of the limits f(a + 0) and f(a – 0) are infinite.

Test the function for continuity

Example 39: $f(x) = \dfrac{1}{1 - e^{-1/x}}$; $x = 0$; $f(0) = 0$ *at* $x = 0$.

Solution: We have

$$f(0+0) = \lim_{h \to 0} f(0+h) = \lim_{h \to 0} f(h)$$

$$= \lim_{h \to 0} \frac{1}{1 - e^{-1/h}}$$

$$= \frac{1}{e^{-\infty}} = \frac{1}{1-0} = 1.$$

Again $f(0-0) = \lim_{h \to 0} f(0-h) = \lim_{h \to 0} f(-h) = \lim_{h \to 0} \dfrac{1}{1 - e^{1/h}}$

$$= \frac{1}{1 - e^{\infty}} = \frac{1}{1 - \infty} = 0.$$

Since $f(0 + 0) \neq f(0 - 0)$, the function f(x) is discontinuous at x = 0.

Example 40: *Test the following functions for continuity:*

$f(x) = x^3$ *at* $x = 2$.

Solution: We have $f(2) = 2^3 = 8$;

$$f(2 + 0) = \lim_{h \to 0} f(2+h) = \lim_{h \to 0} (2+h)^3 = 2^3 = 8;$$

and $\quad f(2 - 0) = \lim_{h \to 0} f(2 - h) = \lim_{h \to 0} (2 - h)^3 = 2^3 = 8.$

Since $f(2) = f(2 + 0) = f(2 - 0)$, therefore f(x) is continuous at x = 2.

Example 41(a): *Test two function for continuity*

$f(x) = x \sin(1/x)$, $x \neq 0$ *and* $f(0) = 0$ *at* $x = 0$.

Solution: We have

$$f(0+0) = \lim_{h\to 0} f(0+h) = \lim_{h\to 0} f(h) = \lim_{h\to 0} h \sin\frac{1}{h} = 0.$$

Again $f(0-0) = \lim_{h\to 0} \lim_{h\to 0} f(0-h) = \lim_{h\to 0} f(-h)$

$$= \lim_{h\to 0}(-h)\sin\left(-\frac{1}{h}\right) = 0.$$

Also f(0) = 0. Thus f(0 + 0) = f(0 – 0) = f(0). Therefore f(x) is continuous at x = 0.

Note: If we are given that f(0) = 2, the above function becomes discontinuous at x = 0.

Example 41(b): *Discuss the points of discontinuity of the function given by:*

$$f(x) = \begin{cases} -x, & for \quad x \le 0 \\ x, & for \quad 0 < x \le 1 \\ 2-x, & for \quad 1 < x \le 2 \\ 1, & for \quad x > 2 \end{cases}$$

Solution: Here we shall test f(x) for continuity at the points x = 0, 1 and 2. Obviously at all other points the function is continuous.

(i) *Continuity at x = 0.* We have f(0) = 0, because f(x) = – x for x = 0.

L.H.L. of f(0) as x → 0 *i.e.*,

$$f(0-0) = \lim_{h\to 0} f(0-h)\ f(-h) = \lim_{h\to 0}\left[-(-h)\right],\ [\because f(x) = -x \text{ or } x < 0]$$

$$= \lim_{h\to 0}(h) = 0;$$

R.H.L. of f(x) as x → 0 *i.e.*,

$$f(0+0) = \lim_{h\to 0} f(0+h) = \lim_{h\to 0} f(h) = \lim_{h\to 0} h,$$

$$[\because f(x) = x \text{ for } 0 < x \le 1]$$

$$= 0.$$

R.H.L. = L.H.L.

Thus, f(0 – 0) = f(0 + 0) = f(0) and so f(x) is continuous at x = 0.

(ii) *Continuity at x = 1.* We have f(1) = 1, because f(x) = x for 0 < x ≤ 1.

Now L.H.L. of f(x) as x → 1 *i.e.*,

$f(1-0) = \lim_{h\to 0} f(1-h) = \lim_{h\to 0} (1-h),$

$[\because f(x) = x \text{ for } 0 < x \leq 1]$

$= 1;$

and R.H.L. $f(1+0) = \lim_{h\to 0} f(1+h) = \lim_{h\to 0} [2-(1+h)],$

$[\because f(x) = 2 - x \text{ for } 0 < x \leq 2]$

$\lim_{h\to 0} (1-h) = 1.$

L.H.L. = R.H.L.

Since $f(1-0) = f(1+0) = f(1)$, therefore f(x) is continuous at $x = 1$.

(iii) *Continuity at x = 2.* We have $f(2) = 2 - 2 = 0$, because f(x) $2 - x$, for $1 < x \leq 2$.

L.H.L. of f(x) as $x \to 2$ *i.e.*,

$f(2-0) = \lim_{h\to 0} f(2-h) = \lim_{h\to 0} [2-(2-h)],$

$[\because f(x) = 2 - x, \text{ for } 1 < x \leq 2]$

$= \lim_{h\to 0} (h) = 0;$

and R.H.L. $= f(2+0) = \lim_{h\to 0} f(2+h) = \lim_{h\to 0} (1),$

$= 1$

$[\because f(x) = 1 \text{ for } x > 2]$

L.H.L. = R.H.L.

Since $f(2+0) \neq f(2-0)$, the function f(x) is discontinuous at $x = 2$.

Example 41(c): *Define continuity of a function f(x) at x = a. Discuss the continuity of the function f(x) defined as:*

f(0) = 0; f9x) = 1 – x, 0 < x < 1; f(1) = 1;

f(x) = 2 – x, 1 < x < 2; f(2) = 0

at the points x = 0, 1, 2,.

Solution: For definition of continuity

(i) *To test f(x) for continuity at x = 0.* We have $f(0) = 0$.

Now $f(0+0) = \lim_{h\to 0} f(1+h) = \lim_{h\to 0} f(h) = \lim_{h\to 0} (1-h)$

$[\because f(x) = 1 - x \text{ for } 0 < x < 1]$

$= 1.$

As $f(0 + 0) \neq f(0)$, therefore f(x) is discontinuous at $x = 0$.

(ii) *To test f(x) for continuity at x = 1.* We have $f(1) = 1$.

Now $f(1 - 0) = \lim_{h \to 0} f(1 - h) = \lim_{h \to 0} [1 - (1 - h)]$

$[\because f(x) = 1 - x \text{ for } 0 < x < 1]$

$= 0,$

and $f(1 + 0) = \lim_{h \to 0} f(1 + h) = \lim_{h \to 0} [2 - (1 + h)],$

$[\because f(x) = 2 - x, \text{ for } 1 < x < 2]$

$= 1.$

$\because$ $f(1 - 0) \neq f(1 + 0)$, therefore f(x) is discontinuous at $x = 1$.

(iii) *To test f(x) for continuity at x = 2.* We have $f(2) = 0$.

Now, $f(2 - 0) = \lim_{h \to 0} f(2 - h) = \lim_{h \to 0} [2 - (2 - h)],$

$[\because f(x) = 2 - x \text{ for } 1 < x < 2]$

$= 0.$

Since $f(2) = f(2 - 0)$, f(x) is continuous at $x = 2$.

Note that here f(x) is not defined for $x < 2$. So the question of finding $f(2 + 0)$ does not arise. Here f(x) will be continuous at $x = 2$ if $f(2) = f(2 - 0)$.

Example 41(d): *Verify Rolle's Theorem for* $f(x) = \sqrt{1 - x^2}$ *in [–1, 1]*

Solution: The given function being an algebraic function of x is continuous in [–1, 1]

Now $f'(x) = \dfrac{-x}{\sqrt{1 - x^2}}$ for $-1 < x < x < 1$ so f(x) is derivable in]–1, 1[such that $f'(c) = 0$

Thus, f(x) satisfies all the conditions of Rolle's Theorem.

So there exist a point $c \in]-1, 1[$ such that $f'(c) = 0$

Clearly $f'(c) = 0$ for $c \in]-1, 1[$

Hence, the given function satisfies the hypotheses and the conclusion of Rolle's Theorem.

Example 42: *Verify Rolle's Theorem in the case of the functions:*

(i) $f(x) = 2x^3 + x^2 - 4x - 2,$

(ii) $f(x) = \sin x$ *in* $[0, \pi]$,

(iii) $f(x) = (x - a)^m (x - b)^n$, *where m and n are +ive integers, and x lies in the interval [a, b].*

Solution:

(i) Here f(x) is a rational integral function of x. We know that rational functions are continuous and differentiable for all real values of x. Thus, the first two conditions of Rolle's Theorem are satisfied in any interval.

Again let $f(x) = 0$. Then $2x^3 + x^2 - 4x - 2 = 0$

or $(x^2 - 2)(2x - 1) = 0$ *i.e.* $x = \pm\sqrt{2}, -\frac{1}{2}$.

Thus $f(\sqrt{2}) = f(-\sqrt{2}) = f\left(-\frac{1}{2}\right) = 0$.

Now, let us consider the interval $[-\sqrt{2}, \sqrt{2}]$. In this interval all the conditions of Rolle's Theorem are satisfied. Therefore there is at least one value of x in the open interval $(-\sqrt{2}, \sqrt{2})$ where $f'(x) = 0$.

Now $f'(x) = 0$ where $6x^2 + 2x - 4 = 0$

or $3x^2 + x - 2 = 0$ or $(3x - 2)(x + 1) = 0$

or $x = -1, 2/3$.

Thus, $f'(x)(-1) = f'\left(\frac{2}{3}\right) = 0$

Since both the points $x = -1$ and $x = \frac{2}{3}$ lie in the open interval $(-\sqrt{2}, \sqrt{2})$, Rolle's Theorem is verified.

(ii) Here $f(0) = \sin 0 = 0$ and $f(\pi) = \sin \pi = 0$. Thus

$$f(0) = f(\pi) = 0.$$

We know that sin x is continuous and differentiable in $[0, \pi]$. Hence all the three conditions of Roll's Theorem are satisfied. Therefore $f'(x) = 0$ for at least one value of x in the open interval $(0, \pi)$.

Now $f'(x) = 0$ gives $\cos x = 0$ or $x = \pm\frac{\pi}{2}, \pm\frac{3\pi}{2}, \pm\frac{5\pi}{2}, \ldots$ Since $x = \pi/2$ lies in the open interval $(0, \pi)$, the Rolle's Theorem is verified.

(iii) Here $f(x) = (x - a)^m (x - b)^n$.

As m and n are positive integers, $(x - a)^m$ and $(x - b)^n$ are polynomials in x on being expanded by binomial theorem. Hence f(x) is also a polynomial in x. Consequently f(x) is continuous and differentiable in the closed interval [a, b]. Also f(x) = f(b) = 0. Thus all the three conditions of Rolle's Theorem are satisfied. So f '(x) = 0 for at least one value of x lying in the open interval (a, b).

Now, $f'(x) = (x - a)^m \cdot n (x - b)^{n-1} + m (x - a)^{m-1} (x - b)^n$.

The equation f '(x) = 0, on being solved, gives

$$x = a, b, \frac{na + mb}{m + n}.$$

Out of these values the value $\frac{na + mb}{m + n}$ is a point lying in the open interval (a, b) as it divides the interval (a, b) internally in the ratio m : n. Thus, the Rolle's Theorem is verified.

Example 43: *Discuss the applicability of Rolle's Theorem for* $f(x) = 2 + (x-1)^{2/3}$ *in the interval [0, 2].*

Solution: Given $f(x) = 2 + (x - 1)^{2/3}$. Obviously f(0) = 3 = f(2), showing that the third condition of Rolle's Theorem is satisfied.

The function f(x), being an algebraic function of x, is continuous in the closed interval [0, 2]. Thus the first condition of Rolle's Theorem is satisfied.

Now $f'(x) = \frac{2}{3} \cdot \left[\frac{1}{(x - 1)^{1/3}}\right]$. We observe that at x = 1, f '(x) is not finite while x = 1 is a point of the open interval 0 < x < 2. Thus, the second condition for Rolle's Theorem is not satisfied.

Hence the Rolle's Theorem is not applicable for the function $2 + (x - 1)^{3/2}$ in the given interval [0, 2].

Example 44: *Discuss the applicability of Rolle's Theorem in the interval [–1, 1] to function f(x) = | x |.*

Solution: Here f(–1) = | –1 | = 1 and f(1) = | 1 | = 1, so that f(–1) = f(1).

Also the function f(x) is continuous throughout the closed interval [–1, 1] but it is not differentiable at x = 0 which is a point of the open interval (–1, 1). Therefore the second condition for Rolle's Theorem is not satisfied, *i.e.*, the Rolle's Theorem is not applicable here.

Example 45: *Are the conditions of Rolle's Theorem satisfied in the case of the following functions?*

(i) $f(x) = x^2$ *in* $2 \le x \le 3$,

(ii) $f(x) = \cos(1/x)$ *in* $-1 \le x \le 1$,

(iii) $f(x) = \tan x$ *in* $0 \le x \le \pi$,

Solution:

(i) The function $f(x) = x^2$ is continuous and differentiable in the interval [2, 3]. Thus, the first two conditions of Rolle's Theorem are satisfied.

Also $f(2) = 4$ and $f(3) = 9$, so that $f(2) \neq f(3)$. Hence the third condition is not satisfied.

(ii) Here $f(-1) = \cos(-1) = \cos 1$ and $f(1) = \cos 1$. Thus, $f(-1) = f(1)$ *i.e.*, the third condition is satisfied.

But the first two conditions of Rolle's Theorem are not satisfied as the function is discontinuous at $x = 0$ and consequently is not differentiable there.

(iii) Here $f(0) = \tan 0 = 0$ and $f(\pi) = \tan \pi = 0$. Thus, $f(0) = f(\pi)$ *i.e.*, the third condition is satisfied.

But the first two conditions of Rolle's Theorem are not satisfied here as the function is not continuous at $x = \pi/2$ and consequently is non-differentiable there.

Example 46(a): *If f(x), $\phi(x)$, $\psi(x)$ have derivatives when $a \leq x \leq b$, show that there is a value c of x lying between a and b such that*

$$\begin{vmatrix} f(a) & \phi(a) & \psi(a) \\ f(b) & \phi(b) & \psi(b) \\ f'(c) & \phi'(c) & \psi'(c) \end{vmatrix} = 0$$

Solution: Consider the following function

$$F(x) = \begin{vmatrix} f(a) & \phi(a) & \psi(a) \\ f(b) & \phi(b) & \psi(b) \\ f(x) & \phi(x) & \psi(x) \end{vmatrix}$$

On expanding the determinant, we observe that the function F(x) is of the form $A\,f(x) + B\,\phi(x) + C\,\psi(x)$, where A, B, C are some real numbers.

Since the functions f(x), $\phi(x)$ and $\psi(x)$ have derivatives when $a \leq x \leq b$, therefore, the function F(x) also possesses derivatives when $a \leq x \leq b$. Consequently F(x) is also continuous when $a \leq x \leq b$. Further $F(a) = F(b) = 0$ because then the two rows of the determinant become identical. Thus, F(x) satisfies all the three conditions of Rolle's Theorem. Hence $F'(x) = 0$ for at least one value of x, say $x = c$, lying between a and b. Thus there is a value c of x lying between a and b such that

$$\begin{vmatrix} f(a) & \phi(a) & \psi(a) \\ f(b) & \phi(b) & \psi(b) \\ f'(c) & \phi'(c) & \psi'(c) \end{vmatrix} = 0.$$

Example 46(b): *Discuss the applicability of Rolle's Theorem to f(x) =* $\log\left[\frac{x^2+ab}{(a+b)x}\right]$, *in the interval [a, b]*

Solution: We have $f(a) = \log\left[\frac{a^2+ab}{(a+b)a}\right] = \log 1 = 0,$

and $$f(b) = \log\left[\frac{b^2+ab}{(a+b)b}\right] = \log 1 = 0.$$

Hence f(a) = f(b) = 0.

Also $$Rf'(x) = \lim_{h\to 0}\frac{f(x+h)-f(x)}{h}$$

$$= \lim_{h\to 0}\frac{1}{h}\left[\log\left\{\frac{(x+h)^2+ab}{(a+b)(x-b)}\right\} - \log\left\{\frac{x^2+ab}{(a+b)x}\right\}\right]$$

$$= \lim_{h\to 0}\frac{1}{h}\left[\log\frac{(x^2+2xh+h^2+ab)(a+b)x}{(a+b)(x-h)(x^2+ab)}\right]$$

$$= \lim_{h\to 0}\frac{1}{h}\left[\log\left\{\frac{(x^2+2xh+h^2+ab)}{(x^2+ab)}\times\frac{x}{x+h}\right\}\right]$$

$$= \lim_{h\to 0}\frac{1}{h}\left[\log\left\{1+\frac{xh+h^2}{x^2+ab}\right\} - \log\left\{1+\frac{h}{x}\right\}\right]$$

$$= \lim_{h\to 0}\frac{1}{h}\left[\frac{2hx+h^2}{x^2+ab} - \frac{h}{x} + \ldots\ldots\right] \quad \ldots(1)$$

[$\because \log(1+y) = y - 1/2\, y^2 + \ldots.$]

$$= \frac{2x}{x^2+ab} - \frac{1}{x}.$$

Again $$Lf'(x) = \lim_{h\to 0}\left[\frac{f(x-h)-f(x)}{-h}\right]$$

$$= \lim_{h\to 0}\frac{1}{(-h)}\left[\frac{-2hx+h^2}{x^2+ab} - \frac{(-h)x}{x} + \ldots\right],$$

[replacing h by – h in (1)]

$$= \frac{2x}{x^2 + ab} - \frac{1}{x}.$$

Thus, Rf '(x) = Lf '(x), showing that f(x) is differentiable for all values of x in [a, b]. Consequently f(x) is also continuous for all values of x in [a, b]. Hence all the three conditions of Rolle's Theorem are satisfied.

$\therefore$ f '(x) = 0 for at least one value of x in the open interval a < x < b.

Now f '(x) = 0 where $\frac{2x}{x^2 + ab} = \frac{1}{x} = 0$

or $2x^2 = (x^2 + ab) = 0$ or $x^2 = ab$

or $x = \sqrt{(ab)}$, which being the geometric mean of a and b lies in the open interval (a, b). Hence Rolle's Theorem is verified.

Example 46(c): *Verify Rolle's Theorem for*

(i) $f(x) = x^3 - 4x$ *in* $[-2, 2]$,

(ii) $f(x) = x(x + 3)\, e^{-x/2}$ *in* $[-3, 0]$,

(iii) $f(x) = e^x (\sin x - \cos x)$ *in* $[\pi/4, 5\pi/4]$,

Solution:

(i) Here $f(x) = x^3 - 4x$. Since f(x) is a polynomial in x, therefore it is continuous and differentiable for every real value of x; Also $f(-2) = 0 = f(2)$.

$\therefore$ f(x) satisfies all the three conditions of Rolle's Theorem.

$\therefore$ in the open interval

(–2, 2) for which f '(c) = 0.

Now f '(x) = 0 gives $3x^2 - 4 = 0$

or $x = \pm \frac{2}{\sqrt{3}} = \pm 1.155$ (aprox).

Both these values lie in the open interval (–2, 2). Thus the theorem is verified.

(ii) Here $f(x) = x(x + 3)\, e^{-x/2} = (x^2 + 3x)\, e^{-x/2}$.

We have $f'(x) = (2x + 3)\, e^{-x/2} + (x^2 + 3x) \, . \, e^{-x/2} \, . \left(-\frac{1}{2}\right)$

$= e^{-x/2}\, [2x + 3 - 1/2\, (x^2 + 3x) = -1/2\, (x^2 - x - 6)\, e^{-x/2}$,

which exists for every value of x in the interval [–3, 0]. Therefore f(x) is differentiable and also continuous in the interval [–3, 0].

Also f(–3) = 0 = f(0). Therefore all the three conditions of Rolle's Theorem are satisfied.

∴ There must exist at least one number, say c, in the open interval (–3, 0) for which $f'(c) = 0$ *i.e.*, $-1/2\ (c^2 - c - 6)\ e^{-c/2} = 0$ or $c^2 - c - 6 = 0$ or $(c - 3)(c + 2) = 0$ or c = 3, –2.

The value c = – 2 lies in the open interval (–3, 0). Hence the theorem is verified.

(iii) Here $f(x) = e^x (\sin x - \cos x)$.

We have $f(\pi/4) = e^{\pi/4} \{\sin(\pi/4) = \cos(\pi/4)\}$

$$= e\pi/4\ [(1/\sqrt{2}) - (1/\sqrt{2})] = 0$$

and $$f\left(\frac{5\pi}{4}\right) = e^{5\pi/4}\left[\sin\frac{5\pi}{4} - \cos\frac{5\pi}{4}\right] = e^{5\pi/4}\left[-\frac{1}{\sqrt{2}} + \frac{1}{\sqrt{2}}\right] = 0.$$

∴ $f(\pi/4 = f(5\pi/4) = 0.$

Further the function f(x) is continuous and differentiable in [π/4, 5π/4]. Therefore all the three conditions of Rolle's Theorem are satisfied.

∴ There must exist at least one number, say c, in the open interval (π/4, 5π/4) for which $f'(c) = 0$.

Now $f'(x) = e^x (\cos x + \sin x) + e^x (\sin x - \cos x) = 2\ e^x \sin x$.

From $f'(x) = 0$ we get $2e^x \sin x = 0$

or $\sin x = 0,$ $[\because e^x \neq 0]$

or $x = 0, \pm\pi, \pm 2\pi, \pm 3\pi, \ldots$

Out of these values x = π lies in the open interval (π/4, 5π/4). Thus the Rolle's Theorem is verified.

Example 47: *State the conditions for the validity for the formula and investigate how far these conditions are satisfied and whether the result is true, when* $f(x) = x \sin(1/x)$ *(being defined to be zero at* $x = 0$*) and* $x < 0 < x + h$.

Solution: The conditions for the validity of the given formula are :

(i) The function f(x) must be continuous in the closed interval [x, x + h].

(ii) The function f(x) must be differentiable in the open interval (x, x + h).

(iii) θ is a real number such that $0 < \theta < 1$.

Now consider the function f(x) defined as:

$$f(x) = x \sin(1/x) \text{ for } x \neq 0,\ f(0) = 0.$$

The first condition is satisfied because f(x) is continuous in the closed interval [x, x + h] for $x < 0 < x + h$. [The students should show here that f(x) is continuous at x = 0].

But the second condition is the satisfied because f(x) is not differentiable at x = 0 which is a point lying in the open interval (x, x + h) for $x < 0 < x + h$. [Show here that f(x) is not differentiable at x = 0].

Hence the result of the given formula is not true for this function f(x).

Example 48: *A function f(x) is continuous in the closed interval $0 \leq x \leq 1$ and differentiable in the open interval $0 < x < 1$, prove that $f'(x_1) = f(1) - f(0)$, where $0 < x_1 < 1$.*

Solution: Here a = a, b = 1. Therefore

$$\frac{f(b) - f(a)}{b - a} = \frac{f(1) - f(0)}{1 - 0} = f(1) - f(0).$$

If we take $c = x_1$ and substitute these values in the result of Lagrange's mean value theorem, we get

$$f(1) - f(0) = f'(x_1) \text{ where } 0 < x_1 < 1.$$

Example 49: *Separate the intervals in which the polynomial $2x^3 - 15x^2 + 36x + 1$ is increasing or decreasing.*

Solution: Let $f(x) = 2x3 = 15x^2 + 36x + 1$.

Then $f'(x) = 6x^2 = 30x + 36 = 6(x - 2)(x - 3)$.

Now $f'(x) > 0$ for $x < 2$; $f'(x) < 0$ for $2 < x < 3$; $f'(x) > 0$ for $x > 3$; $f'(x) = 0$ for x = 2 and 3.

Thus, f(x) is positive in the intervals $(-\infty, 2)$ and $(3, \infty)$ and negative in the interval (2, 3).

Hence f(x) is monotonically increasing in the intervals $(-\infty, 2]$, $[3, \infty)$ and monotonically decreasing in the interval [2, 3].

Example 50: *Find 'c' so that* $f'(c) = \frac{f(b) - f(a)}{b - a}$ *in the following cases:*

(i) $f(x) = x^2 - 3x - 1$; $a = -11/7$, $b = 13/7$

(ii) $f(x) = x^x$; $a = 0$, $b = 1$.

Solution:

(i) Here $f(a) = f\left(-\frac{11}{7}\right) = \frac{121}{49} + \frac{33}{7} - 1 = \frac{303}{49}$

and $f(b) = f\left(-\frac{13}{7}\right) = \frac{169}{49} + \frac{39}{7} - 1 = \frac{153}{49}$

$$\therefore \quad \frac{f(b)-f(a)}{b-a} = \frac{-456/49}{24/7} - \frac{19}{7}$$

Now $\quad f'(x) = 2x - 3; \quad \therefore f'(c) = 2c - 3.$

From Lagrange's mean value theorem, we have

$$2c - 3 = -19/7 \text{ or } c = 1/7.$$

(ii) Here $f(a) = f(0) = e^0 = 1$, and $f(b) = f(1)\ e^1 = e$. Also $f'(x) = e^x$, so that $f'(c) = e^c$

$\therefore$ using Lagrange's mean value theorem, we have

$$\frac{e-1}{1-0} = e^c \text{ or } e^c = e - 1$$

or $\quad c \log_e (e - 1)$

Example 51(a): *Compute the value of* θ *in the first mean value theorem*

$f(x + h) = f(x) + hf'(x + \theta h),$

if $f(x)\ ax^2 + bx + c.$

Solution: We have $f(x) = ax^2 + bx + c$.

$\therefore f(x + h) = a(x + h)^2 + b(x + h) + c,$

$f'(x) = 2ax + b,\ f'(x + \theta h) = 2a(x + \theta h) + b.$

Putting all these values in the Lagrange's mean value theorem, we have

$$a(x + h)^2 + b(x + h) + c = ax^2 + bx + c + h[2a(x + \theta h) + b] \quad ...(1)$$

The relation (1) is identically true for all values of x. So when $x \to 0$, we get

$ah^2 + bh + c = c + h[2a\theta h + b]$

or $ah^2 = 2a\theta h^2$

or $\theta = 1/2$

Example 51(b): *Verify Cauchy's mean value theorem for the functions* x^2 *and* x^3 *in the interval* $[1, 2]$.

Solution: Let $f(x) = x^2$ and $g(x) = x^3$. Both f(x) and g(x) are continuous n the closed interval [1, 2] and differentiable in the open interval (1, 2). Also $g'(x) = 3x^2 \neq 0$ for any point in the open interval (q, 2). Therefore by Cauchy's mean value theorem there exists at least one real number c in the open interval (1, 2), such that

$$\frac{f(2)-f(1)}{g(2)-g(1)} = \frac{f'(c)}{g'(c)} \quad ...(1)$$

Now $\quad \dfrac{f(2)-f(1)}{g(2)-g(1)} = \dfrac{4-1}{8-1} = \dfrac{3}{7}$

Also $f'(x) = 2x,\ g'(x) = 3x^2$.

Therefore $$\frac{f'(c)}{g'(c)} = \frac{2c}{3c^2} = \frac{2}{3c}$$

Substituting these values in (1), we get $\frac{3}{7} = \frac{2}{3c}$ or $c = \frac{14}{9}$ which lies in the open interval (1, 2). This verifies the theorem.

Example 52: *If $f(x) = (x - 1)(x - 2)(x - 3)$ and $a = 0$, $b = 4$, find 'c' using Lagrange's mean value theorem.*

Solution: We have

$$f(x) = (x - 1)(x - 2)(x - 3) = x^3 - 6x^2 + 11x - 6$$

$\therefore f(a) = f(0) = -6$, and $f(b) = f(4) = 6$.

$$\therefore \quad \frac{f(b)-f(a)}{b-a} = \frac{6-(-6)}{4-0} = \frac{12}{4} = 3$$

Also $f'(x) = 3x^2 - 12x + 11$, so that $f'(c) = 3c^2 - 12c + 11$. Substituting these values in Lagrange's mean value theorem.

$$\frac{f(b)-f(a)}{b-a} = f'(c),\ (a < c < b),\ \text{we have}$$

$$3 = 3c^2 - 12c + 11$$

$$\text{or } 3c^2 - 12c + 8 = 0$$

$$\text{or } c = \frac{12 \pm \sqrt{(144-96)}}{6} = 2 \pm \frac{2\sqrt{3}}{3}$$

Both of these values of c lie in the interval (0, 4). Hence both of these are the required values of c.

Example 53: *Find* 'c' of the mean value theorem, if

$$f(x) = x(x - 1)(x - 2);\ a = 0,\ b = 1/2.$$

Solution: Here $f(a) = f(0) = 0$ and

$$f(b)\ f\left(\frac{1}{2}\right) = \frac{1}{2}\left(\frac{1}{2} - 1\right)\left(\frac{1}{2} - 2\right) = \frac{3}{8}.$$

$$\therefore \quad \frac{f(b)-f(a)}{b-a} = \frac{\frac{3}{8} - 0}{\frac{1}{2} - 0} = \frac{3}{4}.$$

Now $f(x) = x^3 - 3x^2 + 2x$.

$\therefore f'(x) = 3x^2 - 6x + 2$, so that $f'(c) = 3c^2 - 6c + 2$.

Substituting these values in Lagrange's mean value theorem

$$\frac{f(b)-f(a)}{b-a} = f'(c), (a < c < b), \text{ we have}$$

$$\frac{3}{4} = 3c^2 - 6c + 2$$

or $12c^2 - 24c + 5 = 0.$

$$c = \frac{24 \pm \sqrt{(24 \times 24 - 4 - \times 12 \times 5}}{24}$$

$$= \frac{24 \pm 4\sqrt{36-15}}{24} = 1 \pm \frac{\sqrt{21}}{6}$$

Out of these two values of c only $1 - \frac{\sqrt{21}}{6}$ lies in the open interval $\left(0, \frac{1}{2}\right)$ which is therefore the required value of c.

Example 54: *If in the Cauchy's mean value theorem, we write f(x) = ex and g(x) = e*$^{-x}$*, show that 'c' is the arithmetic mean between a and b.*

Solution: Here $\frac{f(b)-f(a)}{g(b)-g(a)} = \frac{e^b - e^a}{e^{-b} - e^{-a}} = - e^a e^b = - e^{a+b}.$

Also $\frac{f'(x)}{g'(x)} = \frac{e^x}{-x^{-x}}$

so that $\frac{f'(c)}{g'(c)} = \frac{e^c}{-x^{-c}} = e^{2c}$

Substituting these values in Cauchy's mean value theorem, we get

$- e^{a+b} = e^{2c}$ or $2c = a + b$

or $\frac{1}{2}(a+b)$.

Hence c is the arithmetic mean between a and b.

Example 55: *If, in the Cauchy's mean value theorem, we write*

(i) *f(x) =* $\sqrt{x}$ *and g(x) = 1/*$\sqrt{x}$, *then c is the geometric mean between a and b, and if*

(ii) *f(x) = 1/x*^{2} *and g(x) = 1/x, then c is the harmonic mean between a and b.*

Solution: Here $\dfrac{f(b)-f(a)}{g(b)-g(a)} = \dfrac{\sqrt{b}-\sqrt{a}}{(1/\sqrt{b})-(1/\sqrt{2})} = -\sqrt{(ab)}.$

Also $\dfrac{f'(x)}{g'(x)} = \dfrac{\frac{1}{2}x^{-1/2}}{-\frac{1}{2}x^{-3/2}}$, so that $\dfrac{f'(c)}{g'(c)} = -\dfrac{c^{-1/2}}{c^{-3/2}} = -c.$

Substituting these values in Cauchy's mean value theorem, we get $-\sqrt{(ab)}$ $= -c$ or $c = -\sqrt{(ab)}$ *i.e.*, c is the geometric mean between a and b.

(ii) From the Cauchy's mean value theorem, we have

$$\frac{f(b)-f(a)}{g(b)-g(a)} = \frac{f'(c)}{g'(c)}.$$

Putting f(x) $1/x^2$ and $g(x) = 1/x$, we get

$$\frac{(1/b^2)-(1/a^2)}{(1/b)-(1/a)} = \frac{-2c^{-3}}{-c^{-2}} \text{ or } \frac{a+b}{ab} = \frac{2}{c}$$

or $c = \dfrac{2ab}{a+b}$

i.e., c is the harmonic mean between a and b.

Example 56: Show that $x^3 - 3x^2 + 2$ is monotonically increasing in every interval.

Solution: Let $f(x) = x^3 - 3x^2 + 3x + 2$.

Then $f'(x) = 3x^2 - 6x + 3 = 3(x-1)^2$.

We see that $f'(x) > 0$ for every real value of x except 1 where its value is zero. Hence f(x) is monotonically increasing in every interval.

Example 57: Show that $\dfrac{x}{1+x} < \log(1+x) < x$ for $x > 0$.

Solution: Let $f(x) = \log(1+x) - \dfrac{x}{1+x}$

Then $f'(x) = \dfrac{1}{1+x} - \dfrac{1.(1+x)-x.1}{(1+x)^2}$

$$= \frac{1}{1+x} - \frac{1}{(1+x)^2} = \frac{x}{(1+x)^2}$$

We see that $f'(x) > 0$ for $x > 0$. Therefore, f(x) is monotonically increasing in the interval $[0, \infty)$. But $f(x) = 0$. Therefore, $f(x) > f(0) = 0$ for $x > 0$ *i.e.*,

$$\left[\log(1+x) - \frac{x}{1+x}\right] > 0 \text{ for } x > 0$$

Hence $\log(1 + x) > \dfrac{x}{1+x}$ for $x > 0$. ...(1)

Again let $\phi(x) = x - \log(1 + x)$; then

$$\phi'(x) = 1 - \frac{1}{1+x} = \frac{x}{1+x}$$

We see that $\phi'(x) > 0$ for $x > 0$. Therefore, $\phi(x)$ is monotonically increasing in the interval $[0, \infty)$. But $\phi(0) = 0$. Therefore $\phi(0) > \phi(0) = 0$ for $x > 0$ *i.e.*, $[x - \log(1 + x)] >$ for $x > 0$.

Hence $x > \log(1 + x)$ for $x > 0$. ...(2)

From (1) and (2), we have

$$\frac{x}{1+x} < \log(1 + x) < x \text{ when } x > 0.$$

Example 58: *Find 'θ' if* $f(x + h) = f(x) + hf'(x) + \dfrac{h^2}{2!} f''(x + \theta h)$, $0 < \theta < 1$ *and*

(i) $f(x) = ax^3 + bx^2 = cx + d$,

(ii) $f(x) = x^3$.

Solution: Here $f(x) = ax^3 + bx^2 = cx + d$

$\therefore$ $f(x + h) = a(x + h)^3 + b(x + h)^2 + c(x + h) + d$,

$f'(x) = 3ax^2 + 2bx + c$, $f''(x) = 6ax + 2b$,

$f''(x + \theta h) = 6a(x + \theta h) + 2b$.

The given relation

$$f(x + h) = f(x) + hf'(x) + \left(\frac{h^2}{2!}\right) f''(x + \theta h), \text{ we have}$$

$$a(x + h)^3 + b(x + h)^2 + c(x + h) + d,$$

$$= ax^3 + bx^2 + cx + d + h(3ax^2 + 2bx + c)$$

$$+ \left(\frac{h^2}{2!}\right)\{6a(x + \theta h) + 2b\} \quad ...(1)$$

The relation (1) is an identify in x. Letting $x \to 0$ on both sides of (1), we have

$$ah^3 + bh^2 + ch + d = d + ch + \left(\frac{h^2}{2}\right)(6a\theta h) + 2b)$$

or $\quad ah^3 + bh^2 + ch + d = d + ch + 3a\theta h^3 + bh^2$

or $\quad ah^3 = 3a\theta h^3$

or $\quad \theta = \frac{1}{3}$ $\quad [\because ah^3 \neq 0]$

(ii) Proceed as in part (i) of this question. The required value of θ is 1/3.

Example 59: *Expand the following by Maclaurin's theorem with Lagrange's form of remainder after n terms:*

(i) a^x, (ii) e^x.

Solution:

(i) Here $f(x) = a^x$. ...(1)

$\therefore \quad f^{(n)}(x) = a^x(\log a)^n$. ...(2)

Putting x = 0 in (1) and (2), we get

$$f(0) = a^0 = 1,\ f^{(n)}(0) = a^0(\log a)^n = (\log a)^n.$$

$$\therefore \quad f'(0) = \log a,\ f''(0) = (\log a)^2, \ldots, f^{(n-1)}(0) = (\log a)^{n-1}.$$

Also changing x to θx in (2), we get

$$f^{(n)}(\theta x) = a^{\theta x}(\log a)^n.$$

Now by Maclaurin's theorem with Lagrange's form of remainder after n terms, we have

$$f(x) = f(0) + xf'(0) + \frac{x^2}{2!}f''(0) + \ldots + \frac{x^{n-1}}{(n-1)!}f^{(n-1)}(0)$$

$$+\frac{x^n}{n!}f^{(n)}(\theta x), \text{ where } 0 < \theta < 1. \quad \ldots(3)$$

Substituting values found above in (3), we get

$$a^x = 1 + x \log a \ \frac{x^2}{2!}(\log a)^2 + \ldots + \frac{x^{n-1}}{(n-1)!}(\log a)^{(n-1)}$$

$$+\frac{x^n}{n!}a^{\theta x}(\log a)^n$$

Here Lagrange's form of remainder after n terms

$$+\frac{x^n}{n!}a^{\theta x}(\log a)^n, \text{ where } 0 < \theta < 1.$$

(ii) Here $f(x) = e^x$. Therefore $f^{(n)}(x) = e^x$.

Putting x = 0, in these, we get

$f(0) = e^0 = 1,\ f^{(n)}(0) = e^0 = 1$. Also $f^{(n)} = (\theta x)\ e^{\theta x}$.

Substituting these values in Maclaurin's theorem with Lagrange's form of remainder after n terms, we get

$$e^x = 1 + x + \frac{x^2}{2!} + \frac{x^3}{2!} + \dots + \frac{x^{n-1}}{(n-1)!} + \frac{x^n}{n!} e^{\theta x}$$

Example 60: *Show that*

(i) $\sin x = x - \frac{x^3}{3!} + \frac{x^5}{5!} - \dots + (-1)^{n-1} \frac{x^{2n-1}}{(2n-1)!}$

$$+ (-1)^n \frac{x^{2n}}{(2n)!} \sin \theta x,$$

for every real value of x.

(i) $\log (1 + x) = x - \frac{x^2}{2} + \frac{x^3}{3} - \dots + (-1)^{n-2} \frac{x^{n-1}}{n-1}$

$$+ (-1)^{n-1} \frac{x^n}{n(1+\theta x)^n}, \text{ for } x > -1.$$

Solution:

(i) Here $f(x) = \sin x$.

We know that sin x possesses derivatives of every order for every real number x and

$$f^{(n)}(x) = \sin x\ (x + 1/2n\pi). \quad \dots(1)$$

Putting $x = 0$ in (1) and (2), we get

$f(0) = \sin 0 = 0,\ f^{(n)}(0) = \sin (1/2n\pi)$.

$\therefore\quad f'(0) = \sin (1/2\pi) = 1,\ f''(0) = \sin \pi = 0.$

$f'''(0) = \sin (3\pi/2) = -1.$

$f^{iv}(0) = \sin 2\pi = 0,$

$f^{v}(0) = \sin (5\pi/2) = \sin (2\pi + 1/2\pi) = \sin 1/2\ \pi = 1, \dots,$

$f^{(2n-1)}(0) = \sin (n-1)\ \pi = 0,$

$f^{(2n-1)}(0) = \sin \{1/2\ (2n-1)\ \pi\} = \sin (n\pi = 1/2\pi)$

$= (-1)^n \sin (-\pi/2),$

$[\because \sin (n\pi + \theta) = (-1)^n \sin\theta]$

$= (-1)^n (-1) = (-1)^{n+1} = (-1)^{n-1}.$

Also changing n to 2n and x to θx in (2), we get

$$f^{(2n)}(\theta x) = \sin(\theta x + n\pi) = (-1)^n \sin\theta x.$$

Substituting these values in Maclaurin's theorem with Lagrange's form of remainder after 2n terms *i.e.*,

$$f(x) = f(0) + sf'(0) + \frac{x^2}{2!} f''(0) + \ldots$$

$$+ \frac{x^{2n-1}}{(2n-1)!} f^{(2n-1)}(0) + \frac{x^{2n}}{(2n)!} f^{(2n)}(\theta x),$$

we get $\sin x = x - \frac{x^3}{3!} + \frac{x^5}{5!} - \ldots$

$$+ (-1)^{n-1} \frac{x^{2n-1}}{(2n-1)!} + (-1)^n \frac{x^{2n}}{(2n)!} \sin\theta x.$$

(ii) Here $f(x) = \log(1 + x)$. ...(1)

We know that $\log(1 + x)$ possesses derivatives of every order when $(1 + x) > 0$ *i.e.*, $x > -1$.

Also, $f^{(n)}(x) = (-1)^{n-1}(n-1)!\,(1+x)^{-n}$ ...(2)

Putting $x = 0$ in (1) and (2), we get

$$f(0) = \log 1 = 0,\ f^{(n)}(0) = (-1)^{n-1}(n-1)!.$$

Also changing x to θx in (2), we get

$$f^{(n)}(\theta x) = (-1)^{n-1}(n-1)!\,(1+\theta x)^{-n}.$$

Substituting these values in Maclaurin's theorem with Lagrange's form of remainder after n terms *i.e.*,

$$f(x) = f(0)\frac{x}{1!} f'(0) + \frac{x^2}{2!} f''(0) + \ldots$$

$$+ \frac{x^{n-1}}{(n-1)!} f^{(n-1)}(0) + \frac{x^n}{n!} f^{(n)}(\theta x),$$

we get

$$\log(1 + x) = 0 + \frac{x}{1!}.1 + \frac{x^2}{2!}.(-1).1 + \frac{x^3}{3!}(-1)^2.2! + \ldots$$

$$+ \frac{x^{n-1}}{(n-1)!}(-1)^{n-2}(n-2)! + \frac{x^n}{n!}(-1)^{n-1}(n-1)!(1+\theta x)^{-n}$$

$$= x\frac{x^2}{2} + \frac{x^3}{3} - \ldots + (-1)^{n-2}\frac{x^{n-1}}{n-1} + (-1)^{n-1}\frac{x^n}{n(1+\theta x)}$$

Example 61: *Show that* $f(x) = |x - 1|$, $0 \le x \le 2$ *is not derivable at* $x = 1$. *Is it continuous in [0, 2]?*

Solution:

(i) *To test for continuity at x = 1.* We have f(1) = 0.

Also R.H.L. = f(1 + 0) = $\lim_{h \to 0}$ f(1 + h)

$= \lim_{h \to 0} |1 + h - 1) = 0;$

and L.H.L. = f(1 – 0) = f(1 – 0) = f(1), therefore f(x) is continuous at x = 1.

In fact f(x) is continuous at every point in the interval [0, 2].

(ii) *To test f(x) for differentiability at x = 1.*

We have

$$Rf'(1) = \lim_{h \to 0} \frac{f(1+h) - f(1)}{h} = \lim_{h \to 0} \frac{|1+h-1| - 0}{h}$$

$$= \lim_{h \to 0} \frac{h}{h} = 1;$$

and $$Lf'(1) = \lim_{h \to 0} \frac{f(1-h) - f(1)}{-h} = \lim_{h \to 0} \frac{|1-h-1| - 0}{-h}$$

$$= \lim_{h \to 0} \frac{h}{-h} = -1.$$

Since Rf '(1) ≠ Lf '(1), the function f(x) is not differentiable at x = 1.

Example 62(a): *Show that the function f(x), where*

f(x) = 2 + x if x ≥ 0; f(x) = 2 – x if x < 0 is not derivable at the point x = 0.

Solution: We have

$$Rf'(0) = \lim_{h \to 0} \frac{f(0+h) - f(0)}{h} = \lim_{h \to 0} \frac{f(h) - f(0)}{h}$$

$$= \lim_{h \to 0} \frac{2+h-2}{h} = 1$$

and $$Lf'(0) = \lim_{h \to 0} \frac{f(0-h) - f(0)}{-h} = \lim_{h \to 0} \frac{f(-h) - f(0)}{-h}$$

$$= \lim_{h \to 0} \frac{\{2-(-h)\} - 2}{-h} = -1.$$

Since Rf ' (0) ≠ Lf '(0), the function f(x) is not differentiable at x = 0.

Example 62(b): *A function f(x) is defined as follows:*

$$f(x) = \begin{cases} \left(x^2/a\right) - a, & \text{when} \quad x < 0 \\ 0, & \text{when} \quad x = a \\ a - \left(a^2/x\right), & \text{when} \quad x > 0 \end{cases}$$

Prove that the function f(x) is continuous at x = a.

Solution: We have f(a) = 0.

Also the right hand limit of f(x) at x = a *i.e.*,

$$f(a+0) = \lim_{h\to 0} f(a+h) = \lim_{h\to 0}\left[a - \frac{a^2}{(a+h)}\right],$$

[∵ $f(x) = a - (a^2/x)$ for x > a]

$= a - (a^2/a) = a - a = 0;$

and the left hand limit of f(x) at x = a *i.e.*,

$$f(a-0) = \lim_{h\to 0} f(a-h) = \lim_{h\to 0}\left[\frac{(a-h)^2}{a} - a\right],$$

[∵ $f(x) = (x^2/a) - a$ for x < a]

$= (a^2/a) - a = a - a = 0.$

∵ f (a + 0) = f(a – 0) = f(a), therefore f(x) is continuous at x = a.

Example 62(c): *Give ∈ – δ definition of continuity of a function f at a point and use two definition to show that if f is continuous at c then so is | f | is the converse true? Justify.*

Solution: A function f is said to be continuous at a point c. If for any ∈ > 0. There exist some δ > 0 such that

$|f(x) - f(c)| < \in$ when $|x - c| < \delta$ ...(1)

Now we show that if f is continuous at c then |f| f is continuous at c

We know $||x| - |y|| \le |x - y|$, for all x, y ∈ R

∴ $||f(x)| - f(c)|| \le f(x) - f(c)|$ ...(2)

from (1) and (2) we get

$||f(x) - |f(c)|| < \in$ when $|x - c| < \delta$

Hence δ is continuous at c

However, the converse is not true *i.e.*, if |f| is continuous at a point. Then f may not be continuous at that point

$$\text{Let } f(x) = \begin{cases} \dfrac{|x|}{x} & \text{if } x \neq 0 \\ 1 & \text{if } x = 0 \end{cases}$$

we have $\lim_{x \to 0-} f(x) = \lim_{x \to 0} \frac{-x}{x} = -1$

$$\lim_{x \to 0+} f(x) = \lim_{x \to 0} \frac{x}{x} = 1$$

$f(0) = 1$

Hence f is continuous at x = 0

We have $(|f|)(x) = 1 \ \forall \ x \in R$

i.e. $|f|$ is a constant function and so $|f|$ is continuous at every point of R including x = 0.

Example 63: *Examine the function $f(x) = (x - a) \sin \{1/(x - a)\}$, when $x \neq a$ and $f(a) = 0$ for continuity at $x = a$.*

Solution: We have R.H.L. = f(a + 0)

$$= \lim_{h \to 0} f(a + h)$$

$$= \lim_{h \to 0} (a + h - a) \sin \{1/(a + h - a)\}$$

$$= \lim_{h \to 0} h \sin (1/h) = 0.$$

Again

$$\text{L.H.L.} = f(a - 0) = \lim_{h \to 0} f(a - h)$$

$$= \lim_{h \to 0} (a - h - a) \sin \{1/(a - h - a)\}$$

$$= \lim_{h \to 0} - h \sin (- 1/h) = 0.$$

Also f(a) = 0. Thus, f(a + 0) = f(a – 0) = f(a). Therefore f(x) is continuous at x = a.

Example 64: *Investigate the points of continuity and discontinuity of the following function:*

$$f(x) = \frac{1}{x - a} \operatorname{cosec}\left(\frac{1}{x - a}\right), \ x \neq a;$$

$$f(x) = 0, \qquad x = a.$$

Solution: We have L.H.L. $f(a + 0) = \lim_{h \to 0} f(a + h)$

$$= \lim_{h\to 0} \frac{1}{a+h-a} \operatorname{cosec}\left(\frac{1}{a+h-a}\right) = \lim_{h\to 0} \frac{1}{h\sin(1/h)}$$

$= +\infty$, since h sin (1/h) → 0 as h → 0.

Again,

$$\text{R.H.L.} = f(a-0) = \lim_{h\to 0} f(a-h) = \lim_{h\to 0} \frac{1}{a-h-a} \operatorname{cosec}\left(\frac{1}{a-h-a}\right)$$

$$= \lim_{h\to 0} -\frac{1}{h} \cdot \frac{1}{\sin\{-(1/h)\}} = \lim_{h\to 0} \frac{1}{h\sin(1/h)}$$

$= +\infty$, since h sin (1/h) → 0 as h → 0.

Also as given f(a) = 0.

Since f(a + 0) = f(a – 0) → f(a), the function f(x) is discontinuous at x = a, having an infinite discontinuity of second kind.

Example 65(a): *Examine the function defined below for continuity at x = 0:*

$$f(x) = \frac{\sin^2 ax}{x^2} \text{ for } x \neq 0,\ f(x) = 1 \text{ for } x = 0.$$

Solution: We have f(0) = 1;

$$\text{R.H.L.} = f(0+0) = \lim_{h\to 0} f(0+h) = \lim_{h\to 0} f(h) = \lim_{h\to 0} \frac{\sin^2 ah}{h^2}$$

$$= \lim_{h\to 0} \left(\frac{\sin ah}{ah}\right)^2 . a^2 = 1.a^2 = a^2;$$

$$\text{and L.H.L. } f(0-0) = \lim_{h\to 0} f(0-h) = \lim_{h\to 0} f(-h) = \lim_{h\to 0} \frac{\sin^2(-ah)}{(-h)^2}$$

$$= \lim_{h\to 0} \frac{\sin^2 ah}{h^2} = a^2 .$$

Now f(x) is continuous at x = 0 if and only if

f(0) = f(0 + 0) = f(0 – 0).

Therefore f(x) is discontinuous at x = 0 unless a = ± 1.

Example 65(b): *Discuss the continuity at x = 0, 1, 2 of function f(x) defined as follows:*

$f(x) = -x^2$ for $x \leq 0$, $f(x) = 5x - 4$ for $0 < x \leq 1$.

$f(x) = 4x^2 - 3x$ for $1 < x < 2$ and $f(x) = 3x + 4$ for $x \geq 2$.

Solution:

(i) Continuity at x = 0. We have $f(0) = -0^2 = 0$;

We have

L.H.L. $= f(0-0) = \lim_{h\to 0} f(0-h) = \lim_{h\to 0} f(-h) = \lim_{h\to 0} -(-h)^2$

$= \lim_{h\to 0} -h^2 = 0$;

and R.H.L. $f(0+0) = \lim_{h\to 0} f(0+h)$

$= \lim_{h\to 0} f(h) = \lim_{h\to 0} (5h-4)$. $[\because 0 < h < 1]$

$= 4$.

Since R.H.L. ≠ L.H.L.

$\Rightarrow f(0+0) \neq f(0-0)$, the function f(x) is discontinuous at x = 0.

(ii) Continuity at x = 1. We have $f(1) = 5 \times 1 - 4 = 1$;

R.H.L $= f(1+0) = \lim_{h\to 0} f(1+h)$

$= \lim_{h\to 0} [4(1+h)^2 - 3(1+h)]$, $[\because 1 < 1+h < 2]$

$= \lim_{h\to 0} [4h^2 + 5h + 1] = 1$;

and L.H.L. $f(1-0) = \lim_{h\to 0} f(1-h)$

$= \lim_{h\to 0} [5(1-h) - 4]$, $[\because 0 \leq 1-h < 1]$

$= \lim_{h\to 0} (1-5h) = 1$.

Thus, $f(1+0) = f(1-0) = f(1)$. Therefore f(x) is continuous at x = 1.

(iii) Continuity at x = 2. We have $f(2) = 3 \times 2 + 4 = 10$;

L.H.L. $= f(2-0) = \lim_{h\to 0} f(2-h) = \lim_{h\to 0} [4(2-h)^2 - 3(2-h)]$;

$[\because 1 < 2-h < 2]$

$= \lim_{h\to 0} [4h^2 - 13h + 10] = 10$;

and R.H.L. $= f(2+0) = \lim_{h\to 0} f(2+h)$

$= \lim_{h\to 0} [3(2+h) + 4]$

$= \lim_{h\to 0} (3h + 10) = 10$

$f(2) = f(2-0) = f(2+0)$ therefore f(x) continuous at x = 2.

EXERCISES

1. Show that the function $f(x) = (x - a) \sin \frac{1}{x-a}$.
2. Show that the function $f(x) = |x - 1| + |x + 1|$ is not derivable at $x = -1, 2$ and is deriable at every other point.
3. Determine whether f(x) is continuous and has a derivative at two origin,

   While $$f(x) = 2 + x, \text{ if } x \geq 0$$
   $$= 2 + x, \text{ if } x \geq 0$$
4. Show that function $f(x) = |x - 1| + |x + 1|$ is not derivable at $x = -1, 1$; and is derivable at every other point.
5. Show that $f(x) = \frac{xe^{1/x}}{1+e^{1/x}}$, $x \neq 0$ and $f(x) = 0$ is continuous but not derivable at $x = 0$.
6. Show that $f(x) = x|x|$ is derivable at the origin.
7. Show that x^r, where x is a positive integer, its continuous at $x = a$.
8. Show that $1/x$ is continuous at $x = a$, $(a \neq 0)$.
9. If $f(x) = x$ for $x \leq 1$ and $f(x) = x^2$ for $x < 1$ show that f is continuous at $x = 1$ but not derivable at $x = 1$.
10. Show that $f(x) = |x - a|$ is continuous at $x = a$ but not derivable at $x = a$.
11. Prove that the function f(x) defined by
    $$f(x) = \begin{cases} 2-x & \text{if } x < 2 \\ -2+3x-x^2 & \text{if } x > 2 \end{cases}$$
    is derivable at $x = z$.
12. Construct a function which is continuous in [1, 5] but not derivable at 2, 3, 4.
13. $\lim_{x \to 1} \frac{(x-1)\sqrt{2-x}}{x^2-1}$.
14. $\lim_{x \to 1/2} \frac{8x^3-1}{5x^2-5x+1}$.

2

Applications of Derivatives

INCREASING AND DECREASING FUNCTIONS

Consider the function:

$$y = x^2 + 4x + 5$$

We know an 'increase' in the value of x will make a corresponding 'change' in the value of y. But how does the change look like? Is it an increase or decrease? By substituting different values for x, we can make sure that the 'change' is an increase. Now, consider another function:

$$y = 50 - 4x - x^2$$

Here the 'change' is a decrease. Again consider the function:

$$y = x^2 - 4x + 5$$

We can see that up to the value $x = 2$, the value of y decreases, and from $x = 2$ onwards it increases. *A function $y = f(x)$ is said to be an increasing function, if an increase in x makes an increase in y. A function $y = f(x)$ is said to be a decreasing function, if an increase in x makes a decrease in y.* For example, supply function is increasing while demand function is decreasing. If we draw the tangent to the curve at any point, it will make an angle, which is less than 90° to the positive direction of x–axis.

Hence slope of the curve at any point $= \dfrac{dy}{dx} = \tan\theta > 0$, since $\theta < 90°$.

In other words, an increasing function is positively sloped. If we draw the tangent to the curve at any point, it will make an angle, which is greater than 90° to the positive direction of x–axis. Hence slope of the curve at any point is less than 0. Or, a decreasing function is negatively sloped. A decreasing function for values of $x < 2$, and increasing for values of $x > 2$.

Hence we conclude that: a function $y = f(x)$ is

(i) an *increasing function* at the point x, if $\frac{dy}{dx} \geq 0$, and

(ii) a *decreasing function* at the point x, if $\frac{dy}{dx} \leq 0$

If $\frac{dy}{dx} > 0$, we call $y = f(x)$ a *strictly increasing function* at the point x, and if $\frac{dy}{dx} < 0$, we call $y = f(x)$ a *strictly decreasing function* at the point x.

It is possible that a function f is neither increasing nor decreasing on a given interval. But, we can divide its domain into disjoint union of open intervals, such that the function may be either increasing or decreasing on each of the intervals.

MAXIMA AND MINIMA

In Economics, we frequently need to find the largest profit or the least cost in certain situations. Many of these problems can be solved using the method of differentiation. By the method of differentiation it is possible to find the 'extreme value' of a function in a given domain. This possibility is made use in finding the maxima or minima of functions. (it is discussed in the next chapter in detail).

ECONOMIC APPLICATIONS

(1) Some Economic Concepts

We denote price by p and quantity by q or x.

(i) Total Cost (TC) = Fixed Cost (FC) + Variable Cost (VC)

(ii) FC = TC at zero output

(iii) Average Total Cost (ATC) = $\frac{TC}{x}$

(iv) Marginal Cost (MC) = $\frac{d(TC)}{dx}$

(v) Total Revenue (TR) = price (p) × quantity (x)

(vi) Average Revenue (AR) = $\frac{TR}{x} = \frac{px}{x}$ = p = price

(vii) Marginal Revenue (MR) = $\frac{d(TR)}{dx}$

(viii) Break even point is set up where TR = TC

(ix) Profit (π) = TR –TC

(2) Elasticity of a Function

Elasticity of a function $y = f(x)$ at a point x is defined as the ratio of proportionate change in y to proportionate change in x.

Let dx be the change in x and let dy be the corresponding change in y. Then,

$$\text{Proportionate change in } x = \frac{dx}{x}, \text{ and}$$

$$\text{Proportionate change in } y = \frac{dy}{y}$$

$$\therefore \quad \text{Elasticity of } y \ \frac{dy}{y} \div \frac{dx}{x} = \frac{x}{y}\frac{dy}{dx}$$

The notion of elasticity of a function is very much used in Economics.

Elasticity of Demand

The law of demand, a change in price leads to a change in demand. Other things remaining constant, quantity demanded of a commodity varies inversely with its price. We can measure at what rate demand changes in response to a given change in its price, and this measure is termed as price elasticity of demand, or simply elasticity of demand. Hence, *elasticity of demand is defined as the degree of responsiveness of demand for a commodity to a change in its price* and is denoted by e_p. It is taken as the negative of the ratio of proportionate change in demand (q) to proportionate change in price (p).

$$\text{i.e., } e_p = -\frac{dq}{q} \div \frac{dp}{p} = -\frac{p}{q}\frac{dq}{dp}$$

On the basis of the responsiveness of demand to the change in price, elasticity of demand is generally classified into five categories.

(i) When a slight change in price causes infinitely large change in the quantity demanded, then the demand is said to be *perfectly elastic*, and in that case $e_p = \infty$.

(ii) If quantity demanded does not change at all in response to change in its price, then the demand is said to be *perfectly inelastic*, and in that case $e_p = 0$.

(iii) If a given change in price causes an equal and proportionate change in the quantity demanded, then the demand is said to be *unit elastic*, and in that case, $e_p = 1$.

(iv) If a given change in price leads to a more than proportional change in the quantity demanded, then the demand is said to be *elastic*, and in that case, $e_p > 1$.

(v) On the contrary, if a given change in price leads to a less than proportional change in the quantity demanded, then the demand is said to be *inelastic*, and in that case, $e_p < 1$.

Again if we assume the price of commodity as constant, the quantity demanded (q) of a commodity changes with consumer's income (y). Here $\frac{dq}{dy}$ can be positive or negative according as the type of the commodity. If the commodity is a normal good, then $\frac{dq}{dy} > 0$, and if the commodity is an inferior good, then $\frac{dq}{dy} < 0$. Income elasticity of demand (e_y) may be given by the formula,

$$e_y = \frac{dq}{q} \div \frac{dy}{y} = \frac{y}{q}\frac{dq}{dy}.$$

Hence, for inferior goods $e_y < 0$.

Cross Elasticity

Let there be two commodities. Let q_1 and q_2 are the demands of the first and second commodities respectively and let p_1 and p_2 are their respective prices. Then *cross elasticity of demand*, $e_{1,2}$ is given by :

$$e_{1,2} = \frac{p_2}{q_1}\frac{dq_1}{dp_2}.$$

Cross elasticity of demand is positive for substitutes and negative for complements.

Elasticity of Supply

As in the case of demand, there is relation between price and supply also. Law of supply states that when price increases, more is supplied. *Price elasticity of supply, or simply elasticity of supply is defined as the degree of responsiveness of supply* (q) *of a commodity to a change in its price* (p), and is denoted by e_s.

i.e., $$e_s = \frac{p}{q}\frac{dq}{dp}$$

3. Relation of Elasticity of Demand with Marginal and Average Revenues

We have seen that the elasticity of demand is:

$$e_p = -\frac{dq}{q} \div \frac{1}{e_p} = -\frac{p}{q}\frac{dq}{dp}$$

Taking reciprocal and multiplying by 1 we get,

$$\frac{-1}{e_p} = \frac{q}{p}\frac{dp}{dq} \qquad \ldots(1)$$

We know that, the average revenue (AR) and marginal revenue (MR) are given by:

$$AR = \frac{p.x}{x} = p \qquad \ldots(2)$$

$$MR = \frac{d(TR)}{dq} = \frac{d(pq)}{dq}$$

$$= p \times 1 + q.\frac{dp}{dq} \quad \text{(by product rule of differentiation)}$$

$$= p\left[1 + \frac{q}{p}\frac{dp}{dq}\right]$$

$$= p\left[1 - \frac{1}{e_p}\right] \quad \text{using equation (2)}$$

i.e., $$\frac{MR}{p} = 1 - \frac{1}{e_p}$$

or, $$\frac{1}{e_p} = 1 - \frac{MR}{p}$$

$$= \frac{p\text{-}MR}{p}$$

Hence $$e_p = \frac{p}{p\text{-}MR} = \frac{AR}{AR\text{-}MR} \quad \text{using (1)}$$

DERIVATIVE AS A RATE MEASURE

Differential Calculus has enormous applications in all branches of study. Among Social Sciences and Humanities, its applications are more in the theory of Economics. We use derivatives either to measure the rate of quantities, or to determine the slope of curves or to find the extreme value that a function can attain in a given domain. Application of derivative in determining extreme values is so vast, and hence it is dealt separately in the next Chapter. In this Chapter, we see how the derivatives facilitate the measurement of various

kinds of rates. The determination of slopes of different curves using derivatives is also dealt with. Direct applications of derivatives in Economics are given as a separate session at the end.

A. Rate of Change

The derivative of a function can be considered as a rate of change. Consider a particle, which is moving along a straight line.

Example: *1. Find the rate of change of the area of a circle with respect to its radius*

Solution: Let A be the area of the circle.

Then,

$$A = \pi r^2, \text{ where r is the radius of the circle.}$$

Differentiating with respect to r,

$$\frac{dA}{dr} = 2\pi r$$

i.e., the rate of change of the area of a circle with respect to its radius is 2π times its radius.

B. Relative Rate of Change

Generally we come across with the problems in which the rate of change of one of the quantities involved is required corresponding to the given rate of change of another quantity.

Example: *The radius of a balloon is increasing at the rate of 10 cm/sec. At what rate is the surface area of the balloon increasing when the radius is 15 cm?*

Solution: Let r be the radius and S be the surface area of the balloon at any time t. It is given that $\frac{dr}{dt} = 10$ cm/sec.

We know that $S = 4\pi r^2$

Differentiating with respect to t,

$$\frac{dS}{dt} = 8\pi r \frac{dr}{dt}$$

i.e., $$\frac{dS}{dt} = 8\pi r \times 10 = 80\pi r$$

Rate of increase of the surface area when the radius is 15 cm

$$= \left(\frac{dS}{dt}\right)_{r=15}$$

$$= 80\,\pi \times 15 = 1200\,\pi \text{ cm}^2/\text{sec}$$

C. Errors and Approximations

We have learned that the derivative $f\,'(x)$ of the function $y = f\,(x)$ measures its rate of change of at any point x. If Δx is a small increment in x and Δy is the corresponding change in y, then Δy can be written approximately as, $\Delta y = f\,'(x)$

When Δx is sufficiently small, it is called the differential of x denoted by dx. Since the differentiability of f guarantees its continuity, $f\,'(x)\mathrm{d}x$ is also small and we denote it as dy, which is called the differential of y.

i.e., when Δx is sufficiently small,

$$\mathrm{d}y = f\,'(x)\mathrm{d}x$$

or, $$\frac{\mathrm{d}y}{\mathrm{d}x} = f\,'(x).$$

Hence the notation $\frac{\mathrm{d}y}{\mathrm{d}x}$ will be interpreted as the ratio of the differentials as well as the derivative of $f(x)$. This formula for approximation is useful in the calculation of small changes (errors) in dependent variable corresponding to small changes (errors) in the independent variable and is of great importance in the theory of errors in Engineering, Physics, Statistics and several other branches of Science and in Economics. It can also be used in linear approximation of non–linear functions. For approximating errors, we take the differential dx of the independent variable x as a small amount of error in the measurement of x. Then an approximate value of corresponding error in the dependent variable $y = f(x)$ will be $f\,(x + \mathrm{d}x) - f\,(x)$, which is equal to d$y$. When d$x$ is small, this approximate value will tend to the actual error in y.

SLOPE OF A CURVE

We had studied that slope (or gradient) m of a straight line is the measure of inclination of the line and is measured using the formula,

$$m = \tan\,\theta,$$

where θ is the angle which the line makes with the positive direction of x–axis in anticlockwise sense. Now, what does the slope of a parabola mean? Since it changes its direction continuously, we cannot measure its inclination as a whole. But at any point it possesses a tangent line and it is possible to measure the inclination of the tangent. Slope of a curve at the point (x, y) is defined as the slope of the tangent to the curve at the point. We have already shown that the slope of the tangent to the curve $y = f(x)$ at the point (x, y) is given by $\frac{dy}{dx}$.

SOLVED EXAMPLES

Example 1: *Find the point of inflexion of the curve* $y = x^3 - 3x^2 - 9x + 9$

Solution: $y = x^3 - 3x^2 - 9x + 9$

$$\frac{dy}{dx} = 3x^2 - 6x - 9$$

$$\frac{d^2y}{dx^2} = 6x - 6$$

$$\frac{d^2y}{dx^2} = 0 \Rightarrow 6x - 6 = 0 \Rightarrow x = 1$$

Also, $\left(\frac{d^3y}{dx^3}\right)_{x=1} \neq 0$

When $x = 1$, $y = 1 - 6 = -2$

$\therefore$ The point of inflexion of the function is $(1, -2)$.

Example 2(a): *Find the point of inflexion of the curve:*

$$y = ax^3 + bx^2 + cx + d.$$

Also find the condition for the point of inflexion to be stationary.

Solution: $y = ax^3 + bx^2 + cx + d$

$$\frac{dy}{dx} = 3a\,x^2 + 2b\,x + c$$

$$\frac{d^2y}{dx^2} = 6ax + 2b$$

$$\frac{d^2y}{dx^2} = 0 \quad \Rightarrow 6ax + 2b = 0$$

$$\Rightarrow x = \frac{-b}{3a}$$

Also, $\left(\frac{d^3y}{dx^3}\right)_{x=-b/3a} \neq 0$

$\therefore$ The point of inflexion of the curve is at $x = \frac{-b}{3a}$.

The condition for the point of inflexion to be stationary is that $\frac{dy}{dx} = 0$ at the point.

i.e., $$\left(\frac{dy}{dx}\right)_{x=-b/3a} = 0$$

i.e., $$3a\left(\frac{-b}{3a}\right)^2 + 2b\left(\frac{-b}{3a}\right) + c = 0$$

$$\Rightarrow \quad \frac{b^2}{3a} - \frac{2b^2}{3a} + c = 0$$

$$\Rightarrow \quad b^2 - 2b^2 + 3ac = 0$$

$$\Rightarrow \quad b^2 = 3ac$$

Example 2(b): *Show that the function* $y = \frac{1}{x-1}$ *has a point of inflexion at* $x = 1$.

Solution:

$$y = \frac{1}{x-1}$$

$$\frac{dy}{dx} = \frac{-1}{(x-1)^2}$$

$$\frac{d^2y}{dx^2} = \frac{2}{(x-1)^3}$$

$\frac{d^2y}{dx^2}$ is not defined at $x = 1$. Hence the point of inflexion cannot be determined by applying this method. But we can reach the answer in the following way:

When $x < 1$, $\frac{d^2y}{dx^2} < 0$, and

when $x > 1$, $\frac{d^2y}{dx^2} > 0$

That is, at $x = 1$, the curve changes from concavity to convexity. This ensures that the curve has a point of inflexion at $x = 1$.

Example 2(c): *Find the points of inflection of the standard normal curve:*

$$y = k.e^{\frac{-x^2}{2}}$$

Solution:

Given $y = k.e^{\frac{-x^2}{2}}$

$$\Rightarrow \quad \frac{dy}{dx} = k.e^{\frac{-x^2}{2}} \times \frac{-2x}{2} = -k.xe^{\frac{-x^2}{2}}$$

$$\frac{d^2y}{dx^2} = -k\ x \times -xe^{\frac{-x^2}{2}} + e^{\frac{-x^2}{2}} \times -k$$

$$= (x^2 - 1)\ k.e^{\frac{-x^2}{2}}$$

$$\frac{d^2y}{dx^2} = 0 \Rightarrow x^2 - 1 = 0 \Rightarrow x = 1 \text{ or } x = -1$$

i.e., the points of inflexion of the curve are at $x = 1$ and $x = -1$

Example 3: *Find elasticity of demand of the function:*

$$x = 50 - 2p, \text{ at } p = 10$$

Solution: Elasticity of demand, $e = -\frac{p}{x}\frac{dx}{dp}$; $x = 50 - 2p$; $\frac{dx}{dp} = -2$

Also, when $p = 10$, $x = 50 - 2p = 30$

$$\therefore \quad e = -\frac{10}{30} \times -2 = 0.67$$

Example 4: *Find elasticity of demand with respect to price for the demand function:*

$$q = \frac{4}{p^2 - 1} \text{ when price is Rs. 5.}$$

Solution: Elasticity of demand, $e = -\frac{p}{q}\frac{dq}{dp}$

$$q = \frac{4}{p^2 - 1} \Rightarrow \frac{dq}{dp} = \frac{-4}{(p^2 - 1)^2} \cdot 2p$$

$$= \frac{-8p}{(p^2 - 1)^2}$$

When $p = 5$, $q = \frac{4}{5^2 - 1} = \frac{1}{6}$, and

$$\frac{dq}{dp} = \frac{-8 \times 5}{(5^2 - 1)^2} = \frac{-5}{72}$$

$$\therefore \qquad e = -\frac{5}{1/6} \times \frac{-5}{72} = \frac{25}{12} = 2.08$$

Example 5(a): *Find elasticity of demand of the function $p = -2x^2 + 3x + 150$ at $x = 8$*

Solution: Elasticity of demand, $e = -\dfrac{p}{x}\dfrac{dx}{dp}$

$$p = -2x^2 + 3x + 150$$

$$\frac{dp}{dx} = -4x + 3$$

$$\Rightarrow \qquad \frac{dx}{dp} = \frac{1}{-4x+3}$$

When $x = 8$, $p = -2 \times 8^2 + 3 \times 8 + 150 = 46$, and

$$\frac{dx}{dp} = \frac{1}{-4\times 8+3} = \frac{-1}{29}$$

$$\therefore \qquad e = -\frac{46}{8} \times \frac{-1}{29} = 0.198$$

Example 5(b): *Find the price elasticity of demand if the demand curve is given by $x = \dfrac{9}{\sqrt{p}}$ and show that it is a constant.*

Solution: Elasticity of demand $e = -\dfrac{p}{x}\dfrac{dx}{dp}$

$$x = \frac{9}{\sqrt{p}} = 9\ p^{-1/2}$$

$$\Rightarrow \qquad \frac{dx}{dp} = -\ 9/2\ p^{-3/2}$$

$$\therefore \qquad e = -\frac{p}{9p^{-1/2}} \times -\ 9/2\ p^{-3/2} = \frac{1}{2}$$

Since the value of e does not depend on price, it is a constant.

Example 5(c): *The price elasticity of demand of a commodity when price is Rs. 5 and quantity demanded is 20 units is 1.2. Find the demand function assuming that it is linear.*

Solution: Let $q = f(p)$ is the required demand function where $f(p)$ is linear in p.

Given that p = 5 ; q = 20 and $e = -\frac{p}{q}\frac{dq}{dp} = 1.2$

i.e., $-\frac{5}{20}\frac{dq}{dp} = 1.2$

or $\frac{dq}{dp} = -\frac{1.2 \times 20}{5} = -4.8$

Hence the slope of the line $q = f(p)$ at (5, 20) is –4.8

∴ Equation of the line is:

$$q - 20 = -4.8\,(p - 5)$$

i.e., $q = -4.8p + 44$

∴ The demand function is $q = 44 - 4.8p$

Example 5(d): *Show that the elasticity of demand at all points of the demand function in the form of a rectangular hyperbola given by $xy = c^2$ is unity.*

Solution:

$e = -\frac{x}{y}\frac{dy}{dx}$, where y is the quantity demanded at price x.

Now, $xy = c^2 \Rightarrow y = \frac{c^2}{x}$

$\Rightarrow \frac{dy}{dx} = \frac{-c^2}{x^2}$

∴ $e = -\frac{x}{c^2/x}\frac{-c^2}{x^2} = 1$

i.e., elasticity of demand at all points is unity.

Example 5(e): *Find the price elasticity of demand and marginal revenue when price is Rs 3, if the demand function is $q = 15 - p - p^2$.*

Solution: Price elasticity of demand, $e = -\frac{p}{q}\frac{dq}{dp}$

$$q = 15 - p - p^2$$

$\Rightarrow \frac{dq}{dp} = -1-2p$

When p = 3, q = 3 and $\frac{dq}{dp} = -7$

$$\therefore \quad e = -\frac{3}{3} \times -7 = 7$$

Also we know that $e = \frac{AR}{AR - MR}$, where AR is the average revenue which is equal to p and MR is the marginal revenue.

$$\text{i.e.,} \quad 7 = \frac{3}{3\text{-}MR}$$

$$\Rightarrow \quad 3 = 7(3 - MR)$$

$$\text{or} \quad 7MR = 18$$

$$\text{or,} \quad MR = \frac{18}{7} = 2\frac{4}{7}$$

Example 6: *Find the elasticity of supply when price is Rs. 7, for the supply function: $x = -4p + p^2$.*

Solution: Elasticity of supply $e_s = \frac{p}{x}\frac{dx}{dp}$

$$x = -4p + p^2$$

$$\frac{dx}{dp} = -4 + 2p$$

At $p = 7$, $x = -4'7 + 7^2 = 21$ and

$$\frac{dx}{dp} = -4 + 2'7 = 10$$

$$\therefore \quad e_s = \frac{7}{21} \times 10 = 3.33$$

Example 7: *Find elasticity of supply if the supply function is given as: $q = 2p^2 + 3p + 1$, when $p = 4$*

Solution: Elasticity of supply $e_s = \frac{p}{q}\frac{dq}{dp}$

$$q = 2p^2 + 3p + 1$$

$$\frac{dq}{dp} = 4p + 3$$

At $p = 4$, $q = 45$ and $\frac{dq}{dp} = 19$

$$\therefore \quad e_s = \frac{4}{45} \times 19 = 1.69$$

Example 8: *The supply of certain goods is given by* $x = a\sqrt{p-b}$, *where a and b are positive constants. Find the expression for elasticity of supply and determine the price for unit elasticity.*

Solution:

$$e_s = \frac{p}{x}\frac{dx}{dp}$$

$$x = a\sqrt{p-b}$$

$$\frac{dx}{dp} = \frac{a}{2\sqrt{p-b}}$$

$$\therefore \quad e_s = \frac{p}{a\sqrt{p-b}}\frac{a}{2\sqrt{p-b}} = \frac{p}{2(p-b)}$$

For $e_s = 1, \frac{p}{2(p-b)} = 1$

$\Rightarrow 2(p-b) = p$ or, $p = 2b$

$\therefore$ The price for unit elasticity is $p = 2b$

Example 9: *If the demand law is given by . Show that the marginal revenue MR is given by* $\frac{ab}{(q+b)^2} - c.$

Solution: $(p+c)(q+b) = c \Rightarrow (p+c) = \frac{a}{q+b}$

$$\Rightarrow p = \frac{a}{q+b} - c$$

$$\therefore \quad AR = \frac{a}{q+b} - c$$

$$TR = AR \times q = \frac{aq}{q+b} - cq$$

$$MR = \frac{d}{dq}TR = \frac{(q+b).a - aq}{(q+b)^2} - c$$

$$= \frac{aq + ab - aq}{(q+b)^2} - c$$

$$= \frac{ab}{(q+b)^2} - c$$

Example 10: *If the demand law of a commodity is* $p = \dfrac{a}{q^b}$ *; a, b constants, show that elasticity of demand is constant for all values of p and q.*

Solution:

$$p = \frac{a}{q^b} = aq^{-b}$$

$$\Rightarrow \quad \frac{dp}{dq} = a.-b.q^{-b-1}$$

$$\text{Elasticity of demand} = e = \frac{-q}{p}\frac{dp}{dq} = \frac{q}{a/q^b}ab.q^{-b-1}$$

$$= \frac{q}{a}.q^b ab.q^{-b-1} = b\text{, a constant}$$

Example 11(a): *If* p_1, p_2 *are the prices of two commodities* q_1, q_2 *respectively, and the price line is* $M = p_1q_1 + p_2q_2$, *show that* $d^2q_1 = 0$.

Solution: $M = p_1q_1 + p_2q_2$

$$\Rightarrow \quad p_1q_1 = M - p_2q_2$$

$$\Rightarrow \quad q_1 = \frac{M - p_2q_2}{p_1}$$

$$\Rightarrow \quad \frac{dq_1}{dq_2} = \frac{d}{dq_2}\left(\frac{M - p_2q_2}{p_1}\right)$$

$$\Rightarrow \quad dq_1 = \frac{d}{dq_2}\left(\frac{M - p_2q_2}{p_1}\right)dq_2$$

$$= \frac{-p_2}{p_1}dq_2$$

$$\Rightarrow \quad d^2q_1 = \frac{d}{dq_2}\left(\frac{-p_2}{p_1}dq_2\right)dq_2 = (0)\,dq_2 = 0.$$

Example 11(b): *The distance traveled by a particle in t seconds is given by,*

$$s = a\,e^{t} + b\,e^{t}$$

Show that the acceleration is always equal to the distance passed over.

Solution: We know that velocity of the body is the rate of change of its displacement and acceleration of a moving body is the rate of change of its velocity.

That is if s is the distance traveled by a particle in time t, then

velocity , $v = \dfrac{ds}{dt}$,

and acceleration, $a = \dfrac{dv}{dt}\left(= \dfrac{d^2s}{dt^2}\right)$

Here, $s = a\,e^{t} + b\,e^{-t}$

$$\frac{ds}{dt} = a\,e^{t} - b\,e^{-t}$$

$$\frac{d^2s}{dt^2} = a\,e^{t} + b\,e^{-t} = s$$

i.e., the acceleration is always equal to the distance passed over.

Example 11(c) *Find out MC if TC = 5 x^2 + 2 x –1, when x = 3*

Solution: TC = 5 x^2 + 2 x –1

$$MC = \frac{d(TC)}{dx} = 10\,x + 2 = 32 \text{ when } x = 3$$

Example 12 *Find out AR and MR if TR = –3q^2 + 30q –12, when q=4*

Solution: TR = –3q^2 +30q –12

$$AR = \frac{TR}{q}$$

$$= -3q + 30 - \frac{12}{q} = 15 \text{ when } q = 4$$

$$MR = \frac{d(TR)}{dq}$$

$$= -6q + 30 = 6 \text{ when } x = 3$$

Example 13: *A manufacturer produces spherical ice–crème bowls with radius 10 cms. If there is a possibility of an error of 0.1cm in its measurement, what will be the error in the volume?*

Solution: We know that volume V of a sphere is related to its radius r by the formula,

$$V = \frac{4}{3}\pi\,r^3$$

Taking differentials on both sides,

$$dV = \frac{4}{3}\pi .3r^2\ dr = 4\pi\ r^2\ dr$$

Since dr = 0.1 ,when r = 10,

Error in the volume = dV

$$= 4\pi\ 10^2\ (0.1) = 125.7\ cm^3$$

For linear approximation of a non–linear function $y = f(x)$ when x is very close to a value a we assume $x–a$ = dx. Hence, the value $f(x)– f(a)$ will approximately be equal to $f'(a)\ (x - a\ )$,

or, $f(x)$ is approximately equal to $f(a) + f'(a)\ (x - a)$

i.e., the non–linear function $f(x)$ can be approximated by the linear function:

$$f(a) + f'(a)\ (x - a).$$

Example 14(a) *Approximate the function* $f(x) = x^2 + 2x - 3$ *by a linear function when x is very close to 2.*

Solution:

Given $f(x) = x^2 + 2x - 3,$

$$a = 2 \Rightarrow f(a) = 5$$

$$f'(x) = 2x + 2,$$

$$f'(a) = 6$$

We know that,

$$f(x) \approx f(a) + f'(a)\ (x - a\ )$$

$$f(x) \approx 5 + 6\ (x - 2) = 6x - 7$$

Remark:

Note that $f(x) = 6x - 7$ is the tangent to the curve $f(x) = x^2 + 2x - 3$ at $x = 2$. Thus the linear approximation of a non–linear function $y = f(x)$ when x is very close to a value a is the tangent to the curve $y = f(x)$ at $x = a$.

Example 14(b): *A particle moves along a straight line such that its distance* s_1 *at any time t is given by* $s_1 = t^3 + 3t^2 + 5t$. *Another particle moves such that its distance* s_2 *at any time t is given by* $s_2 = 3\ t^3 - 3t^2 - 7t + 8$. *Find their positions and speeds when their accelerations are equal.*

Solution: Let v_1 and v_2 are the velocities and a_1 and a_2 are the accelerations respectively of the two particles at time t.

Given, $s_1 = t^3 + 3t^2 + 5t$,

$$s_2 = 3t^3 - 3t^2 - 7t + 8$$

$$v_1 = \frac{ds_1}{dt} = 3t^2 + 6t + 5 \, ;$$

$$v_2 = \frac{ds_2}{dt} = 9t^2 - 6t - 7$$

$$a_1 = \frac{d^2s_1}{dt^2} = 6t + 6;$$

$$a_2 = \frac{d^2s_2}{dt^2} = 18t - 6$$

$$a_1 = a_2 \Rightarrow 6t + 6 = 18t - 6$$

$$\Rightarrow t = 1.$$

When t = 1,

$$s_1 = 1 + 3 + 5 = 9,$$

$$s_2 = 3 - 3 - 7 + 8 = 1,$$

$$v_1 = 3 + 6 + 5 = 14,$$

$$v_2 = 9 - 6 - 7 = -4$$

∴ The distance traveled by the first particle is 9 units and by the second particle is 1 unit. Also, the speed of the first particle is 14 units and the second particle is (–1) unit.

Example 14(c) *Find the slope of the tangent to the curve* $y = x^3 + 3x^2 + 2x - 3$ *at* $x = 1$.

Solution: Given curve is $y = x^3 + 3x^2 + 2x - 3$

Differentiating with respect to x,

$$\frac{dy}{dx} = 3x^2 + 6x + 2$$

Slope of the tangent to the curve at $(x = 1) = \left(\frac{dy}{dx}\right)_{x=1} = 11.$

Example 15: *If the area of a circle increases at a uniform rate, show that the rate of increase of its circumference varies inversely as the radius.*

Solution: Let r be the radius, A be the area and S be the circumference of the circle at time t. Since area increases at a uniform rate,

$$\frac{dA}{dt} = k, \text{ where k is a constant.}$$

We have, $A = \pi r^2 \Rightarrow \frac{dA}{dt} = 2\pi r$

$$\Rightarrow 2\pi r \frac{dr}{dt} = k$$

$$\Rightarrow \frac{dr}{dt} = \frac{k}{2\pi r} \qquad ...(1)$$

Now, $S = 2\pi r \Rightarrow \frac{dS}{dt} = 2\pi \frac{dr}{dt} = 2\pi \frac{k}{2\pi r}$ from (1)

$$\Rightarrow \frac{dS}{dt} = \frac{k}{r}$$

i.e., rate of increase of its circumference varies inversely as its radius.

Example 16(a): *A man 1.8 meters height walks at the rate of 7.2 km/hr from a lamp-post 6 meters high. Find the rate at which the length of his shadow increases.*

Solution: Let the man be x meters away from the lamp–post at time t. Let the length of his shadow at that time be y meters.

Let AB be the position of the man at time t (see figure 5.5). Let PQ be the lamp–post. Then AO is the shadow of the man so that AO = y and AP = x. In figure, OAB and OPQ are similar triangles.

Therefore, $\frac{OA}{OP} = \frac{AB}{PQ}$.

i.e., $\frac{y}{y+x} = \frac{1.8}{6} = \frac{3}{10} \Rightarrow 10y = 3(y + x)$

$$\Rightarrow y = \frac{3}{7}x$$

Differentiating with respect to t, $\frac{dy}{dt} = \frac{3}{7}\frac{dx}{dt}$ ...(1)

Given that $\frac{dx}{dt} = 7.2$ km/hr $= \frac{7200}{3600} = 2$m/sec

Then (1) becomes, $\frac{dy}{dt} = \frac{3}{7} \times 2 = \frac{6}{7}$

∴ The rate at which the length of his shadow increases 6/7 = m/sec.

Example 16(b): *Water is poured into an inverted conical vessel at the rate of 10 cc/min. if the radius of the base is 10 cm and height is 15 cm, find the rate at which the water level rising at the instant when the depth is 5 cm.*

Solution: Let V be the volume, r be the base radius and h be the height of the cone formed by the water in the conical vessel at any time t. Given that $\frac{dV}{dt} = 10$ cc/min.

$$\frac{h}{r} = \frac{15}{10} \Rightarrow r = \frac{2}{3}h \qquad \ldots(1)$$

We know that, $V = \frac{1}{3}\pi r^2 h = \frac{1}{3}\pi(\frac{2}{3}h)^2 h = \frac{4}{27}\pi h^3$, from (1)

$$\therefore \qquad \frac{dV}{dt} = \frac{12}{27} = \pi h^2 \frac{dh}{dt} \Rightarrow 10 = \frac{4}{9}\pi h^2 \frac{dh}{dt}$$

$$\Rightarrow \qquad \frac{dh}{dt} = \frac{90}{4\pi h^2} = \frac{90}{4\pi \times 5^2} = \frac{9}{10\pi} \text{ when } h = 5 \text{ cm.}$$

$\therefore$ The rate at which the water level rising at the instant when the depth is 5 cm = $\frac{9}{10\pi}$ cm/min.

Example 16(c): *The diameter and altitude of a can in the shape of a right circular cylinder are measured as 4cm and 6cm respectively. The possible error in each measurement is 0.1cm. Find the possible error in the compounded volume.*

Solution: Let V be the volume, r be the base radius and h be the height of the right circular cylinder.

Given that, r = 2cm and h = 6cm

We know that,

$$V = \pi r^2 h$$

Taking differentials on both sides,

$$dV = \pi r^2 \, dh + 2\pi rh \, dr$$

Since dr = dh = 0.1,

Error in the volume = dV

$$= \pi \times 2^2 \times .1 + 2\pi \times 2 \times 6 \times .1 = 2.8\pi \text{ cc.}$$

Example 16(d): *If q is calculated from the formula* $q = k r^2 \sqrt{h}$ *; where k is a constant, show that a small percentage error in r is four times as serious as the same percentage error in h.*

Solution: [If dx is the error in calculating x, then $\frac{dx}{x}$ is called the relative error and $\frac{dx}{x} \times 100$ is called the percentage error in calculating x.]

Given that $q = k\, r^2 \sqrt{h}$

Taking logarithm through out,

$$\log q = \log k + 2 \log r + ½ \log h$$

Taking differentials,

$$\frac{dq}{q} = 0 + \frac{2dr}{r} + \frac{dh}{2h}$$

Multiplying through out by 100,

$$\frac{dq}{q} \times 100 = \frac{2dr}{r} \times 100 + \frac{dh}{2h} \times 100 \qquad \ldots(1)$$

Suppose that a small percentage error e is occurred in the measurement of r alone.

i.e., $\frac{dr}{r} \times 100 = \varepsilon$, and $dh = 0$

Then (1) becomes,

$$\frac{dq}{q} \times 100 = 2\varepsilon + 0 = 2\varepsilon.$$

Now suppose that the same small percentage error e is occurred in the measurement of h alone.

i.e., $\frac{dh}{h} \times 100 = \varepsilon$, and $dr = 0$

Then (1) becomes,

$$\frac{dq}{q} \times 100 = 0 + \frac{\varepsilon}{2} = \frac{\varepsilon}{2}.$$

Hence a small percentage error in r is four times as serious as the same percentage error in h.

Example 17: *If x is small show that* $\frac{1}{1+3x}$ *can be approximated by* $1 - 3x$.

Solution: We know that the non–linear function $f(x)$ can be approximated by the linear function:

$$f(a) + f'(a)\,(x - a)$$

When x is small,

$$f(x) \approx f(0) + x\, f'(0)$$

Here, $f(x) = \dfrac{1}{1+3x}$; $f(0) = 1$

$$f'(x) = \frac{-3}{(1+3x)^2}\ ;\ f'(0) = -3$$

$$\therefore \qquad f(x) \approx 1 + x \times -3 = 1-3x$$

Example 18: *If x is nearly equal to 1, show that* $f(x) = \dfrac{1}{x^2}$ *can be approximated by 3–2x.*

Solution: When x is nearly equal to 1, $f(x) \approx f(1) + (x-1)f'(1)$.

Here, $f(x) = \dfrac{1}{x^2}$; $f(1) = 1$

$$f'(x) = \frac{-2}{x^3}\ ;\ f'(1) = -2$$

$$\therefore \qquad f(x) \approx 1 + (x-1) \times -2 = 3 - 2x.$$

Example 19(a): *Find the slope of the tangent to the curve:*

$$y^2 = \frac{x^3}{4-x} \text{ at the point } (2, -2).$$

Solution: Given curve is $y^2 = \dfrac{x^3}{4-x}$

Differentiating with respect to x,

$$2y\frac{dy}{dx} = \frac{(4-x)\,3x^2 - x^3(-1)}{(4-x)^2}$$

$$= \frac{12x^2 - 2x^3}{(4-x)^2}$$

$$\Rightarrow \qquad \frac{dy}{dx} = \frac{6x^2 - x^3}{y(4-x)^2}$$

$\therefore$ Slope of the tangent to the curve at

$$(2, -2) = \left(\frac{dy}{dx}\right)_{(2,-2)} = -2$$

Example 19(b): *Find the range of values of x for which the function:*

$y = x^3 - 6x^2 - 36x + 7$ increases with x.

Solution: Given $y = x^3 - 6x^2 - 36x + 7$;

$$\frac{dy}{dx} = 3x^2 - 12x - 36 = 3(x-6)(x+2)$$

$$\frac{dy}{dx} > 0 \Rightarrow x > 6 \text{ or } x < -2.$$

Hence the function is increasing for all values of $x > 6$ and $x < -2$.

Example 19(c): *Show that $(x-1)\,e^x + 1$; $x > 0$, is strictly increasing.*

Solution:

$$f(x) = (x-1)\,e^x + 1;\ x > 0$$

$$f'(x) = (x-1)\,e^x + e^x = x\,e^x > 0 \text{ since } x > 0\ .$$

$\therefore$ $f(x)$ is strictly increasing.

Example 20: *Show that $y = 2x^2 - 1/3\ x^3 - 6x - 2$ is strictly decreasing.*

Solution:

$$y = 2x^2 - 1/3\ x^3 - 6x - 2$$

$$\Rightarrow \frac{dy}{dx} = 4x - x^2 - 6 = -(x^2 - 4x + 6)$$

$$= -[(x-2)^2 + 2] < 0; \text{ since } (x-2)^2 + 2 > 0$$

$\therefore$ y is strictly decreasing.

Example 21: *Find the greatest value of k such that the function:*

$y = 3x^2 - k\,x - 2$ is increasing in $[0, 1]$.

Solution:

$$y = 3x^2 - k\,x - 2$$

$$\Rightarrow \frac{dy}{dx} = 6x - k \text{ and } \frac{d^2y}{dx^2} = 6.$$

Since $\frac{d^2y}{dx^2} > 0$, we get $\frac{dy}{dx}$ as a strictly increasing function in [0, 1].

Hence the least value of $\frac{dy}{dx}$ is at $x = 0$.

i.e., the least value of $\frac{dy}{dx} = -k$.

Also $\frac{dy}{dx} \geq 0\ \forall\ x \in [0, 1]$, since y is given to be increasing in [0, 1].

Therefore the least value of $\frac{dy}{dx}$ is also ≥ 0.

i.e., $-k \geq 0$; or $k \leq 0$

Hence, the greatest value of k such that y is increasing in [0, 1] is 0.

Example 22: *Test whether $y = 4x^3 - 3x^2 - 2x - 1$ is concave or convex at $x = 3$.*

Solution: $y = 4x^3 - 3x^2 - 2x - 1$;

$$\frac{dy}{dx} = 12x^2 - 6x - 2 ;$$

$$\frac{d^2y}{dx^2} = 24x - 6$$

$$\left(\frac{d^2y}{dx^2}\right)_{x=3} = 66 > 0$$

$\Rightarrow$ the function is convex at $x = 3$.

Example 23: *Show that the curve $y = 2x - 3 + 1/x$ is convex from below for all positive values of x.*

Solution:

$$y = 2x - 3 + \frac{1}{x};$$

$$\frac{dy}{dx} = 2 - \frac{1}{x^2};$$

$$\frac{d^2y}{dx^2} = \frac{2}{x^3} \geq 0 \text{ , when } x \geq 0.$$

$\therefore$ The curve is convex from below for all positive values of x.

Example 24: *Find the points of inflection, if any, of $y = x^4 - 6x^2$*

Solution:

$$y = x^4 - 6x^2$$

$$\frac{dy}{dx} = 4x^3 - 12x$$

$$\frac{d^2y}{dx^2} = 12x^2 - 12$$

$$\frac{d^2y}{dx^2} = 0 \quad \Rightarrow 12x^2 - 12 = 0 \Rightarrow x^2 - 1 = 0$$

$$\Rightarrow (x - 1)(x + 1) = 0$$

$$\Rightarrow x = 1, x = -1$$

Also, $\left(\frac{d^3y}{dx^3}\right)_{x=1} \neq 0$ and $\left(\frac{d^3y}{dx^3}\right)_{x=-1} \neq 0$

when $x = 1, y = 1-6 = -5$;

when $x = -1, y = 1-6 = -5$

$\therefore$ The points of inflexion of the function are (1, –5) and (–1, –5).

Example 25: *Find whether the average cost function $y = x^2 - 4x + 5$ increases or decreases at each of the points; x = 1, 2 and 3.*

Solution:

Given $y = x^2 - 4x + 5$

Differentiating with respect to x,

$$\frac{dy}{dx} = 2x - 4$$

$\left(\frac{dy}{dx}\right)_{x=1} = -2 < 0 \Rightarrow$ the cost function is decreasing at $x = 1$.

$\left(\frac{dy}{dx}\right)_{x=2} = 0$

$\Rightarrow$ the cost function is neither decreasing nor increasing at $x = 2$.

$\left(\frac{dy}{dx}\right)_{x=3} = +2 > 0 \Rightarrow$ the cost function is increasing at $x = 3$.

Example 26: *Show that the curve $y = x^3 - 3x^2$ is concave at $x = 1/2$, and convex at $x = 4$.*

Solution:

Given $y = x^3 - 3x^2 \Rightarrow \frac{dy}{dx} = 3x^2 - 6x \frac{d^2y}{dx^2} = 6x - 6$

$$\left(\frac{d^2y}{dx^2}\right)_{x=\frac{1}{2}} = -3 < 0$$

$\Rightarrow$ the curve is concave at $x = \frac{1}{2}$

$$\left(\frac{d^2y}{dx^2}\right)_{x=4} = 18 > 0$$

$\Rightarrow$ the curve is convex at $x = 4$

EXERCISES

1. Show that the function $y = x^{1/3}$ is convex when $x < 0$ and concave when $x > 0$. Find its point of inflexion.
2. The radius of a sphere is found to be 10cms., with a possible error of 0.02 cms. What is the relative error in the computed volume?
3. Find the range of values of a for which the function:

   $y = x^3 - 2x^2 + ax + 9$ increases.
4. Show that the function $y = -2x^3 + 4x^2 + 9x - 15$ is concave at $x = 3$.
5. The height of a right circular cylinder is exactly equal to the diameter of the base. The height is measured as 12 cms, with a possible error of 0.1 cm. Find the approximate possible error in the calculated volume.
6. Show that the average rate of change of the function $y = \frac{3x+2}{x+1}$ as x increases from 2 to 2.1 is 10/3.
7. Show that the slope of the curve y = b at the point where it crosses the y–axis is $\frac{-b}{a}$.

   [Hint: the curve crosses the y–axis at the point where $x = 0$; i.e., at (0, b)]
8. A body is projected vertically upwards and travels a distance of s cms. Above the point of projection in time t seconds, where $s = 140t - 10t^2$. When does the velocity vanish? What is the greatest height attained by the body?
9. The volume of a sphere is increasing at the rate of 30 cubic cms per second. Find the rate of change of the surface area at the instant when the radius is 12 cms.
10. Show that the function $y = x^3 - 7x^2 + 6x - 2$ decreases at $x = 4$
11. Show that the function $y = (x - 1)\, e^x + 1$ is increasing for all values of x.
12. Show that for all x less than k, the function $y = x^2 - 2kx + 1$ decreases.
13. Show that the function $y = 2x^3 - 3x^2 + 5x - 10$ is convex when $x > ½$ and concave when $x < ½$. What happens when $x = ½$?
14. If the demand function is given by q = 36 – 2p, find the marginal revenue for q = 10 units.

15. A rod 20 meters long has one end on the ground and the other in contact with a vertical wall. The lower end slips along the ground. Show that when the foot of the rod is 16 meters away from the wall, the upper end is moving 4/3 times as fast as the lower end.

16. An inverted cone has a depth of 10cms. and the base radius 5cms. Water is poured into it at the rate of 1.5 c.c per minute. Find the rate at which the level of the water in the cone is rising when the depth of water in the cone is 4cms.

17. A man is walking at the rate of 6.5km/hr. towards the foot of a tower 120m high. At what rate is he approaching the top of the tower when he is 50m away from the tower?

18. Show that $y = \frac{1}{x^2}$ can be approximated as $3 - 2x$ when x is nearly equal to 1. What percentage error is involved in the use o f the expression when $x = 1.1$?

19. Find the slope of the curve $y = 2x^2 - 3x + 5$ at the point where its x–coordinate is 2.

20. Find the percentage error in calculating the volume of a cubical box if an error of 1% is made in measuring the length of edges of the cube?

21. Find the slope of the tangent to the parabola $y^2 = 4ax$ at the point $(at^2, 2at)$

22. Find the slope of the tangent to the graph $f(x) = 3x^2 - 5x + 7$ at the point (2, 9). Also find the equation of the tangent through the point.

3

Maxima and Minima

INTRODUCTION

Let f(x, y, z, ...) be any function of several independent variables z, y, z... supposed to be continuous for all values of these variables in the neighbourhood of their values a, b, c,... respectively. Then (a, b, c,...) is said to be a *maximum or a minimum* value of f(x, y, z,...) for all sufficiently small independent values of h, k, *l*..., positive or negative, provided they are not all zero. We use the method of finding maxima and minima in these type of problems. For maximizing a firm's profit, or for maximizing a consumer's utility, or for minimizing the cost of production we use the method of maxima and minima. The most common problems arise in Economics concern with the choice of best or the optimal. This in turn warrants determining the maximum or minimum possible solution of a problem. A rational consumer thinks always to maximize his utility, while a prudent producer aims to attain maximum profit or minimum cost.

FIRST AND SECOND ORDER CONDITIONS FOR EXTREME VALUES

From the above figures, it is clear that a function has a maximum at a point if its curve is concave from below and has a minimum at a point if its curve is convex from below; and consequently,

i. a sufficient condition for maxima is $\frac{dy}{dx} = 0$ and $\frac{d^2y}{dx^2} < 0$

ii. and for minima is $\frac{dy}{dx} = 0$ and $\frac{d^2y}{dx^2} > 0$ at the point

In Economics, the conditions $\frac{dy}{dx} = 0$ and $\frac{d^2y}{dx^2} < 0$ (or $\frac{d^2y}{dx^2} > 0$) are usually called as the *first–order condition* and *second order condition* respectively for the occurrence of extreme values. Problems such as maximum utility, maximum profits, minimum cost may be effectively solved with these conditions.

Remark 1:

The set of conditions for the function $y = f(x)$ to attain a maximum at $x = a$ are:

(i) $\frac{dy}{dx} = 0$ at $x = a$; and

(ii) $\frac{d^2y}{dx^2} < 0$ at $x = a$

Similarly, the set of conditions for the function $y = f(x)$ to attain a minimum at $x = a$ are:

(i) $\frac{dy}{dx} = 0$ at $x = a$; and

(ii) $\frac{d^2y}{dx^2} > 0$ at $x = a$

Remark 2:

For a function $y = f(x)$, if both $\frac{dy}{dx}$ and $\frac{d^2y}{dx^2}$ are zeros at $x = a$, then the point $x = a$ may be a point of inflexion of the curve; provided $\frac{d^2y}{dx^2} \neq 0$.

MAXIMA AND MINIMA SUBJECT TO CONSTRAINTS

Suppose we have to find the maximum or minimum of a function,

$$u = f(x, y) \quad \text{...(1)}$$

where x and y are connected by an equation:

$$g(x, y) = 0 \quad \text{...(2)}$$

The subsidiary equation $g(x, y) = 0$ is called the *constraint*. Applying the total differential formula in (1) we get,

$$du = f_x\, dx + f_y\, dy = 0 \quad \text{...(3)}$$

which gives a necessary condition for the occurrence of an extreme point for $u = f(x, y)$. From (2) we get,

$$g_x\, dx + g_y\, dy = 0 \quad \text{...(4)}$$

Assuming that at the extreme points, g_y 0; multiply (4) by $\frac{f_y}{g_y}$, we get,

$$\frac{f_y}{g_y} g_x\, dx + f_y\, dy = 0 \qquad \ldots(5)$$

Substituting (5) in (3) we get,

$$du = f_x\, dx - \frac{f_y}{g_y} g_x\, dx = 0$$

i.e.,
$$du = (f_x - \frac{f_y}{g_y} g_x)dx = 0$$

i.e.,
$$f_x - \frac{f_y}{g_y} g_x = 0$$, since dx is an independent variable

i.e.,
$$\frac{f_x}{g_x} = \frac{f_y}{g_y} \qquad \ldots(6)$$

(6) is the necessary condition for the occurrence of an extreme value, when there is given a constraint. Now, let each ratio in (6) to be equals some constants, say $(-\lambda)$.

i.e.,
$$\frac{f_x}{g_x} = \frac{f_y}{g_y} = -\lambda$$

Hence we get,

$$f_x + g_x \;= 0 \qquad \ldots(A)$$

$$f_y + g_y \;= 0 \qquad \ldots(B)$$

Solving (A)& (B) together with (2) gives us the extreme value.

In some cases we can also solve these kinds of problems by substituting the subsidiary equation in the function, even though it is not feasible always. This reduces the problem to one involving unconstrained function, and then it can be done by the method described in the previous section.

NECESSARY CONDITIONS FOR THE EXISTENCE OF MAXIMA OR MINIMA

From the definition it is obvious that we shall have a maximum or a minimum of f(x, y, z,...) for those values of x, y, z,... for which the expression f(x + h, y + k, z + *l*, ...) – f(x, y, z, ...) is of *invariable sign* for all sufficiently small independent values of h, k, *l*, ... provided they are not all equal to zero. There will be a maximum or a minimum according as this negative or positive.

Expanding by Taylor's Theorem for several variables, we have

$$f(x + h, y + k, z + l,...)$$

$$\left[1+\frac{1}{1!}\left(h\frac{\partial}{\partial x}+k\frac{\partial}{\partial y}+l\frac{\partial}{\partial z}+...\right)+\frac{1}{2!}\left(h\frac{\partial}{\partial x}+k\frac{\partial}{\partial y}+l\frac{\partial}{\partial x}+...\right)^2\right]f(x,y,z,...)$$

$\therefore$ f(x + h, y + k, z + l, ...) − f(x, y, z)

$$=\left(h\frac{\partial f}{\partial x}+k\frac{\partial f}{\partial y}+l\frac{\partial f}{\partial z}+...\right) + \text{terms of the second and}$$

higher orders in h, k, l,... . ...(1)

Now by taking h, k, l,... sufficiently small, the first degree terms in h, k, l,... can be made to govern the sign of the right hand side and therefore of the left hand side of (1). But $h\frac{\partial f}{\partial x}+k\frac{\partial f}{\partial y}+l\frac{\partial f}{\partial z}+...$ is not equal to zero, the sign of this expression will change by changing the sign of each of h, k, l, ... Hence as a necessary condition for the occurrence of a maximum or a minimum or f(x, y, z, ...), we must have

$$h\frac{\partial f}{\partial x}+k\frac{\partial f}{\partial y}+l\frac{\partial f}{\partial z}+...=0. \qquad ...(2)$$

Since (2) is true whatever be the values of h, k, l,... independent of each other, we must have as a necessary consequence

$$\frac{\partial f}{\partial x}=0, \frac{\partial f}{\partial y}=0, \frac{\partial f}{\partial z}=0,...$$

If there are n independent variables, we have then obtained n simultaneous equations to give us the values a, b, c,... of the n variables x, y, z,... for which f(x, y, z,...) may have a maximum or minimum value.

The conditions $\frac{\partial f}{\partial x}=0, \frac{\partial f}{\partial y}=0, \frac{\partial f}{\partial z}=0$,... *are necessary but not sufficient for the existence of maxima and minima.*

STATIONARY AND EXTREME POINTS

A point $(a_1, a_2, ..., a_n)$ is called a **stationary point,** if all the first order partial derivatives of the function $f(x_1, x_2, ..., x_n)$ vanish at that point. Also then the value of the function $f(x_1, x_2,..., x_n)$ is said to be stationary at that point. A stationary point which is either a maximum or a minimum is called an **extreme point** and the value of the function at that point is called an **extreme value.** A stationary point is not necessarily an extreme point. Thus, a stationary value may be a maximum or a minimum or neither of these two. To decide whether a stationary point is really an extreme point, a further investigation is required.

LAGRANGE'S NECESSARY AND SUFFICIENT CONDITION FOR THE MAXIMA OR MINIMA OF A FUNCTION OF THREE INDEPENDENT VARIABLES

Necessary Conditions: Let f(x, y, z) be a function of three independent variables x, y and z. Then as derived in for f(x, y, z) to be a maximum or a minimum at any point (a, b, c), it is necessary that

$$\frac{\partial f}{\partial x}=0, \frac{\partial f}{\partial y}=0 \text{ and } \frac{\partial f}{\partial z}=0$$

at that point.

Hence, the points where the value of the function f(x, y, z) is stationary (*i.e.*, may be a maximum or a minimum) are obtained by solving the simultaneous equations

$$\frac{\partial f}{\partial x}=0, \frac{\partial f}{\partial y}=0, \frac{\partial f}{\partial z}=0 .$$

Sufficient Conditions: Before deriving the sufficient conditions for the existence of a maximum or a minimum of a function of three independent variables, we obtain the following two algebraic lemmas regarding the signs of quadratic expressions.

Lemma 1: Let I_2 $ax^2 + 2hxy + by^2$ be a quadratic expression in two variables x and y. We can write

$$I_2 = \frac{1}{a}\left[a^2x^2 + 2ahxy + aby^2\right], \text{ if } a \neq 0$$

$$= \frac{1}{a}\left[(ax+hy)^2 + \left(ab-h^2\right)y^2\right].$$

The expression within the square brackets will be positive if $ab - h^2$ is positive and in that case the sign of the expression I_2 will be the same as that of a.

In case $ab - h^2$ is not positive, we can say nothing about the sign of the expression within the square brackets and hence nothing about the sign of the given quadratic expression I_2.

Lemma 2: In three variables x, y and z,

$$I_3 = ax^2 + by^2 + cz^2 + 2fyz + 2gzx + 2hxy$$

$$= \frac{1}{a}[a^2x^2 + aby^2 - acz^2 + 2fayz + 2gazx + 2haxy], \text{ if } a \neq 0$$

$$= \frac{1}{a}[a^2x^2 + 2\,ax\,(gz + hy) + aby^2 + acz^2 + 2fayz]$$

$$= \frac{1}{a}[(ax + hy + gz)^2 + aby^2 + acz^2 + 2fayz - (gz - hy)^2]$$

$$= \frac{1}{a}[(ax + hy + gz)^2 + (ab - h^2)\,y^2 + 2yz\,(fa - gh) + (ac - g^2)z^2].$$

Now, I_3 will be of the same sign as a provided the expression within the square brackets is positive which will of course be so if

$$ab - h^2 \text{ and } \{(ab - h^2)(ac - g^2) - (fa - gh)^2\}$$

are both positive *i.e.*, if

$$ab - h^2 \text{ and } a\,(abc + 2fgh - af^2 - bg^2 - ch^2)$$

are both positive.

This I_3 will be positive if

$$a, \begin{vmatrix} a & h \\ h & b \end{vmatrix}, \begin{vmatrix} a & h & g \\ h & b & f \\ g & f & c \end{vmatrix}$$

be all positive and will be negative if these three expression are alternately negative and positive.

Now, we are in a position to derive Lagrange's sufficient conditions for the existence of a maximum or a minimum of a function of three independent variables at a stationary point.

Let a set of the values of x, y, z obtained by solving the equations

$$\frac{\partial f}{\partial x} = \frac{\partial f}{\partial y} = \frac{\partial f}{\partial z} = 0$$

be a, b, c.

Let the values of the six second order partial derivatives

$$\frac{\partial^2 f}{\partial x^2}, \frac{\partial^2 f}{\partial y^2}, \frac{\partial^2 f}{\partial z^2}, \frac{\partial^2 f}{\partial y\,\partial z}, \frac{\partial^2 f}{\partial z\,\partial x} \text{ and } \frac{\partial^2 f}{\partial x\,\partial y}$$

at the point (a, b, c) be denoted by A, B, C, F, G and H respectively.

Then, we have

$$f(a + h, b + k, c + l) - f(a, b, c)$$

$$= \frac{1}{2!}\,(Ah^2 + Bk^2 + Cl^2 + 2Fkl + 2Glh + 2Hhk) + R_3, \qquad ...(1)$$

where R_3 consists of terms of third and higher orders of small quantities h, k and l. By taking h, k and l sufficiently small, the second degree terms in h, k and l can be made to govern the sign of the right hand side and therefore of the left hand side of (1). If this group of terms forms an expression of invariable sign for all such values of h, k and l, we shall have a maximum or a minimum value of f(x, y, z) at (a, b, c) according as that sign is negative or positive.

Hence by our lemma 2, if the expressions

$$A, \begin{vmatrix} A & H \\ H & B \end{vmatrix}, \begin{vmatrix} A & H & G \\ H & B & F \\ G & F & C \end{vmatrix}$$

be all positive, we shall have a minimum of f(x, y, z) at (a, b, c) and if these expressions be alternately negative and positive, we shall have a maximum of f(x, y, z) at (a, b, c), whilst if these conditions are not satisfied, we shall in general have neither a maximum nor a minimum of f(x, y, z) at (a, b, c).

WORKING RULE FOR FINDING THE MAXIMA AND MINIMA OF A FUNCTION THREE INDEPENDENT VARIABLES

Suppose f(x, y, z) is a given function of three independent variables x, y and z. Find $\partial f/\partial x$, $\partial f/\partial y$ and $\partial f/\partial z$ and solve the simultaneous equations $\partial f/\partial x = 0$, $\partial f/\partial y = 0$ and $\partial f/\partial z = 0$. All the triads (a, b, c) of the values of x, y and z obtained on solving these equations will give the stationary values of f(x, y, z) *i.e.*, will give the points at which the function f(x, y, z) may be a maximum or a minimum.

To discuss the maximum or minimum of f(x, y, z) at any point (a, b, c) obtained on solving the equations $\partial f/\partial x = 0$, $\partial f/\partial y = 0$ and $\partial f/\partial z = 0$, we find the values at this point of the six partial derivatives of second order of f(x, y, z) symbolically denoted as follows:

$$A = \frac{\partial^2 f}{\partial x^2}, B = \frac{\partial^2 f}{\partial y^2}, C = \frac{\partial^2 f}{\partial z^2}, F = \frac{\partial^2 f}{\partial y \partial z}, G = \frac{\partial^2 f}{\partial z \partial x} \text{ and } H = \frac{\partial^2 f}{\partial x \partial y}.$$

If the expressions

$$A, \begin{vmatrix} A & H \\ H & B \end{vmatrix}, \begin{vmatrix} A & H & G \\ H & B & F \\ G & F & C \end{vmatrix}$$

be all positive, we shall have a minimum of f(x, y, z) at (a, b, c) and if these expressions be alternately negative and positive, we shall have a maximum of f(x, y, z) at (a, b, c), whilst if these conditions are not satisfied, we shall in general have neither a maximum nor a minimum of f(x, y, z) at (a, b, c).

CONDITIONS FOR MAXIMUM OR MINIMUM

Consider a function $u = f(x, y)$ of two independent variables x and y. instead of getting a curve as in functions of one independent variable, here we get a surface as the graph of the function. The function is said to have a *relative maximum* value at $(x = a, y = b)$, if $f(a, b) > f(x, y)$ for all values x and y close to $(x = a, y = b)$; and a *relative minimum* value at the point, if $f(a, b) < f(x, y)$ for all values x and y close to $(x = a, y = b)$. If $f(x, y)$ has a maximum at $(x = a, y = b)$, it follows that the function $f(x, b)$ has a maximum at $x = a$ and $f(a, y)$ has a minimum at $x = b$.

Thus for the function $f(x, y)$ to have maximum at $(x = a, y = b)$, it is necessary to have the condition:

$$\frac{\partial}{\partial x} f(x,y) = \frac{\partial}{\partial y} f(x,y) = 0 \text{ at } (x = a,\ y = b)$$

Similar is the case with minimum also. This condition indicates only that at the point (a, b) the function is neither increasing, nor decreasing with respect to the principal axes, but is at a relative plateau. This point is called a *critical point* of the function.

As in the one-variable case, the first derivative condition gives only a necessary condition for an extreme value. i.e., at an extreme point the function will have its first partial derivatives zero, but being the first partial derivatives of a function at a point zero, we can not conclude that the point is an extreme point. To determine whether the critical point is in fact a point of maximum or minimum, it should satisfy certain second order conditions. The second order direct partial derivatives must both be negative for a maximum and positive for a minimum, when evaluated at the critical points.

i.e.,f_{xx} , f_{yy} <0, for maximum; and

f_{xx} , f_{yy} >0, for minimum

This condition ensures that the function is moving downward from the critical point in case of a maximum and upward in case of a minimum. More over, the product of the second order direct partial derivatives at the critical points should exceed the product of the cross partial derivatives at the critical points.

i.e.,$f_{xx} \cdot f_{yy} > f_{xy} \cdot f_{yx}$ $(= f_{xy}{}^2$, by Young's Theorem)

These conditions can be summarized as:

	Maximum	Minimum
i.	$f_x = 0, \quad f_y = 0$	$f_x = 0, \quad f_y = 0$
ii.	$f_{xx}, \quad f_{yy} < 0$	$f_{xx}, \quad f_{yy} > 0$
iii.	$f_{xx} \cdot f_{yy} > f_{xy}^2$	$f_{xx} \cdot f_{yy} > f_{xy}^2$

LAGRANGE'S METHOD

Let $u = f(x_1, x_2, \ldots, x_n)$

be a function of n variables $x_1, x_2, \ldots, x_n$. Let these variables be connected by m equations

$\phi_1(x_1, x_2, \ldots, x_n) = 0, \phi_2(x_1, x_2, \ldots, x_n) = 0, \ldots, \phi_m(x_1, x_2, \ldots x_n) = 0$

so that only n − m of the n variables are independent.

For a maximum or a minimum of u, we have

$$du = \frac{\partial u}{\partial x_1}dx_1 + \frac{\partial u}{\partial x_2}dx_2 + \frac{\partial u}{\partial x_3} + \ldots + \frac{\partial u}{\partial x_n} = 0.$$

Also differentiating the m given equations connecting the variables, we get

$$d\phi_1 = \frac{\partial \phi_1}{\partial x_1}dx_1 + \frac{\partial \phi_1}{\partial x_2}dx_2 + \frac{\partial \phi_1}{\partial x_3}dx_3 + \ldots + \frac{\partial \phi_1}{\partial x_n}dx_n = 0,$$

$$d\phi_2 = \frac{\partial \phi_2}{\partial x_1}dx_1 + \frac{\partial \phi_2}{\partial x_2}dx_2 + \frac{\partial \phi_2}{\partial x_3}dx_3 + \ldots + \frac{\partial \phi_2}{\partial x_n}dx_n = 0,$$

... ...

... ...

$$d\phi_m = \frac{d\phi_m}{dx_1}dx_1 + \frac{d\phi_m}{dx_2}dx_2 + \frac{d\phi_m}{dx_3}dx_3 + \ldots + \frac{d\phi_m}{dx_n}dx_n = 0.$$

Multiplying the above m + 1 equations obtained on differentiation by 1, $\lambda_1, \lambda_2, \ldots \lambda_m$ respectively and adding, we get an equation which may be written as

$$P_1dx_1 + P_2dx_2 + P_3dx_3 + \ldots + P_ndx_n = 0, \qquad \ldots(1)$$

where $P_r = \frac{\partial u}{\partial x_r} + \lambda_1 \frac{\partial \phi_1}{\partial x_r} + \lambda_2 \frac{\partial \phi_2}{\partial x_r} + \ldots + \lambda_m \frac{\partial \phi_m}{\partial x_r}$.

Now the m multiplies $\lambda_1, \lambda_2, \ldots \lambda_m$ are at our choice. We choose them such that they satisfy the m linear equations

$$P_1 = 0, P_2 = 0, \ldots, P_m = 0.$$

Then the equation (1) reduces to

$$P_{m+1}dx_{m+1} + P_{m+2}dx_{m+2} + \ldots + P_n dx_n = 0. \quad \ldots(2)$$

It is immaterial which of the n – m of the n variables $x_1, x_2, \ldots, x_n$ are regarded as independent. Let us regard the n – m variables $x_{m+1}, x_{m+2}, \ldots, x_n$ as independent. Then since the n – m quantities $dx_{m+1}, dx_{m+2}, \ldots, dx_n$ are all independent of one another, their coefficients must be separately zero in the relation (2). Hence we must have

$$P_{m+1} = 0, P_{m+2} = 0, \ldots, P_n = 0.$$

Thus, we get m + n equations

$$P_1 = 0, P_2 = 0, \ldots, P_n = 0$$

and $$\phi_1 = 0, \phi_2 = 0, \ldots, \phi_m = 0,$$

which together with the relation $u = f(x_1, x_2, \ldots, x_n)$ determine the m multiplies $\lambda_1, \lambda_2, \ldots, \lambda_m$, the values of $x_1, x_2, \ldots, x_n$ and u at the stationary point. This method is known s *Lagrange's Method of Undetermined Multipliers.*

It is very convenient to apply and it often gives us the maximum or minimum values of u without actually determining the values of the multipliers $\lambda_1, \ldots, \lambda_m$. The only drawback of this method is that it does not determine the nature of the stationary point.

The most widely used method of obtaining maxima or minima of functions subject to equality constraints is that of Lagrange's Multiplier. For that we write the Lagrangean function *L*, given by:

$$L(x, y,) = f(x, y) + \lambda\, g(x, y) \quad \ldots(6)$$

where $f(x, y) = 0$ is the function to be maximized or minimized subject to the constraint $g(x, y) = 0$ and λ is an unknown, called the *Lagrange's multiplier*.

Differentiating (6) partially with respect to x, y and λ and setting the results equals zero, we get, three equations in x, y and λ; and we can solve for the three. (In many cases, the value of λ is not of interest and is not found).

SOLVED EXAMPLES

Example 1: *The utility function of a consumer on two goods x and y is U= x y + y. Price of x is Rs 20 per unit and that of y is Rs 5 per unit. If the consumer has only Rs 100 to spend on the two goods, find the units of goods, which give him maximum utility.*

Solution: Since price of x is Rs 20 per unit and that of y is Rs 5 per unit, and since the consumer has only Rs 100 to spend on the two goods, the budget constraint can be written as:

$$20x + 5y = 100.$$

$$\therefore \quad y = 20 - 4x$$

Hence the utility function can be written as

$$U = x(20 - 4x) + 20 - 4x = -4x^2 + 16x + 20$$

$$\frac{dU}{dx} = -8x + 16$$

$$\frac{dU}{dx} = 0 \quad \Rightarrow -8x + 16 = 0 \Rightarrow x = 2$$

$$\frac{d^2U}{dx^2} = -8 < 0 \Rightarrow x = 2 \text{ maximises utility.}$$

Also, when $x = 2$, $y = 20 - 4x = 20 - 8 = 12$.

Hence for getting maximum utility, the consumer has to purchase 2 units of good x and 12 units of good y.

Example 2: *A monopolist with cost function $C(x) = x^2 + 1$ faces a demand curve $x = 12 - p$. What will be the equilibrium price and quantity?*

Solution: $x = 12 - p$

$$\Rightarrow \quad p = 12 - x$$

$$TR = px = (12 - x)\,x$$

$$= 12x - x^2$$

Profit, $\pi = TR - TC = 12x - x^2 - (x^2 + 1)$

$$= -2x^2 + 12x - 1$$

$$\frac{d\pi}{dx} = -4x + 12$$

$$\frac{d\pi}{dx} = 0$$

$$\Rightarrow -4x + 12 = 0 \Rightarrow x = 3$$

$$\frac{d^2\pi}{dx^2} = -4$$

$\Rightarrow$ profit is maximum at $x = 3$

$\therefore$ Equilibrium quantity = 3; and

Equilibrium price = 12 – 3 = 9.

Example 3(a): *For a firm under perfect competition, it is given that:*

$$p = 3 \text{ and } C = 100 + \frac{15x^2}{1000}$$

Find the number of items to be produced to attain maximum profit. Also find the maximum profit.

Solution: Let be the profit.

Total revenue, TR = price × quantity = $3x$

Then, π = TR − TC

$$= 3x - \left[100 + \frac{15x^2}{1000}\right]$$

$$= 3x - 100 - \frac{15x^2}{1000}$$

$$= 3 - \frac{30x}{1000}$$

$$\frac{d\pi}{dx} = 0$$

$$\Rightarrow 3 - \frac{30x}{1000} = 0$$

$$\Rightarrow x = 100$$

$$\frac{d^2\pi}{dx^2} = -\frac{30}{1000} < 0$$

$\Rightarrow$ $x = 100$ maximizes

In other words, the number of items to be produced to attain maximum profit is 100.

$$\text{Maximum profit} = 3\times100 - 100 - \frac{15\times100^2}{1000} = 50$$

Example 3(b): *A monopolist produces x sets of radios per week at a total cost C* = $x^2 + 78x + 2500$. *The demand function is given by* $x = \dfrac{600-p}{8}$, *where p is the price per set. Find the number of sets to be produced to get a maximum profit. What is the monopoly price?*

Solution: Let be the profit.

$$x = \frac{600-p}{8}$$

$$\Rightarrow p = 600 - 8x.$$

$\therefore \quad TR = p\,x = 600\,x - 8x^2$

Then, $\quad \pi = TR - TC$

$= 600\,x - 8x^2 - [x^2 + 78x + 2500]$

$= -9\,x^2 + 522x - 2500$

$$\frac{d\pi}{dx} = -18x + 522$$

$$\frac{d\pi}{dx} = 0 \qquad \Rightarrow -18x + 522 = 0$$

$\Rightarrow \quad x = 29$

$$\frac{d^2\pi}{dx^2} = -18 < 0 \Rightarrow x = 29 \text{ maximizes.}$$

In other words, the number of sets to be produced to attain maximum profit is 29.

Monopoly price = $600 - 8x = 600 - 8 \,'\, 29$ = Rs. 368

Example 3(c): *For a firm under perfect competition, total cost function is given by: $C = 0.04x^3 - 0.9\,x^2 + 10x + 12$. If the price is Rs. 4 per unit, will the firm continue production?*

Solution: Let be the profit.

$$\pi = TR - TC = 4x - [0.04x^3 - 0.9\,x^2 + 10x + 12]$$

$$= -0.04x^3 + 0.9\,x^2 - 6x - 12$$

$$\frac{d\pi}{dx} = -0.12\,x^2 + 1.8\,x - 6$$

$$\frac{d\pi}{dx} = 0 \qquad \Rightarrow -0.12\,x^2 + 1.8\,x - 6 = 0$$

$$\Rightarrow 12\,x^2 - 180\,x + 600 = 0$$

$$\Rightarrow x^2 - 15\,x + 50 = 0$$

$$\Rightarrow (x - 5)(x - 10) = 0$$

$$\Rightarrow x = 5,\ x = 10$$

$$\frac{d^2\pi}{dx^2} = -0.24\,x + 1.8$$

$$\left(\frac{d^2\pi}{dx^2}\right)_{x=5} = 0.6 > 0 \Rightarrow x = 5 \text{ does not maximises.}$$

$$\left(\frac{d^2\pi}{dx^2}\right)_{x=10} = -0.6 > 0 \qquad \Rightarrow x = 10 \text{ maximises.}$$

Maximum profit $= -0.04 \times 10^3 + 0.9 \times 10^2 - 6 \times 10 - 12$

$= -40 + 90 - 60 - 12 = -22.$

Hence the maximum profit is a loss of Rs. 22. Since the fixed cost (i.e., the cost for $x = 0$) is Rs. 12, the loss is more than the fixed cost. So the firm will not continue production.

(Also we know that for shut down point, price = min AVC.

Variable Cost, $VC = 0.04x^3 - 0.9x^2 + 10x$

$\therefore$ Average Variable Cost, $AVC = \dfrac{VC}{x}$

$= 0.04x^2 - 0.9x + 1$

$$\frac{d(AVC)}{dx} = 0.08x - 0.9$$

$$\frac{d(AVC)}{dx} = 0 \quad \Rightarrow x = \frac{0.9}{.08} = 11.25$$

$$\frac{d^2(AVC)}{dx^2} = 0.08 > 0 \Rightarrow x = 11.25 \text{ minimizes AVC.}$$

$\therefore$ Minimum value of AVC

$= 0.04 \times (11.25)^2 - 0.9 \times 11.25 + 10 = 4.9375$

Example 4: *If the demand law is:* $x = \alpha e^{-\beta p}$; $\alpha, \beta > 0$, *express marginal revenue as a function of x. At what levels of output and price the total revenue is maximum? Find the maximum total revenue.*

Solution: Given $x = \alpha e^{-\beta p}$

$$\Rightarrow \quad e^{-\beta p} = \frac{x}{\alpha}$$

$$\Rightarrow \quad -\beta p = \log \frac{x}{\alpha}$$

$$= \log x - \log \alpha$$

$$\therefore \quad p = \frac{-1}{\beta}[\log x - \log \alpha]$$

$$= \frac{-1}{\beta}[\log \alpha - \log x]$$

Then, $\quad TR = p\,x = \dfrac{x}{\beta}[\log \alpha - \log x]$

$$MR = \frac{d(TR)}{dx}$$

$$= \frac{x}{\beta}[-\frac{1}{x}] + \frac{1}{\beta}[\log a - \log x]$$

$$= \frac{1}{\beta}[\log \alpha - \log x - 1]$$

$$\frac{d(TR)}{dx} = 0 \qquad \Rightarrow \frac{1}{\beta}[\log \alpha - \log x - 1] = 0$$

$$\Rightarrow \log \alpha - \log x - 1 = 0$$

$$\Rightarrow \log x = \log \alpha - 1 = \log \alpha - \log e$$

(since log e =1)

$$\Rightarrow \qquad \log x = \log \frac{\alpha}{e}$$

$$\Rightarrow \qquad x = \frac{\alpha}{e}$$

$$\frac{d^2(TR)}{dx^2} = \frac{1}{\beta}[-\frac{1}{x}] < 0, \text{ when } x = \frac{\alpha}{e}.$$

Therefore $x = \frac{\alpha}{e}$ maximises TR.

Also, when $x = \frac{\alpha}{e}$,

$$p = \frac{1}{\beta}[\log \alpha - \log \frac{\alpha}{e}]$$

$$= \frac{1}{\beta}[\log \alpha - \log \alpha + \log e] = \frac{1}{\beta}$$

Hence the level of output for which the total revenue is maximum is $\frac{\alpha}{e}$ and the corresponding price is $\frac{1}{\beta}$.

Also maximum total revenue = $\frac{\alpha}{\beta e}$.

Example 5: *Find the maxima and minima (if any) of :*

i. $x^4 - 8x^3 + 22x^2 - 24x + 1$

ii. $x^5 - 5x^4 + 5x^3 + 10$

Solution:

i. Given $y = x^4 - 8x^3 + 22x^2 - 24x + 1$

$$\Rightarrow \qquad \frac{dy}{dx} = 4x^3 - 24x^2 + 44x - 24$$

$$\frac{dy}{dx} = 0 \quad \Rightarrow 4x^3 - 24x^2 + 44x - 24 = 0$$

$$\Rightarrow 4(x-1)(x-2)(x-3) = 0$$

$$\Rightarrow x = 1,\ x = 2,\ x = 3$$

$\therefore$ The function has extreme values at $x = 1$, $x = 2$ and $x = 3$.

$$\frac{d^2y}{dx^2} = 12x^2 - 48x + 44$$

$$\left(\frac{d^2y}{dx^2}\right)_{x=1} = 8 > 0 \Rightarrow x = 1 \text{ is a minimum point.}$$

$$\left(\frac{d^2y}{dx^2}\right)_{x=2} = -6 < 0 \Rightarrow x = 2 \text{ is a maximum point.}$$

$$\left(\frac{d^2y}{dx^2}\right)_{x=3} = 8 > 0 \Rightarrow x = 3 \text{ is a minimum point.}$$

$\therefore$ Maximum value $= -7$

Minimum value corresponding to $(x = 1) = -8$, and

Minimum value corresponding to $(x = 3) = -8$.

ii. Given $y = x^5 - 5x^4 + 5x^3 + 10$

$$\Rightarrow \quad \frac{dy}{dx} = 5x^4 - 20x^3 + 15x^2$$

$$\frac{dy}{dx} = 0 \quad \Rightarrow 5x^4 - 20x^3 + 15x^2 = 0$$

$$\Rightarrow 5x(x-1)(x-3) = 0$$

$$\Rightarrow x = 0,\ x = 1,\ x = 3$$

$\therefore$ The function has extreme values at $x = 0$, $x = 1$ and $x = 3$.

$$\frac{d^2y}{dx^2} = 20x^3 - 60x^2 + 30x$$

$$\left(\frac{d^2y}{dx^2}\right)_{x=3} = 0$$

$\Rightarrow x = 0$ is neither a maximum point, nor a minimum point

[$x = 0$ will be a point of inflexion]

$$\left(\frac{d^2y}{dx^2}\right)_{x=1} = -10 < 0 \Rightarrow x = 1 \text{ is a maximum point.}$$

$$\left(\frac{d^2y}{dx^2}\right)_{x=3} = 90 > 0 \Rightarrow x = 3 \text{ is a minimum point.}$$

$\therefore$ Maximum value = 11, and Minimum value = –17.

Example 6(a): *Show that the function $y = x + 1/x$ has one maximum value and one minimum value and the latter is larger than the former.*

Solution: Given $y = x + \frac{1}{x}$

$$\Rightarrow \quad \frac{dy}{dx} = 1 - \frac{1}{x^2}$$

$$\frac{dy}{dx} = 0 \Rightarrow 1 - \frac{1}{x^2} = 0$$

$$\Rightarrow \quad \frac{x^2 - 1}{x^2} = 0$$

$$\Rightarrow \quad x^2 - 1 = 0$$

$$\Rightarrow \quad (x + 1)(x - 1) = 0$$

$$\Rightarrow \quad x = 1, x = -1.$$

$\therefore$ The function has extreme values at $x = 1$ and $x = -1$.

$$\frac{d^2y}{dx^2} = \frac{2}{x^3}$$

$$\left(\frac{d^2y}{dx^2}\right)_{x=1} = 2 > 0 \Rightarrow x = 1 \text{ is a minimum point.}$$

$$\left(\frac{d^2y}{dx^2}\right)_{x=-1} = -2 < 0 \Rightarrow x = -1 \text{ is a maximum point.}$$

$\therefore$ Maximum value = –2 , and

Minimum value = 2

Hence the minimum value is larger than the maximum value.

Example 6(b): *Find the point of inflexion of the curve:*

$$y = x^3 - 15x^2 + 20x + 10$$

Solution:

Given $y = x^3 - 15x^2 + 20x + 10$

$$\Rightarrow \quad \frac{dy}{dx} = 3x^2 - 30x + 20$$

$$\frac{d^2y}{dx^2} = 6x - 30$$

$$\frac{d^2y}{dx^2} = 0 \Rightarrow x = 5$$

$$\frac{d^3y}{dx^3} = 6 \Rightarrow \left(\frac{d^3y}{dx^3}\right)_{x=5} \neq 0.$$

$\therefore$ $x = 5$ is a point of inflexion of the function.

Example 6(c): *Find the maxima and minima (if any) of the function:*

$$\frac{1-x+x^2}{1+x+x^2}$$

Solution:

Let $$y = \frac{1-x+x^2}{1+x+x^2}$$

$$\frac{dy}{dx} = \frac{(1+x+x^2)(-1+2x)-(1-x+x^2)(1+2x)}{(1+x+x^2)^2}$$

$$= \frac{2(x^2-1)}{(1+x+x^2)^2}$$

$$\frac{dy}{dx} = 0 \Rightarrow x^2 - 1 = 0$$

i.e., $x = 1, x = -1$

$$\frac{d^2y}{dx^2} = \frac{(1+x+x^2)^2 4x - 2(x^2-1)2(1+x+x^2)(1+2x)}{(1+x+x^2)^4}$$

$$= \frac{4(1+3x-x^3)}{(1+x+x^2)^3}$$

$$\left(\frac{d^2y}{dx^2}\right)_{x=1} = \frac{4}{9} > 0$$

$\Rightarrow$ $x = 1$ is a minimum point.

$$\left(\frac{d^2y}{dx^2}\right)_{x=-1} = \frac{-4}{27} < 0$$

$\Rightarrow$ $x = -1$ is a maximum point.

$\therefore$ Maximum value = 3 and

Minimum value = 1/3

Example 7: Show that y = sin x (1+ cos x) has a maximum at $x = \frac{\pi}{3}$

Solution:

Given

$$y = \sin x\,(1 + \cos x)$$

$$\frac{dy}{dx} = \sin x\,(-\sin x) + (1 + \cos x)\cos x$$

$$= -\sin^2 x + \cos x + \cos^2 x$$

$$\frac{dy}{dx} = 0 \quad \Rightarrow -\sin^2 x + \cos x + \cos^2 x = 0$$

$$\Rightarrow \cos^2 x - 1 + \cos x + \cos^2 x = 0$$

$$\Rightarrow 2\cos^2 x - 1 + \cos x = 0$$

$$\Rightarrow (\cos x + 1)(2\cos x - 1) = 0$$

$$\Rightarrow \cos x = -1\ ,\ \cos x = 1/2$$

$$\Rightarrow x = \pi,\ x = \pm\frac{\pi}{3}$$

$$\frac{d^2y}{dx^2} = -4\cos x \sin x - \sin x$$

$$= \sin x\,(-4\cos x - 1)$$

$$\left(\frac{d^2y}{dx^2}\right)_{x=\frac{\pi}{3}} = \sin\frac{\pi}{3}\,\left(-4\cos\frac{\pi}{3} - 1\right)$$

$$= \frac{\sqrt{3}}{2}\left(-4.\frac{1}{2} - 1\right) = \frac{-3\sqrt{3}}{2} < 0$$

$\Rightarrow y$ is maximum at $x = \frac{\pi}{3}$.

Example 8(a): *Show that of all rectangles of given perimeter, the square has the greatest area.*

Solution: Let x be the length and y be the breadth of the rectangle. Since the perimeter is given, we have $2x + 2y = 2c$, where c is a constant.

$$\therefore \quad y = c - x$$

The area A of the rectangle is:

$$A = x\,y = x\,(c - x) = cx - x^2$$

Differentiating with respect to x,

$$\frac{dA}{dx} = c - 2x$$

$$\frac{dA}{dx} = 0 \Rightarrow c - 2x = 0$$

$$\Rightarrow \qquad x = \frac{c}{2}$$

Differentiating again with respect to x,

$$\frac{d^2A}{dx^2} = -2$$

$$\left(\frac{d^2A}{dx^2}\right)_{x=\frac{c}{2}} < 0 \Rightarrow \text{A is maximum when } x = \frac{c}{2}$$

Now, $y = c - x = c - \frac{c}{2} = \frac{c}{2}$

i.e., for maximum area, the rectangle should have its length and breadth equal. That means, it should be a square. In other words, of all rectangles of given perimeter, the square has the greatest area.

Example 8(b): *An open rectangular tank, with square base and vertical sides is to be constructed with sheet metal so as to contain a given quantity of water. Show that the cost of material is least when depth is half the width.*

Solution: Let x be the side of the square base and y be the depth of the tank. Since the tank contains a given quantity of water, its volume is constant.

Volume of the tank = base area × depth

$= x^2y = c$, where c is a constant.

$$\therefore \qquad y = \frac{c}{x^2}$$

Let S be the area of the sheet metal with which the tank is to be constructed.

Then, S = base area + lateral area

$$= x^2 + 4xy = x^2 + 4x\frac{c}{x^2}$$

$$= x^2 + \frac{4c}{x}$$

Differentiating with respect to x,

$$\frac{dS}{dx} = 2x - \frac{4c}{x^2}$$

$$\frac{dS}{dx} = 0 \qquad \Rightarrow 2x - \frac{4c}{x^2} = 0$$

$$\Rightarrow \frac{2x^3 - 4c}{x^2} = 0$$

$$\Rightarrow \quad x^3 = 2c \text{ or, } x = (2c)^{1/3}$$

Differentiating again with respect to x,

$$\frac{d^2S}{dx^2} = 2 + \frac{8c}{x^3}$$

$$\left(\frac{d^2S}{dx^2}\right)_{x=(2c)^{1/3}} = 2 + \frac{8c}{2c} = 6 > 0$$

$\Rightarrow$ S is minimum when $x = (2c)^{1/3}$

Now, $$y = \frac{c}{x^2} = \frac{(x^3/2)}{x^2} = x/2 \text{ [since } x^3 = 2c \Rightarrow c = x^3/2]$$

i.e., Area of the sheet metal is minimum when $y = x/2$. Or, cost of material is least when depth is half the width.

Example 8(c): *The short–run cost function of a manufacturer is given by:*

$$C = x^3 - 10x^2 + 100x + 1000. \text{ Find}$$

i. *Find AC, AVC and MC functions.*
ii. *Show that MC = min. of AC*
iii. *Show that MC = min. of AVC*
iv. *Show that total cost function has a point of inflexion at a level of output where MC is minimum. Find minimum of MC.*

Solution:

Given $\quad C = x^3 - 10x^2 + 100x + 1000$

i. $$AC = \frac{C}{x} = x^2 - 10x + 100 + \frac{1000}{x}$$

$$VC = x^3 - 10x^2 + 100x$$

$$AVC = \frac{VC}{x} = x^2 - 10x + 100$$

$$MC = \frac{d(C)}{dx} = 3x^2 - 20x + 100$$

ii. $$AC = x^2 - 10x + 100 + \frac{1000}{x}$$

$$\frac{d(AC)}{dx} = 2x - 10 - \frac{1000}{x^2}$$

$$\frac{d(AC)}{dx} = 0 \quad \Rightarrow \quad 2x - 10 - \frac{1000}{x^2} = 0$$

$$\Rightarrow \quad 2x^3 - 10x^2 - 1000 = 0$$

$$\Rightarrow \quad x^3 - 5x^2 - 500 = 0$$

$$\Rightarrow \quad (x - 10)(x^2 + 5x + 50) = 0$$

$$\Rightarrow \quad x = 10$$

[neglecting the imaginary values of x, which are the roots corresponding to the factor $(x^2 + 5x + 50)$].

$$\frac{d^2(AC)}{dx^2} = 2 + \frac{2000}{x^3} > 0 \text{ when } x = 10.$$

$\therefore$ AC is minimum when $x = 10$.

Also, at $x = 10$,

$$AC = 10^2 - 10 \times 10 + 100 + \frac{1000}{10} = 200;$$

and $\quad MC = 3 \times 10^2 - 20 \times 10 + 100 = 200$

$\therefore \quad MC = \text{min. of } AC$

iii. $\quad AVC = x^2 - 10x + 100$

$$\frac{d(AVC)}{dx} = 2x - 10$$

$$\frac{d(AVC)}{dx} = 0, \Rightarrow 2x - 10 = 0, \Rightarrow x = 5$$

$$\frac{d^2(AVC)}{dx^2} = 2 > 0 \text{ when } x = 5$$

$\therefore$ AC is minimum when $x = 5$.

Also, at $x = 5$,

$$AVC = 5^2 - 10 \times 5 + 100 = 75 \text{ ; and}$$

$$MC = 3 \times 5^2 - 20 \times 5 + 100 = 75$$

$\therefore \quad MC = \text{min. of } AVC$

iv. $C = x^3 - 10x^2 + 100x + 1000$

$$\frac{d(C)}{dx} = 3x^2 - 20x + 100$$

$$\frac{d^2(C)}{dx^2} = 6x - 20$$

$$\frac{d^2(C)}{dx^2} = 0 \text{ at } x = \frac{10}{3}$$

Hence $x = \frac{10}{3}$ is a point of inflection of the total cost function.

Now, $\frac{d(MC)}{dx} = 6x - 20,$

since $\frac{d(MC)}{dx} = \frac{d^2(C)}{dx^2}$

Also, $\frac{d^2(MC)}{dx^2} = 6 > 0,$

when $x = \frac{10}{3}$.

i.e., MC is minimum at $x = \frac{10}{3}$.

Hence the total cost function has a point of inflexion at a level of output where MC is minimum.

$$\text{Minimum of MC} = 3\left(\frac{10}{3}\right)^2 - 20\times\frac{10}{3} + 100$$

$$= \frac{100-200+300}{3} = \frac{200}{3}.$$

Example 9: *Find the maxima and minima (if any) of :*

$x^3 - 6x^2 + 9x - 8$

Solution:

Given $\quad y = x^3 - 6x^2 + 9x - 8$

$$\Rightarrow \quad \frac{dy}{dx} = 3x^2 - 12x + 9$$

$$\frac{dy}{dx} = 0, \Rightarrow 3x^2 - 12x + 9 = 0, \Rightarrow 3(x-1)(x-3) = 0$$

$$\Rightarrow x = 1, x = 3$$

$\therefore$ The function has extreme values at $x = 1$ and $x = 3$.

$$\frac{d^2y}{dx^2} = 6x - 12$$

$$\left(\frac{d^2y}{dx^2}\right)_{x=1} = -6 < 0 \Rightarrow x = 1 \text{ is a maximum point.}$$

$$\left(\frac{d^2y}{dx^2}\right)_{x=1} = 6 > 0 \Rightarrow x = 3 \text{ is a minimum point.}$$

$\therefore$ maximum value $= -4$, and

minimum value $= -62$

Example 10: *Find the point of inflexion of* $f(x) = x^3 - 3x + 1$

Solution:

$$f(x) = x^3 - 3x + 1$$
$$f'(x) = 3x^2 - 3$$
$$f''(x) = 6x$$
$$f''(x) = 0$$
$$\Rightarrow \quad 6x = 0, \text{ or } x = 0.$$

Now, $x = 0$

$\Rightarrow$ $f(x) = 1$

$\therefore$ The point of inflexion $= (0, 1)$.

Example 11: *Find the maxima and minima of* $f(x) = (x + 1)^2 (x - 1)^3$.

Solution: Given $f(x) = (x + 1)^2 (x - 1)^3$

$$\Rightarrow \quad f'(x) = 3(x + 1)^2 (x - 1)^2 + 2(x + 1)(x - 1)^3$$
$$f'(x) = 0 \Rightarrow (x + 1)(x - 1)^2 [3(x + 1) + 2(x - 1)] = 0$$
$$\Rightarrow (x + 1)(x - 1)^2 (5x + 1) = 0$$
$$\Rightarrow \quad x = -1, x = 1, x = \frac{-1}{5}$$

$\therefore$ The function has extreme values at $x = -1$, $x = 1$,

$$x = \frac{-1}{5}$$

$$f''(x) = (x - 1)^2 (5x + 1) + 2(x + 1)(x - 1)(5x + 1) + 5(x + 1)(x - 1)^2$$

$[f''(x)]_{(x=-1)} = -16 < 0 \Rightarrow x = -1$ is a maximum point.

$[f''(x)]_{(x=1)} = 0 \Rightarrow x = 1$ is neither a maximum point, nor a minimum point but a point of inflexion.

$[f''(x)]_{(x=-1/5)} = > 0$

$\Rightarrow x = \frac{-1}{5}$ is a minimum point.

$\therefore$ Maximum value $= 0$, and Minimum value $= \frac{-4^2 \times 6^3}{5^5}$

Example 12(a): *Find the maximum value of* $f(x) = \frac{\log x}{x}$, $0 < x < \infty$

Solution:

Given $f(x) = \frac{\log x}{x}$

$$\Rightarrow \quad f'(x) = \frac{x.\frac{1}{x} - \log x}{x^2} = \frac{1-\log x}{x^2}$$

$$f'(x) = 0 \Rightarrow \frac{1-\log x}{x^2} = 0$$

$\Rightarrow 1-\log x = 0$

$\Rightarrow \quad \log x = 1$

$\Rightarrow \quad \log_e x = \log_e e, \Rightarrow x = e$

$\therefore$ The function has extreme value at $x = e$

$$f''(x) = \frac{x^2.\frac{-1}{x} - (1-\log x).2x}{x^4}$$

$$= \frac{x^2.\frac{-1}{x} - (1-\log x).2x}{x^4}$$

$$= \frac{-3x + 2x\log x}{x^4}$$

$$= \frac{-3 + 2\log x}{x^3}$$

$$[f''(x)]_{(x=e)} = \frac{-3+2\log e}{e^3} = \frac{-3+2}{e^3}$$

$$= \frac{-1}{e^3} < 0$$

$\Rightarrow \quad x = e$ is a maximum point.

Maximum value $= \frac{\log e}{e} = \frac{1}{e}$.

Example 12(b): *Show that the function* $y = k\ e^{-\frac{x^2}{2}}$; $k > 0$, *has one maximum value at* $x = 0$.

Solution:

Given $\quad y = ke^{-\frac{x^2}{2}}$

$$\Rightarrow \quad \frac{dy}{dx} = k\ e^{-\frac{x^2}{2}}.\frac{-2x}{2} = -k\,x\,e^{-\frac{x^2}{2}}$$

$$\frac{dy}{dx} = 0 \Rightarrow -k\,x\,e^{-\frac{x^2}{2}} = 0$$

$\Rightarrow \qquad x = 0$; since $e^{-\frac{x^2}{2}} \neq 0$ for finite values of x

∴ The function has extreme value at $x = 0$

$$\frac{d^2y}{dx^2} = k\,x^2\, e^{-\frac{x^2}{2}} - k e^{-\frac{x^2}{2}}$$

$$= k e^{-\frac{x^2}{2}}\,(x^2 - 1)$$

$$\left(\frac{d^2y}{dx^2}\right)_{x=0} = -\,k e^{-\frac{x^2}{2}} < 0 \Rightarrow x = 0 \text{ is a maximum point.}$$

∴ The function has one maximum value at $x = 0$.

Example 12(c): *Find the minimum value of* $x^4 + y^4 + z^4$, *where* $xyz = c^3$.

Solution: Let $u = x^4 + y^4 + z^4$, ...(1)

where the variables x, y, z are connected by the relation

$$xyz = c^3. \qquad ...(2)$$

For a maximum or a minimum of u, we have

$$du = 0 \Rightarrow 4x^3\,dx + 4y^3\,dy + 4z^3\,dz = 0$$

$$\Rightarrow \qquad x^3\,dx + y^3\,dy + z^3\,dz = 0. \qquad ...(3)$$

Also from the given relation (2), we have

$$\log x + \log y + \log z = \log c^3.$$

Differentiating this, we get

$$\left(\frac{1}{x}\right)dx + \left(\frac{1}{y}\right)dy + \left(\frac{1}{z}\right)dz = 0. \qquad ...(4)$$

Multiplying eqns. (3) by 1 and (4) by λ, and adding and then equating the coefficients of dx, dy, dz to zero, we get

$$x^3 + \frac{\lambda}{x} = 0,\; y^3 + \frac{\lambda}{y} = 0,\; z^3 + \frac{\lambda}{z} = 0.$$

From these, we get $x^4 = y^4 = z^4 = -\lambda$.

Now from (2), $x^4y^4z^4 = c^{12}$.

$$\therefore \qquad -\lambda^3 = c^{12} \text{ or } \lambda = -c^4.$$

∴ u is stationary when $x^4 = y^4 = z^4 = c^4$ *i.e.*, when $x = y = z = c$. Now, regard x and y as independent variables and z as a function of x and y given by (2).

From (1), we have $\dfrac{\partial u}{\partial x} = 4x^3 + 4z^3\,\dfrac{\partial z}{\partial x}$.

Now from (2), we have log x + log y + log z = log c^3.

Differentiating this partially w.r.t. x taking y as constant, we get

$$\frac{1}{x}+\frac{1}{z}\frac{\partial z}{\partial x}=0 \text{ or } \frac{\partial z}{\partial x}=-\frac{z}{x}.$$

$$\therefore \quad \frac{\partial u}{\partial x}=4x^3-4z^3\cdot\frac{z}{x}=4x^3-4\frac{z^4}{x},$$

so that $$\frac{\partial^2 u}{\partial x^2}=12x^2+\frac{4z^4}{x^2}-\frac{16}{x}z^3\frac{\partial z}{\partial x}$$

$$=12x^2+\frac{4z^4}{x^2}-\frac{16z^3}{x}\left(-\frac{z}{x}\right)=12x^2+\frac{4z^4}{x^2}+\frac{16z^4}{x^2}.$$

At the stationary point (c, c, c) found above, we have

$$\frac{\partial^2 u}{\partial x^2} = 12c^2 + 4c^2 + 16c^2 = 32c^2, \text{ which is +ive.}$$

∴ u is minimum at the point x = y = z = c and the minimum value of

$$u = c^4 + c^4 + c^4 = 3c^4.$$

Example 13: *A monopolist sells two products x and y for which the demand functions are* $x=50-0.5p_x$ *and* $y=76-p_y$*; and the combined cost function is* $C=3x^2+2xy+2y^2+55$. *Find :*

i. *profit maximizing level of output for each product*

ii. *the profit maximizing price for each product; and*

iii. *the maximum profit.*

Solution:

Profit, $\pi = TR-TC=(TR_x+TR_y)-TC$; where

TR_x = TR from commodity x and

TR_y = TR from commodity y

Given, $x = 50-0.5p_x$

$\Rightarrow$ $p_x = 100-2x$

Also, $y = 76-p_y$

$\Rightarrow$ $p_y = 76-y$

Hence, TR_x = price of x quantity

$$=(100-2x)x=100x-2x^2$$

and $\quad TR_y = \text{price of } y \text{ quantity}$

$$= (76-y)y = 76y - y^2$$

Given $\quad TC = 3x^2 + 2xy + 2y^2 + 55$

$$\therefore \quad \pi = (100x - 2x^2 + 76y - y^2) - (3x^2 + 2xy + 2y^2 + 55)$$

$$= -5x^2 - 3y^2 - 2xy + 100x + 76y - 55$$

$$\frac{\partial \pi}{\partial x} = -10x - 2y + 100$$

$$\frac{\partial \pi}{\partial y} = -2x - 6y + 76$$

$$\frac{\partial \pi}{\partial x} = 0, \frac{\partial \pi}{\partial y} = 0 \Rightarrow x = 8, y = 10$$

$$\frac{\partial^2 \pi}{\partial x^2} = -10 < 0, \frac{\partial^2 \pi}{\partial y^2} = -6 < 0, \frac{\partial^2 \pi}{\partial x \partial y} = -2$$

$$\frac{\partial^2 \pi}{\partial x^2} \cdot \frac{\partial^2 \pi}{\partial y^2} - \frac{\partial^2 \pi}{\partial x \partial y} = -10 \times -6 - 4 = 56 > 0$$

Hence π is maximum at $x = 8, \; y = 10$.

Maximum $\quad \pi = -5(8)^2 - 3(10)^2 - 2(8 \times 10) + 100(8) + 76(10) - 55$

$$= -320 - 300 - 160 + 800 + 760 - 55 = 725$$

Hence,

i. profit maximizing level of output $x = 8$

profit maximizing level of output $y = 10$

ii. the profit maximizing price for product $x = 100 - 2 \times 8 = 84$

iii. profit maximizing price for product $y = 76 - 10 = 66$

iv. the maximum profit = 725.

Examples 14(a): $f_x = 0, f_y = 0$

$$\Rightarrow \quad 2x + y - 3 = 0 \qquad ...(1)$$

$$2y + x = 0 \qquad ...(2)$$

Solving (1) and (2) we get, $x = 2, y = -1$.

$\therefore$ *(2, –1) is a critical point.*

Conditon – 1:

$$f_{xx} = 2, f_{yy} = 2, f_{xy} = 1$$

Hence both f_{xx} , $f_{yy} > 0$

Conditon – 2:

$$f_{xx} \cdot f_{yy} = 2.2 = 4$$

$$f_{xy}{}^2 = 1$$

Hence $f_{xx} \cdot f_{yy} > f_{xy}{}^2$

(2, –1) is a minimum point.

Minimum value = $2^2 + (-1)^2 + 2.(-1) - 3.2 + 2 = -1$.

Example 14(b): *Examine the function $f(x, y) = 5x^2 + 6y^2 - xy$ for maxima and minima, if any, subject to $x + 2y = 24$.*

Solution:

$$f(x, y) = 5x^2 + 6y^2 - xy$$

$$g(x, y) = x + 2y - 24$$

$$L(x, y, \lambda) = f(x, y) + \lambda\, g(x, y)$$

$$= 5x^2 + 6y^2 - xy + \lambda(x + 2y - 24)$$

$$\frac{\partial L}{\partial x} = 10x - y + \lambda$$

$$\frac{\partial L}{\partial y} = 12y - x + 2\lambda$$

$$\frac{\partial L}{\partial \lambda} = x + 2y - 24$$

Conditon – 1:

Solving $\frac{\partial L}{\partial x} = 0$, $\frac{\partial L}{\partial y} = 0$ and $\frac{\partial L}{\partial \lambda} = 0$, we get:

$$10x - y + \lambda = 0 \quad \text{...(1)}$$

$$12y - x + 2\lambda = 0 \quad \text{...(2)}$$

$$x + 2y - 24 = 0 \quad \text{...(3)}$$

(To solve these three equations, we use the following method:

Make the magnitude of the coefficients of λ in the first two equations the same. Then add or subtract the two equations, as the case may be, to eliminate λ. The new equations contain only *x* and *y*. This, together with the third equation gives a pair of simultaneous equations, which can be solved for *x* and *y*)

From the first two equations,

$$20x - 2y + 2\lambda = 0,$$

$$-x + 12y + 2\lambda = 0$$

(the first equation is multiplied by 2)

Subtracting,

$$21x - 14y = 0$$

i.e., $3x - 2y = 0$ ...(4)

From equations (3) and (4) we get:

$$x + 2y = 24$$

$$3x - 2y = 0$$

Solving we get,

$$x = 6, y = 9$$

Putting the values of x and y in equation (1) , we get $\lambda = 5$

$\therefore$ (6, 9) is a critical point.

Conditon – 2:

$$L_{xx} = 10, L_{yy} = 12, L_{xy} = -1$$

Hence both $L_{xx}, L_{yy} > 0$

Conditon – 3:

$$L_{xx} . L_{yy} = 10. 12 = 120$$

$$L_{xy}^2 = 1$$

Hence $L_{xx} . L_{yy} > L_{xy}^2$

$\therefore$ (6, 9) is a minimum point

Minimum value of $f(x, y) = 5. 6^2 + 6. 9^2 - 6. 9 = 612$.

Example. 14(c): *Use Lagrange's multiplier method to find the values of x and y which maximizes or minimizes the given functions:*

i. $f(x, y) = x^2 + y^2$; *subject to* $x + y = 1$

ii. $f(x, y) = 6x^2 + y^2$; *subject to* $4x - y = 1$

iii. $u(x, y) = 12xy - 3y^2 - x^2$; *subject to* $x + y = 16$

iv. $f(x,y,z) = x^2 + y^2 + z^2$, *subject to* $x + y + z = 1$.

Solution:

i. $f(x, y) = x^2 + y^2$

$$g(x, y) = x + y - 1$$

$$L(x, y,) = f(x, y) + g(x, y) = x^2 + y^2 + (x + y - 1)$$

$$\frac{\partial L}{\partial x} = 2x + \lambda$$

$$\frac{\partial L}{\partial y} = 2y + \lambda$$

$$\frac{\partial L}{\partial \lambda} = x + y - 1$$

Conditon – 1:

Solving $\frac{\partial L}{\partial x} = 0$, $\frac{\partial L}{\partial y} = 0$ and $\frac{\partial L}{\partial \lambda} = 0$, we get:

$$2x + \lambda = 0 \qquad \ldots(1)$$

$$2y + \lambda = 0 \qquad \ldots(2)$$

$$x + y - 1 = 0 \qquad \ldots(3)$$

Eliminating λ from (1) and (2) we get:

$$2x - 2y = 0$$

Or, $$x - y = 0 \qquad \ldots(4)$$

Solving (3) and (4) , we get,

$$x = \frac{1}{2}, y = \frac{1}{2}$$

Putting $x = \frac{1}{2}, y = \frac{1}{2}$ in (1), we get, $\lambda = -1$

$\therefore (\frac{1}{2}, \frac{1}{2})$ is a critical point.

Conditon – 2:

$$L_{xx} = 2, L_{yy} = 2, L_{xy} = 0$$

Hence both L_{xx} , $L_{yy} > 0$

Conditon – 3:

$$L_{xx} . L_{yy} = 4 > L_{xy}{}^2 = 0$$

$(\frac{1}{2}, \frac{1}{2})$ is a minimum point

Minimum value of $f(x, y) = \frac{1}{4} + \frac{1}{4} = \frac{1}{2}$

ii. $f(x, y) = 6x^2 + y^2$; $g(x, y) = 4x - y - 1$

$$L(x, y, \lambda) = f(x, y) + \lambda\, g(x, y) = 6x^2 + y^2 + \lambda(4x - y - 1)$$

$$\frac{\partial L}{\partial x} = 12x + 4\lambda$$

$$\frac{\partial L}{\partial y} = 2y - \lambda$$

$$\frac{\partial L}{\partial \lambda} = 4x - y - 1$$

Conditon – 1:

Solving $\frac{\partial L}{\partial x} = 0$, $\frac{\partial L}{\partial y} = 0$ and $\frac{\partial L}{\partial \lambda} = 0$, we get:

$$12x + 4\lambda = 0 \quad \text{...(1)}$$

$$2y - \lambda = 0 \quad \text{...(2)}$$

$$4x - y - 1 = 0 \quad \text{...(3)}$$

Eliminating from (1) and (2) we get:

$$3x + 2y = 0$$

Or, $x - y = 0$ ...(4)

Solving (3) and (4) , we get, $x = \frac{2}{11}$, $y = \frac{-3}{11}$

Putting $x = \frac{2}{11}$, $y = \frac{-3}{11}$ in (1), we get, $\lambda = \frac{-6}{11}$

$\therefore$ $(\frac{2}{11}, \frac{-3}{11})$ is a critical point.

Conditon – 2:

$$L_{xx} = 12,\ L_{yy} = 2,\ L_{xy} = 0$$

Hence both L_{xx} , $L_{yy} > 0$

Conditon – 3:

$$L_{xx} \cdot L_{yy} = 24 > L_{xy}{}^2 = 0$$

$\therefore$ $(\frac{2}{11}, \frac{-3}{11})$ is a minimum point

Minimum value of $f(x, y) = 6. \left(\frac{2}{11}\right)^2 + \left(\frac{-3}{11}\right)^2 = \frac{33}{121}$

iii. $u(x, y) = 12\,xy - 3y^2 - x^2$

$$g(x, y) = x + y - 16$$

$$L(x, y, \lambda) = u(x, y) + \lambda\, g(x, y)$$

$$= 12\,xy - 3y^2 - x^2 + \lambda(x + y - 16)$$

$$\frac{\partial L}{\partial x} = 12y - 2x + \lambda$$

$$\frac{\partial L}{\partial y} = 12x - 6y + \lambda$$

$$\frac{\partial L}{\partial \lambda} = x + y - 16$$

Conditon – 1:

Solving $\frac{\partial L}{\partial x} = 0$, $\frac{\partial L}{\partial y} = 0$ and $\frac{\partial L}{\partial \lambda} = 0$, we get:

$$12y - 2x + \lambda = 0 \quad \text{...(1)}$$

$$12x - 6y + \lambda = 0 \quad \text{...(2)}$$

$$x + y - 16 = 0 \quad \text{...(3)}$$

Eliminating λ from (1) and (2) we get:

$$7x - 9y = 0 \quad \text{...(4)}$$

Solving (3) and (4) , we get, $x = 9$, $y = 7$

$\therefore$ (9, 7) is a critical point.

Conditon – 2:

$$L_{xx} = -2,\ L_{yy} = -6,\ L_{xy} = 12$$

Hence both L_{xx} , $L_{yy} < 0$

Conditon – 3:

$$L_{xx} \cdot L_{yy} = 12 < L_{xy}{}^2 = 144$$

The test fails. So we use an alternate method to find the maximum or minimum of the function. Here, it is easy to substitute for x or y in $u(x, y)$ so that u becomes a function of x or y alone.

Alternate Method:

We have $u(x, y) = 12\,xy - 3y^2 - x^2$ ...(A)

and $x + y - 16 = 0$ ...(B)

From (B) we get, $y = 16 - x$, and substituting this in (A) we get:

$$u(x) = 12\,x\,(16 - x) - 3(16 - x)^2 - x^2$$

$$= -16\,x^2 + 288\,x - 768$$

$$\frac{du}{dx} = -32\,x + 288$$

$$\frac{du}{dx} = 0 \Rightarrow x = 9$$

$$\frac{d^2u}{dx^2} = -32 < 0$$

$\therefore$ u attains maximum when $x = 9$.

Also, when $x = 9$, $y = 16 - 9 = 7$.

Hence (9, 7) is a maximum point.

Maximum value of u subject to the constraint $x + y = 16$ is 528.

iv.
$$f(x,y,z) = x^2 + y^2 + z^2$$

$$g(x,y,z) = x + y + z - 1$$

$$L(x, y,) = f(x,y,z) + \lambda g(x,y,z)$$

$$= x^2 + y^2 + z^2 + \lambda(x + y + z - 1)$$

$$\frac{\partial L}{\partial x} = 2x + \lambda$$

$$\frac{\partial L}{\partial y} = 2y + \lambda$$

$$\frac{\partial L}{\partial z} = 2z + \lambda$$

$$\frac{\partial L}{\partial \lambda} = x + y + z - 1$$

Solving $\frac{\partial L}{\partial x} = 0$, $\frac{\partial L}{\partial y} = 0$, $\frac{\partial L}{\partial z}$ and $\frac{\partial L}{\partial \lambda} = 0$, we get:

$$2x + \lambda = 0,\ 2y + \lambda = 0,\ 2z + \lambda = 0,\ x + y + z = 1$$

Solving we get,

$$x = \frac{1}{3},\ y = \frac{1}{3},\ z = \frac{1}{3}$$

$\therefore$ $\left(\frac{1}{3}, \frac{1}{3}, \frac{1}{3}\right)$ is a critical point.

For the second order condition we use the formulae:

$d^2L < 0$, for maximum and

$d^2L > 0$, for minimum, subject to $g(x,y,z)=0$.

$$d^2L = f_x d^2x + f_{xx}dx^2 + f_{yy}dy^2 + f_{zz}dz^2 + 2f_{xy}dxdy + 2f_{xz}dxdz + 2f_{yz}dydz$$

$$= 2x.0 + 2.dx^2 + 2.dy^2 + 2.dz^2 + 2.0.dxdy + 2.0.dxdz + 2.0.dydz$$

$$= 2.dx^2 + 2.dy^2 + 2.dz^2 > 0 \text{ , subject to}$$

(note that here $d^2L > 0$, for all values of x, y and z.)

Hence $\left(\frac{1}{3}, \frac{1}{3}, \frac{1}{3}\right)$ is a minimum point.

Example 15: *Find the maximum value of*

$$(ax + by + cz)\, e^{-}(\alpha^2x^2 + \beta^2y^2 + \gamma^2z^2).$$

Solution: Let $u = (ax + by + cz)\, e^{-}(\alpha^2x^2 + \beta^2y^2 + \gamma^2z^2)$.

Then $\log u = \log (ax + by + cz) - (\alpha^2x^2 + \beta^2y^2 + \gamma^2z^2)$.

$$\therefore \quad \frac{1}{u}\frac{\partial u}{\partial x} = \frac{a}{ax+by+cz} - 2\alpha^2 x, \quad \frac{1}{u}\frac{\partial u}{\partial y} = \frac{b}{ax+by+cz} - 2\beta^2 y,$$

$$\frac{1}{u}\frac{\partial u}{\partial z} = \frac{c}{ax+by+cz} - 2\gamma^2 z.$$

For a maximum or a minimum of u, we must have

$$\frac{\partial u}{\partial x} = 0, \quad \frac{\partial u}{\partial y} = 0, \quad \frac{\partial u}{\partial z} = 0.$$

These give $x(ax + by + cz) = \dfrac{a}{2\alpha^2}$ ...(1)

$$y(ax + by + cz) = \frac{b}{2\beta^2} \quad \text{...(2)}$$

and $z(ax + by + cz) = \dfrac{c}{2\gamma^2}$. ...(3)

Multiplying (1), (2), (3) by a, b, c respectively and adding, we get

$$(ax + by + cz)^2 = \frac{1}{2}\left(\frac{a^2}{\alpha^2} + \frac{b^2}{\beta^2} + \frac{c^2}{\gamma^2}\right),$$

so that $(ax + by + cz) = \sqrt{\left\{\frac{1}{2}\left(\frac{a^2}{\alpha^2} + \frac{b^2}{\beta^2} + \frac{c^2}{\gamma^2}\right)\right\}} = R$, say.

Then $x = \dfrac{a}{2\alpha^2 R}$, $y = \dfrac{b}{2\beta^2 R}$, $z = \dfrac{c}{2\gamma^2 R}$.

$\therefore$ u is stationary when $x = \dfrac{a}{2\alpha^2 R}$, $y = \dfrac{b}{2\beta^2 R}$, $z = \dfrac{c}{2\gamma^2 R}$.

Again $\dfrac{1}{u}\dfrac{\partial^2 u}{\partial x^2} - \dfrac{1}{u^2}\left(\dfrac{\partial u}{\partial x}\right)^2 = -\dfrac{a^2}{(ax+by+cz)^2} - 2\alpha^2$.

Now at a stationary point, we have $\dfrac{\partial u}{\partial x} = 0$.

$\therefore$ at the stationary point found above, we have

$\dfrac{\partial^2 u}{\partial x^2} = -u\left[\dfrac{a^2}{(ax+by+cz)^2} + 2\alpha^2\right]$, which is –ive because u is positive for the values of x, y, z found above.

$\therefore$ u is maximum at the stationary point found above.

Also putting the values of x, y, z found above in the value of u, the maximum value of u

$$= R.e^{-\frac{1}{4R^2}\left(\frac{a^2}{\alpha^2}+\frac{b^2}{\beta^2}+\frac{c^2}{\gamma^2}\right)} = R.e^{-\frac{1}{4R^2}.2R^2}$$

$$= R.e^{-1/2} = \frac{R}{\sqrt{e}} = \frac{1}{\sqrt{e}} \cdot \sqrt{\left\{\frac{1}{2}\left(\frac{a^2}{\alpha^2}+\frac{b^2}{\beta^2}+\frac{c^2}{\gamma^2}\right)\right\}}$$

$$= \sqrt{\left\{\frac{1}{2e}\left(\frac{a^2}{\alpha^2}+\frac{b^2}{\beta^2}+\frac{c^2}{\gamma^2}\right)\right\}}.$$

Example 16: *Discuss the maximum or minimum values of u where*

$u = x^2 + y^2 + z^2 + x - 2z - xy.$

Solution: For a maximum or a minimum of u, we must have

$$\frac{\partial u}{\partial x} = 2x - y + 1 = 0,$$

$$\frac{\partial u}{\partial y} = -x + 2y = 0,$$

and $$\frac{\partial u}{\partial z} = 2z - 2 = 0.$$

These equations give $x = -\dfrac{2}{3}$,

$y = -\dfrac{1}{3}$, $z = 1$.

$\therefore \left(-\frac{2}{3}, -\frac{1}{3}, 1\right)$ is the only point at which u is stationary *i.e.*, at which u may have a maximum or a minimum.

Now $\quad \frac{\partial^2 u}{\partial x^2} = 2, \frac{\partial^2 u}{\partial y^2} = 2, \frac{\partial^2 u}{\partial z^2} = 2,$

$$\frac{\partial^2 u}{\partial y \partial z} = 0, \frac{\partial^2 u}{\partial z \partial x} = 0$$

and $\quad \frac{\partial^2 u}{\partial x \partial y} = -1.$

If A, B, C, F, G and G denote the respective values of these six partial derivatives of second order at the point $\left(-\frac{2}{3}, -\frac{1}{3}, 1\right)$, then

$$A = 2, B = 2, C = 2, F = 0, G = 0, H = -1.$$

Now we have

$$A = 2, \begin{vmatrix} A & H \\ H & B \end{vmatrix} = \begin{vmatrix} 2 & -1 \\ -1 & 2 \end{vmatrix} = 3$$

and $\quad \begin{vmatrix} A & H & G \\ H & B & F \\ G & F & C \end{vmatrix} = \begin{vmatrix} 2 & -1 & 0 \\ -1 & 2 & 0 \\ 0 & 0 & 2 \end{vmatrix} = 6.$

Since these three expressions are all positive, we have a minimum of u when $x = -\frac{2}{3}, y = -\frac{1}{3}, z = 1.$

Example 17: *Find the maximum or minimum values of u where*

$$u = axy^2z^3 - x^2y^2z^3 - xy^3z^3 - xy^2z^4.$$

Solution: We have $\frac{\partial u}{\partial x} = ay^2z^3 - 2xy^2z^3 - y^3z^3 - y^2z^4$

$$= y^2z^3 (a - 2x - y - z),$$

$$\frac{\partial u}{\partial y} = 2a\, xyz^3 - 2x^2yz^3 - 3xy^2z^3 - 2xyz^4$$

$$= xyz^3 (2a - 2x - 3y - 2z)$$

and $\quad \frac{\partial u}{\partial z} = 2axy^2z^2 - 3x^2y^2z^2 - 3xy^3z^3 - 4xy^2z^3$

$$= xy^2z^2(3a - 3x - 3y - 4z).$$

For a maximum or a minimum of u, we must have

$$\frac{\partial u}{\partial x}=0,\ \frac{\partial u}{\partial y}=0,\ \frac{\partial u}{\partial z}=0.$$

Now one solution of the equations $\frac{\partial u}{\partial x}=0,\ \frac{\partial u}{\partial y}=0,\ \frac{\partial u}{\partial z}=0$ is given by the equations

$$2x + y + z = a,\ 2x + 3y + 2z = 2a,\ 3x + 3y + 4z = 3a.$$

Solving these equations, we get $x=\frac{a}{7},\ y=\frac{2a}{7},\ z=\frac{3a}{7}$.

∴ u is stationary at the point $\left(\frac{a}{7},\frac{2a}{7},\frac{3a}{7}\right)$.

Now $\quad A=\frac{\partial^2 u}{\partial x^2}=y^2z^3.(-2)$.

At the stationary point $\left(\frac{a}{7},\frac{2a}{7},\frac{3a}{7}\right)$, the value of A is –ive.

∴ u has a maximum value at the point

$$\left(\frac{a}{7},\frac{2a}{7},\frac{3a}{7}\right).$$

Putting $x=\frac{a}{7},\ y=\frac{2a}{7},\ z=\frac{3a}{7}$ in the value of u, the maximum value of u at the point $\left(\frac{a}{7},\frac{2a}{7},\frac{3a}{7}\right)$.

Example 18(a): *Find the maximum value of u where*

$$u=\frac{xyz}{(a+x)(x+y)(y+z)(z+b)}.$$

Solution: We have

$$\log u = \log x + \log y + \log z - \log (a + x) - \log (x + y) - \log (y + z) - \log (z + b).$$

$$\therefore\ \frac{1}{u}\frac{\partial u}{\partial x}=\frac{1}{x}-\frac{1}{a+x}-\frac{1}{x+y}=\frac{ay-x^2}{x(a+x)(x+y)}$$

$$\frac{\partial u}{\partial x}=\frac{(ay-x^2)u}{x(a+x)(x+y)}.$$

Similarly $\frac{\partial u}{\partial x}=\frac{(xz-y^2)u}{y(x+y)(y+z)}$

and $$\frac{\partial u}{\partial z} = \frac{(by - z^2)u}{z(y+z)(z+b)}.$$

Now for a maximum or a minimum of u, we must have

$$\frac{\partial u}{\partial x} = 0 \text{ i.e., } ay - x^2 = 0$$

$$\frac{\partial u}{\partial y} = 0 \text{ i.e., } xz - y^2 = 0$$

and $$\frac{\partial u}{\partial z} = 0 \text{ i.e., } by - z^2 = 0.$$

From the above equations, it follows that a, x, y, z and b are in geometrical progression. Let r be the common ratio of this geometrical progression. Then

$$ar^4 = b \text{ or } r = \left(\frac{b}{a}\right)^{1/4}$$

Also $x = ar$, $y = ar^2$, $z = ar^3$.

Substituting these values, we get

$$u = \frac{ar.ar^2.ar^3}{a(1+r).ar(1+r).ar^2(1+r).ar^3(1+r)}$$

$$= \frac{1}{a(1+r)^4} = \frac{1}{a\left[1+\left(\frac{b}{a}\right)^{1/4}\right]^4} = \frac{1}{\left(a^{1/4}+b^{1/4}\right)^4}$$

This gives a stationary value of u. To decide whether this value of u is a maximum or a minimum we proceed to find the second order partial derivatives of u.

We have

$$\frac{\partial^2 u}{\partial x^2} = \frac{-2xu}{x(a+x)(x+y)} + \left(ay - x^2\right)\frac{\partial}{\partial x}\left[\frac{u}{x(a+x)(x+y)}\right].$$

$\therefore$ when $x = ar$, $y = ar^2$, $z = ar^3$, we have

$$A = \frac{\partial^2 u}{\partial x^2} = \frac{-2.ar.u}{ar.a(1+r).ar(1+r)} = \frac{-2u}{a^2 r(1+r)^2},$$

which is negative.

Hence the above stationary value of u is a maximum.

Ans. Maximum value of $u = \dfrac{1}{\left(a^{1/4} - b^{1/4}\right)^4}$.

Note: In the complicated problems in order to find whether a stationary value of u is a maximum or a minimum, it is sufficient to find the value of a second partial differential coefficient of u with respect to any of the independent variables. The value of u will be maximum or minimum according as the value of this second partial derivative at the stationary point under consideration is –ive or +ive.

Example 18(b): *Show that the point such that the sum of the squares of its distances from n given points shall be minimum, is the centre of the mean position of the given points.*

Solution: Let the n given points be (a_1, b_1, c_1), (a_2, b_2, c_2), ..., (a_n, b_n, c_n) and let (x, y, z) be the coordinates of the required point.

If u denotes the sum of the squares of the distances of (x, y, z) from the n given points, then

$$u = \Sigma[(x - a_1)^2 + (y - b_1)^2 + (z - c_1)^2]$$

$$= \Sigma(x - a_1)^2 + S(y - b_1)^2 + S(z - c_1)^2.$$

For a maximum or a minimum of u, we must have

$$\frac{\partial u}{\partial x} = 2\Sigma(x - a_1) = 2nx - 2\Sigma a_1 = 0,$$

$$\frac{\partial u}{\partial y} = 2\Sigma(y - b_1) = 2ny - 2\Sigma b_1 = 0,$$

and $$\frac{\partial u}{\partial z} = 2\Sigma(z - c_1) = 2nz - 2\Sigma c_1 = 0.$$

Solving these equations, we get

$$x = \frac{\Sigma a_1}{n},\ y = \frac{\Sigma b_1}{n},\ z = \frac{\Sigma c_1}{n}.$$

Now $$A = \frac{\partial^2 u}{\partial x^2} = 2n,\ B = \frac{\partial^2 u}{\partial y^2} = 2n,\ C = \frac{\partial^2 u}{\partial z^2} = 2n,$$

$$F = \frac{\partial^2 u}{\partial y\,\partial z} = 0,\ G = \frac{\partial^2 u}{\partial z\,\partial x} = 0,\ H = \frac{\partial^2 u}{\partial x\,\partial y} = 0.$$

We have $A = 2n$, $\begin{vmatrix} A & H \\ H & B \end{vmatrix} = \begin{vmatrix} 2n & 0 \\ 0 & 2n \end{vmatrix} = 4n^2$, and $$\begin{vmatrix} A & H & G \\ H & B & F \\ G & F & C \end{vmatrix} = \begin{vmatrix} 2n & 0 & 0 \\ 0 & 2n & 0 \\ 0 & 0 & 2n \end{vmatrix} = 8n^3.$$

Since these three expressions are all positive, u is minimum when

$$x = \frac{\Sigma a_1}{n},\ y = \frac{\Sigma b_1}{n} \text{ and } z = \frac{\Sigma c_1}{n}$$

i.e., u is minimum when the point (x, y, z) is the centre of the mean position of the n given points.

Example 18(c): *Show that* $u = (x + y + z)^3 - 3(x + y + z) - 24xyz + a^3$ *has minimum at (1, 1, 1) and maximum at (–1, –1, –1).*

Solution: For a maximum or a minimum of u, we must have

$$\frac{\partial u}{\partial x} = 3\,(x + y + z)^2 - 3 - 24yz = 0, \qquad ...(1)$$

$$\frac{\partial u}{\partial y} = 3\,(x + y + z)^2 - 3 - 24zx = 0 \qquad ...(2)$$

and $$\frac{\partial u}{\partial z} = 3\,(x + y + z)^2 - 3 - 24xy = 0. \qquad ...(3)$$

Subtracting (2) from (1), we get

$$24z\,(x - y) = 0$$

which has x = z for one of its solutions.

Thus, the equations (1), (2) and (3) are satisfied when

$$x = y = z$$

Putting y = x and z = x (1), we get

$$27x^2 - 3 - 24x^2 = 0 \text{ or } 3x^2 = 3 \text{ or } x^2 = 1 \text{ or } x = \pm 1.$$

∴ x = y = z = 1 and x = y = z = – 1 are solutions of the equations (1), (2) and (3).

Hence u is stationary at the points (1, 1, 1)

and (– 1, – 1, – 1).

Now $$A = \frac{\partial^2 u}{\partial x^2} = 6(x + y + z),$$

$$B = \frac{\partial^2 u}{\partial y^2} = 6(x + y + z),$$

$$B=\frac{\partial^2 u}{\partial y^2}=6(x+y+z),$$

$$F=\frac{\partial^2 u}{\partial y\,\partial z}=6(x+y+z)-24x,$$

$$G=\frac{\partial^2 u}{\partial z\,\partial x}=6(x+y+z)-24y,$$

$$H=\frac{\partial^2 u}{\partial x\,\partial y}=6(x+y+z)=-24z.$$

Nature of u at (1, 1, 1): At the stationary point (1, 1, 1), we have A = 18, B = 18, C = 18, F = – 6, G = – 6, H = – 6.

∴ at the point (1, 1, 1), we have

$$A = 18, \begin{vmatrix} A & H \\ H & B \end{vmatrix} = \begin{vmatrix} 18 & -6 \\ -6 & 18 \end{vmatrix} = 288$$

and $$\begin{vmatrix} A & H & G \\ H & B & F \\ G & F & C \end{vmatrix} = \begin{vmatrix} 18 & -6 & -6 \\ -6 & 18 & -6 \\ -6 & -6 & 18 \end{vmatrix}$$

$$=6^3.\begin{vmatrix} 3 & -1 & -1 \\ -1 & 3 & -1 \\ -1 & -1 & 3 \end{vmatrix}$$

$$=6^3.\begin{vmatrix} 1 & 1 & 1 \\ -1 & 3 & -1 \\ -1 & -1 & 3 \end{vmatrix}, \text{ by } R_1+R_2+R_3$$

$$=6^3.\begin{vmatrix} 1 & 0 & 0 \\ -1 & 4 & 0 \\ -1 & 0 & 4 \end{vmatrix}, \text{ by } R_2-R_1 \text{ and } R_3-R_1$$

$= 6^3.16.$

Since these three expressions are all positive, we have a minimum of u at the point (1, 1, 1).

Nature of u at the stationary point (–1, –1, –1).

At the stationary point (–1, –1, –1), we have

$$A = -18 = B = C,\ F = 6 = G = H.$$

$\therefore$ at the point (–1, –1, –1), we have

$$A = -18, \begin{vmatrix} A & H \\ H & B \end{vmatrix} = \begin{vmatrix} -18 & 6 \\ 6 & -18 \end{vmatrix} = 288$$

and
$$\begin{vmatrix} A & H & G \\ H & B & F \\ G & F & C \end{vmatrix} = \begin{vmatrix} -18 & 6 & 6 \\ 6 & -18 & 6 \\ 6 & 6 & -18 \end{vmatrix} = -6^3.16.$$

Since these three expressions are alternately negative and positive, we have a maximum of u at the point (–1, –1, –1).

Example 19: *If two variables x and y are connected by the relation* $ax^2 + by^2 = ab$, *show that the maximum and minimum values of the function* $u = x^2 + y^2 + xy$ *will be the roots of the equation*

$$4(u - a)(u - b) = ab.$$

Solution: We have $u = x^2 + y^2 + xy$, ...(1)

where the variables x and y are connected by the relation

$$ax^2 + by^2 = ab. \quad ...(2)$$

For a maximum or a minimum of u, we have

$$du = 0 \Rightarrow 2x\,dx + 2y\,dy + y\,dx + x\,dy = 0$$

$$\Rightarrow \quad (2x + y)\,dx + (2y + x)\,dy = 0. \quad ...(3)$$

Also differentiating (2), we have

$$2ax\,dx + 2by\,dy = 0 \text{ or } ax\,dx + by\,dy = 0. \quad ...(4)$$

Multiplying (3) by 1 and (4) by λ, and adding and then equating to zero the coefficients of dx and dy, we get

$$(2x + y) + \lambda\,ax = 0 \quad ...(5)$$

and
$$(2y + x) + \lambda\,by = 0. \quad ...(6)$$

Multiplying (5) by x and (6) by y and adding, we get

$$2(x^2 + y^2 + xy) + \lambda\,(ax^2 + by^2) = 0$$

or
$$2u + \lambda\,ab = 0 \text{ or } \lambda = -\frac{2u}{ab}.$$

Putting $\lambda = -\frac{2u}{ab}$ in (5), we get

$$(2x+y)-\frac{2u}{b}x=0$$

or $\quad 2(b - u)\ x + by = 0.$ ...(7)

Similarly putting $\lambda=-\frac{2u}{ab}$ in (6), we get

$$(2y+x)-\frac{2u}{a}y=0$$

or $\quad ax + 2\ (a - u)\ y = 0.$ ...(8)

Eliminating x and y from (7) and (8), we get

$$\begin{vmatrix} 2(b-u) & b \\ a & 2(a-u) \end{vmatrix}=0$$

or $\quad 4(b - u)\ (a - u) - ab = 0$

or $4(u - a)\ (u - b) = ab$, which gives the maximum and minimum values of u.

Example 20: *Find the stationary values of $x^2 + y^2 + z^2$ subject to the conditions*

$$ax^2 + by^2 + cz^2 = 1 \text{ and } lx + my + nz = 0.$$

Interpret the result geometrically.

Solution: Let $u = x^2 + y^2 + z^2$, ...(1)

where the variables x, y and z are connected by the relations

$$ax^2 + by^2 + cz^2 = 1, \quad ..(2)$$

and $\quad lx + my + nz = 0.$

For a stationary value of u, we have

$$du = 0$$

$\Rightarrow \quad 2x\ dx + dy\ dy + 2z\ dz = 0$

$\Rightarrow \quad x\ dx + y\ dy + z\ dz = 0.$ ...(4)

Also differentiating the given relations (2) and (3), we get

$$2ax\ dx + 2by\ dy + 2cz\ dz = 0$$

i.e., $\quad ax\ dx + by\ dy + cz\ dz = 0$ ...(5)

and $\quad l dx + m\ dy + n\ dz = 0.$ ...(6)

Multiplying equations (4) by 1, (5) by λ and (6) by μ and adding, and then equating the coefficients of dx, dy, dz to zero, we get

$$x + \lambda ax + \mu l = 0 \quad ...(7)$$

$$y + \lambda by + \mu m = 0 \qquad ...(8)$$

$$z + \lambda cz + \mu n = 0 \qquad ...(9)$$

Multiplying the equations (7), (8) and (9) by x, y and z respectively we get

$$x^2 + y^2 + z^2 + \lambda (ax^2 + by^2 + cz^2) + \mu (lx + my + nz) = 0,$$

or $u = \lambda.1 + \mu.0 = 0$, using (1), (2) and (3)

or $\lambda = -u$.

Substituting for λ in the equations (7), (8) and (9), we get

$$x = \frac{\mu l}{au - 1}, y = \frac{\mu m}{bu - 1}, z = \frac{\mu n}{cu - 1}.$$

Substituting these values of x, y, z in (3), we get

$$\frac{\mu l^2}{au - 1} + \frac{\mu m^2}{bu - 1} + \frac{\mu n^2}{cu - 1} = 0$$

or
$$\frac{l^2}{au - 1} + \frac{m^2}{bu - 1} + \frac{n^2}{cu - 1} = 0. \qquad ...(10)$$

Hence the stationary (*i.e.*, maximum or minimum) values of u are given by the equation (10). The equation (10) is a quadratic in u and so it gives two stationary values of u.

Geometrical Interpretation : The surface $ax^2 + by^2 + cz^2 = 1$ represents an ellipsoid (or a hyperboloid) whose centre is origin, and $lx + my + nz = 0$ is a plane passing through the origin. Therefore the point (x, y, z) satisfying both the conditions (2) and (3) lies on the conic in which (2) and (3) intersect. Also $x^2 + y^2 + z^2$ gives the square of the distance of (x, y, z) from the origin which is also the centre of the conic of intersection. The maximum and minimum values of this distance are the major and minor semi-axes of the conic. So the equation (10) gives the squares of the lengths of the semi-axes of the conic of intersection.

Example 21: *Find the maximum and minimum values of*

$$u = a^2x^2 + b^2y^2 + c^2z^2,$$

where $x^2 + y^2 + z^2 = 1$ *and* $lx + my + nz = 0$.

Solution: We have $u = a^2x^2 + b^2y^2 + c^2z^2$, ...(1)

where the variables x, y, z are connected by relations

$$x^2 + y^2 + z^2 = 1 \qquad ...(2)$$

and
$$lx + my + nz = 0 \qquad ...(3)$$

For a maximum or a minimum of u, we have

$$du = 0$$

$\Rightarrow \quad 2a^2x\, dx + 2b^2y\, dy + 2c^2z\, dz = 0$

$\Rightarrow \quad a^2x\, dx + b^2y\, dy + c^2z\, dz = 0.$ ...(4)

Also differentiating the two given equations (2) and (3) connecting the variables x, y and z, we get

$$2x\, dx + 2y\, dy + 2z\, dz = 0$$

i.e., $\quad x\, dx + y\, dy + z\, dz = 0$ ...(5)

and $\quad l dx + m\, dy + n\, dz = 0$ ...(6)

Multiplying (4), (5) and (6) by 1, λ and μ respectively and adding, and then equating to zero the coefficients of dx, dy and dz, we get

$a^2x + \lambda x + \mu l = 0,$ ...(7)

$b^2y + \lambda y + \mu m = 0,$ ...(8)

and $\quad c^2z + \lambda z + \mu m = 0.$ ...(9)

Multiplying the equations (7), (8) and (9) by x, y and z respectively and adding, we get

$$u + \lambda.1 + \mu.0 = 0 \text{ or } \lambda = -u.$$

Putting $\lambda = -u$ in (7), we get

$$a^2x - ux + \mu l = 0$$

or $\quad x = \dfrac{\mu l}{u - a^2}.$

Similarly, from (8) and (9), we get

$$y = \frac{\mu m}{u - b^2} \text{ and } z = \frac{\mu m}{u - c^2}.$$

Substituting these values of x, y, z in $lx + my + nz = 0$, we get

$$\frac{ul^2}{u - a^2} + \frac{\mu m^2}{u - b^2} + \frac{\mu n}{u - c^2} = 0.$$

or $\quad \dfrac{l^2}{u - a^2} + \dfrac{m^2}{u - b^2} + \dfrac{n}{u - c^2} = 0.$ ...(10)

These equation (10) gives the required maximum or minimum values of u.

Example 22: *Find the points where*

$$u = ax^p + by^q + cz^r$$

has extreme values subject to the condition

$$x^l + y^m + z^n = k.$$

Solution: We have $u = ax^p + by^q + cz^r$, ...(1)

where the variables x, y, z are connected by the relation

$$x^l + y^m + z^n = k. \qquad ...(2)$$

For a maximum or a minimum of u, we have

$$du = 0 \Rightarrow apx^{p-1}\,dx + bqy^{q-1}\,dy + crz^{r-1}\,dz = 0. \qquad ..(3)$$

Also differentiating (2), we get

$$lx^{l-1}\,dx + my^{m-1}\,dy + nz^{n-1}\,dz = 0. \qquad ...(4)$$

Multiplying (3) by 1 and (4) by λ, and adding and then equating to zero the coefficients of dx, dy, dz, we get

$$apx^{p-1} + \lambda l x^{l-1} = 0, \qquad ...(5)$$

$$bqy^{q-1} + \lambda m y^{m-1} = 0, \qquad ...(6)$$

and $$cr\, z^{r-1} + \lambda n\, z^{n-1} = 0. \qquad ...(7)$$

From (5), we have

$$ap\, x^{p-1} = -\lambda l\, x^{l-1} \text{ or } ap\, x^{p-l} = -\lambda l$$

or $$\frac{x^{p-l}}{l/pa} = -\lambda.$$

Similarly from (6) and (7), we have

$$\frac{y^{q-m}}{\frac{m}{qb}} = -\lambda \text{ and } \frac{z^{r-n}}{\frac{n}{rc}} = -\lambda.$$

Hence, the values of x, y, z for which u has extreme values are given by

$$\frac{x^{p-l}}{\frac{l}{pa}} = \frac{y^{q-m}}{\frac{m}{qb}} = \frac{z^{r-m}}{\frac{n}{rc}}.$$

Example 23: *Prove that of all rectangular parallelopipeds of the same volume the cube has the least surface.*

Solution:

Let x, y, z be the dimensions of the rectangular parallelopiped, V be its volume and S be its surface. Then

$$S = 2xy + 2yz + 2zx \qquad ...(1)$$

and xyz = V = some constant. ...(2)

For a maximum or minimum of S, we have

$$dS = 2(y + z)\,dx + 2(z + x)\,dy + 2(x + y)\,dz = 0$$

i.e., $$(y + z)\,dx + (z + x)\,dy + (x + y)\,dz = 0 \qquad ...(3)$$

Also differentiating (2) and observing that V is constant, we have

$$yz\,dx + zx\,dy + xy\,dz = 0. \qquad ...(4)$$

Multiplying (3) by 1 and (4) by λ, and adding and then equating to zero the coefficients of dx, dy and dz, we get

$$(y + z) + \lambda yz = 0, \qquad ...(5)$$

$$(z + x) + \lambda zx + 0, \qquad ...(6)$$

and $$(x + y) + \lambda xy = 0.. \qquad ...(7)$$

These give $-\lambda = \frac{1}{y} + \frac{1}{z} = \frac{1}{z} + \frac{1}{x} = \frac{1}{x} + \frac{1}{y}$.

$$\therefore \qquad \frac{1}{y} - \frac{1}{x} = 0 \text{ or } x = y.$$

Similarly y = z.

Hence for a stationary value of S, we have

$$x = y = z = V^{1/3}, \text{ from (2).}$$

Thus S is stationary when the rectangular parallelopiped is a cube.

Let us now find the nature of this stationary value of S.

Here S is a function of three variables x, y, z which are connected by the relation (2) so that only two variables are independent. Let us regard x and y as independent variables and z to be dependent on them.

Then from (1), $\frac{\partial S}{\partial x} = 2y + 2y\frac{\partial z}{\partial x} + 2z + 2x\frac{\partial z}{\partial x}$.

Also from (2), $yz + xy\frac{\partial z}{\partial x} = 0$ *i.e.*, $\frac{\partial z}{\partial x} = -\frac{z}{x}$.

$$\therefore \quad \frac{\partial S}{\partial x} = 2y = \frac{2yz}{x} + 2z - 2z = 2y - \frac{2yz}{x}$$

and $\frac{\partial^2 S}{\partial x^2} = \frac{2yz}{x^2} - \frac{2y}{x} \cdot \frac{\partial z}{\partial x} = \frac{2yz}{x^2} + \frac{2yz}{x^2} = \frac{4yz}{x^2} = 4$ at x = y = z.

Similarly by symmetry $\frac{\partial^2 S}{\partial y^2} = 4$ at x = y = z.

Also $$\frac{\partial^2 S}{\partial x\,\partial y} = 2 - \frac{2z}{x} - \frac{2y}{x}\frac{\partial z}{\partial y}.$$

But differentiating (2) partially w.r.t. y taking x as constant, we get

$$xz + xy\frac{\partial z}{\partial y} = 0 \text{ or } \frac{\partial z}{\partial y} = -\frac{z}{y}.$$

$$\therefore \quad \frac{\partial^2 S}{\partial x \partial y} = 2 - \frac{2z}{x} - \frac{2y}{x}\left(-\frac{z}{y}\right) = 2 - \frac{2z}{x} + \frac{2z}{x} = 2.$$

That at the stationary point $x = y = V^{1/3}$, we have

$$r = \frac{\partial^2 S}{\partial x^2} = 4, \; s = \frac{\partial^2 S}{\partial x \partial y} = 2 \text{ and } t = \frac{\partial^2 S}{\partial y^2} = 4.$$

$\therefore$ $rt - s^2 = 4 \times 4 - 2^2 = 12$ which is > 0.

Also $r = 4$ which is > 0.

Hence the stationary value of S given by $x = y = z = V^{1/3}$ is a minimum.

Thus of all rectangular parallelopipeds of the same volume, the cube has the least surface.

Example 24(a): *Find the maximum or minimum value of $x^2 + y^2 + z^2$, subject to the conditions*

$$lx + my + nz = 1, \; l'x + m'y\, n'z = 1.$$

Solution: Let $u = x^2 + y^2 + z^2$, ...(1)

where the variables x, y and z connected by the relations

$$lx + my + nz = 1, \quad ...(2)$$

and $l'x + m'y + n'z = 1$. ...(3)

For a maximum or a minimum of u, we have $du = 0$

$\Rightarrow$ $2xdx + 2ydy + 2\,zdz = 0$

$\Rightarrow$ $xdx + ydy + zdz = 0$. ...(4)

Also differentiating (2) and (3), we get

$$ldx + n'dy + ndz = 0 \quad ...(5)$$

and $l'dx + m'dy + n'dz = 0$. ...(6)

Multiplying (4) by 1, (5) by λ and (6) μ, and adding and then equating to zero the coefficients of dx, dy and dz, we get

$$x + l\lambda + l'\mu = 0, \quad ...(7)$$

$$y + m\lambda + m'\mu = 0, \quad ...(8)$$

and $z + n\lambda + n'\mu = 0$. ...(9)

Multiplying the equations (7), (8) and (9) by x, y and z respectively and adding, we get

$$u + \lambda.z + \mu.1 = 0. \quad ...(10)$$

Again multiplying the equations (7), (8) and (9) by l, m and n respectively and adding, we get

$$1 + \lambda\Sigma l^2 + \mu\ \Sigma ll'^2 = 0. \quad ...(11)$$

Next multiplying the equations (7), (8) and (9) by l', m' and n' respectively and adding, we get

$$1 + \lambda\ \Sigma ll' + \mu\ \Sigma l'^2 = 0. \quad ...(12)$$

Now eliminating λ and μ from (10), (11) and (12), we get

$$\begin{vmatrix} u & 1 & 1 \\ 1 & \Sigma l^2 & \Sigma ll' \\ 1 & \Sigma ll' & \Sigma l'^2 \end{vmatrix} = 0.$$

The above equation gives the maximum or minimum value of u.

Note: If we wish to find an explicit expression for the extreme value of u and also wish to say whether it is maximum or minimum we proceed as follows:

Solving the equations (11) and (12) for λ and μ, we get

$$\frac{1}{\Sigma ll' - \Sigma l'^2} = \frac{\mu}{\Sigma ll' - \Sigma l^2} = \frac{1}{\Sigma l^2 \Sigma l'^2 - (\Sigma ll')^2}$$

But when a, b, c are all +ive, r is –ive and so the stationary value is a true maximum in this case. Also when a, b, c are all –ive, r is +ive and so the stationary value is a true minimum in this case.

or $\lambda = \dfrac{\Sigma ll' - \Sigma l'^2}{\Sigma(mn' - m'n)^2}$ and $\mu = \dfrac{\Sigma ll' - \Sigma l'^2}{\Sigma(mn' - m'n)^2}$

[$\because \Sigma l^2\ \Sigma l'^2 - (\Sigma ll')^2 = \Sigma\ (mn' - m'n)^2$, by Lagrange's identity]

Putting these values of λ and μ in (10), the maximum or minimum value of u is given by

$$u = -\lambda - \mu = \frac{\Sigma l^2 + \Sigma l'^2 - 2\Sigma ll'}{\Sigma(mn' - m'n)^2} = \frac{\Sigma(l - l')^2}{\Sigma(mn' - m'n)^2}.$$

Example 24(b): *Show that the maximum and minimum values of* $u = ax^2 + by^2 + cz^2 + 2fyz + 2gzx + 2hxy$ *subject to the conditions* $lx + my + nz = 0$ *and* $x^2 + y^2\ z^2 = 1$ *are given by the equation*

$$\begin{vmatrix} a-u, & h & g & l \\ h & b-u & f & m \\ g & f & c-u & n \\ l & m & n & 0 \end{vmatrix} = 0.$$

Solution: We have $u = ax^2 + by^2 + cz^2 + 2fyz + 2gzx + 2hxy$, ...(1)

where the variables x, y and z are connected by the relations

$$lx + my + nz = 0 \quad ...(2)$$

and $$x^2 + y^2 + z^2 = 1. \quad ...(3)$$

For a maximum or a minimum of u, we have du = 0

$\Rightarrow 2(ax + gz + hy) + 2(by + fz + hx)dy + 2(cz + fy + gx)dz = 0$

$\Rightarrow (ax + hy + gz)dx + (hx + by + fz)dy + (gx + fy + cz)\, dz = 0.$ ...(4)

Also differentiating (2) and (3), we get

$$l\, dx + m\, dy + n\, dz = 0 \quad ...(5)$$

and $$x\, dx + y\, dy + z\, dz = 0. \quad ...(6)$$

Multiplying (4), by 1, (5) by λ and (6) by μ, and adding and then equating to zero the coefficient of dx, dy and dz, we get

$$(ax + hy + gz)\, \lambda l + \mu x = 0 \quad ...(7)$$

$$(hx + by + fz)\, \lambda m + \mu y = 0 \quad (8)$$

and $$(gx + fy + cz)\, \lambda n + \mu z = 0 \quad ...(9)$$

Multiplying eqns. (7), (8) and (9) by x, y and z respectively and adding, we get

$$u + \lambda.0 + \mu.1 = 0 \text{ or } m = -u.$$

Putting $\mu = -u$ in (7), (8) and (9), we get

$$(a - u\, x + hy + gz + \lambda l = 0, \quad ....(10)$$

$$hx + (b - y) + fz + \lambda m = 0, \quad ...(11)$$

and $$gx + fy + (c - u)\, z + ln = 0. \quad ...(12)$$

Also the relation (2) can be written as

$$lx + my + nz + \lambda.0 = 0. \quad ...(13)$$

Eliminating x, y, z and λ from (10), (11), (12) and (13), we get

$$\begin{vmatrix} a-u, & h & g & l \\ h & b-u & f & m \\ g & f & c-u & n \\ l & m & n & 0 \end{vmatrix} = 0,$$

which gives the maximum and minimum values of u.

To Find the Nature of This Stationary Value of u : Since there are two relations amongst the variables x, y and z, therefore only one variable will be independent. Let it be x. Then

$$\frac{du}{dx}=\frac{\partial x}{\partial x}\cdot\frac{dy}{dx}+\frac{\partial u}{\partial z}\cdot\frac{dz}{dx}$$

$$=2x+2y\cdot\frac{dy}{dx}+2z\cdot\frac{dz}{dx}.$$

Now differentiating (2) and (3) w.r.t. x, we get

$$l+m\frac{dy}{dx}+n\frac{dz}{dx}=0$$

and $$l'+m'\frac{dy}{dx}+n'\frac{dz}{dx}=0.$$

Solving these, we get

$$\frac{\frac{dy}{dx}}{nl'-n'l}=\frac{\frac{dz}{dx}}{m'l-l'm}=\frac{1}{mn'-m'n}.$$

$$\therefore \qquad \frac{dy}{dx}=\frac{nl'-n'l}{mn'-m'n} \text{ and } \frac{dz}{dx}=\frac{m'l-l'm}{mn'-m'n}.$$

$$\therefore \qquad \frac{du}{dx}=2x+2y\cdot\frac{nl'-n'l}{mn'-m'n}+2z\cdot\frac{m'l-l'm}{mn'-m'n}$$

and so $$\frac{d^2u}{dx^2}=2+2\frac{dy}{dx}\cdot\frac{nl'-n'l}{mn'-m'n}+2\frac{dz}{dx}\cdot\frac{m'l-l'm}{mn'-m'n}$$

$$=2+2\left(\frac{nl'-n'l}{mn'-m'n}\right)^2+2\left(\frac{m'l-l'm}{mn'-m'n}\right)^2,$$

which is +ive.

Hence, the stationary value of u found above is the minimum value.

Example 25: *Prove that if $x+y+z=1$, $ayz+bzx+cxy$ has an extreme value equal to*

$$\frac{abc}{2abc+2ca+2ab-a^2-b^2-c^2}.$$

Prove also that if a, b, c are all positive and c lies between $a+b-2\sqrt{(ab)}$ and $a+b+2\sqrt{(ab)}$ this value is true maximum and that if a, b, c are all negative and c lies between $a+b\pm 2\sqrt{(ab)}$, it is true minimum.

Solution: Let u = ayz + bzx + cxy, ...(1)

where the variables x, y and z are connected by the relation

x + y + z = 1. ...(2)

For a maximum or a minimum of u, we have du = 0

$\Rightarrow$ $(bz + cy)\,dx + (cx + az)\,dy + (ay + bx)\,dz = 0.$ ...(3)

Also differentiating (2), we get

$$dx + dy + dz = 0. \qquad ...(4)$$

Multiplying (3) by 1 and (4) by λ, and adding and then equating to zero the coefficients of dx, dy and dz, we get

$$bz + cy + \lambda = 0,\ cx + az + \lambda = 0,\ ay + bx + \lambda = 0.$$

$\therefore$ $-\lambda = bz + cy = cx + az = ay + bx.$

From these, we have

$$z = \frac{ay + bx - cx}{a} = \frac{ay + bx - cy}{b}.$$

$\therefore$ $bx(a + c - b) = ay(b + c - a)$

or $\dfrac{x}{a(b+c-a)} = \dfrac{y}{b(a+c-b)} = \dfrac{z}{c(a+b-c)},$ (by symmetry)

$$= \frac{x+y+z}{2\Sigma bc - \Sigma a^2} = \frac{1}{2\Sigma bc - \Sigma a^2}. \qquad ...(5)$$

Hence u is stationary for the values of x, y and z given by (5).

Also the stationary value of u

$$= ayz + bzx + cxy$$

$$= \frac{abc\left(2bc + 2ca + 2ab - a^2 - b^2 - c^2\right)}{\left(2bc + 2ca + 2ab - a^2 - b^2 - c^2\right)^2}$$

$$= \frac{abc}{2bc + 2ca + 2ab - a^2 - b^2 - c^2}.$$

Now let us regard x and y as independent variables and z as a function of x and y given by (2).

From (1), $\dfrac{\partial u}{\partial x} = (bz + cy) + (ay + bx)\dfrac{\partial z}{\partial x}$, regarding y as a constant.

Also from (2), $1 + \left(\dfrac{\partial z}{\partial x}\right) = 0$ or $\dfrac{\partial z}{\partial x} = -1.$

$\therefore$ $\dfrac{\partial u}{\partial x} = bz + cy - ay - bx,$

so that $r = \dfrac{\partial^2 u}{\partial x^2} = b\dfrac{\partial z}{\partial x} - b = -b - b = -2b,$

$$s = \frac{\partial^2 u}{\partial x \partial y} = b\frac{\partial z}{\partial y} + c - a = c - a - b. \qquad \left[\because \text{ from (2), } \frac{\partial z}{\partial y} = -1\right]$$

Similarly $t = \frac{\partial^2 u}{\partial y^2} = -2a.$

$\therefore$ $rt - s^2 = 4ab - (c - a - b)^2$

$$= \left\{2\sqrt{(ab)} + c - a - b\right\}\left\{2\sqrt{(ab)} - c + a + b\right\}$$

$$= \left[c - \left\{a + b - 2\sqrt{(ab)}\right\}\right]\left[\left\{a + b + 2\sqrt{(ab)}\right\} - c\right].$$

Hence $rt - s^2$ will be positive when $c > a + b - 2\sqrt{(ab)}$ and $< a + b + 2\sqrt{(ab)}$ whether a, b, c are all +ive or –ive.

Example 26: *If* $u = x^2 + y^2 + z^2$,

where $ax^2 + by^2 + cz^2 + 2fy + 2gzx + 2hxy = 1$, *find the maximum or minimum values of u.*

Solution: We have $u = x^2 + y^2 + z^2$, ...(1)

where the variables x, y, z are connected by the relation

$$ax^2 + by^2 + cz^2 + 2fyz + 2gzx + 2hxy = 1. \qquad ...(2)$$

For a maximum or a minimum of u, we have

$$du = 0$$

$$\Rightarrow \qquad 2x\,dx + 2y\,dy + 2z\,dz = 0$$

$$x\,dx + y\,dy + z\,dz = 0. \qquad ...(3)$$

Also differentiating the given relation (2), we get

$$2ax\,dx + 2by\,dy + 2cz\,dz + 2fy\,dz + 2fz\,dy + 2gz\,dx + 2gx\,dz + 2hx\,dy + 2hy\,dx = 0$$

or $(ax + hy + gz)\,dx + (hx + by + fz)\,dy + (gx + fy + cz)\,dz = 0.$ ...(4)

Multiplying (3) by 1, (4) by λ and adding, and then equating the coefficients of dx, dy, dz to zero, we have

$$x + \lambda(ax + hy + gz) = 0, \qquad ...(5)$$

$$y + \lambda(hx + by + fz) = 0, \qquad ...(6)$$

and $$z + \lambda(gx + fy + cz) = 0. \qquad ...(7)$$

Multiplying (5) by x, (6) by y, (7) by z and adding, we get

$$x^2 + y^2 + z^2 + \lambda(ax^2 + by^2 + cz^2 + 2fyz + 2gzx + 2hxy) = 0$$

or $u + \lambda.1 = 0$, using eqns. (1) and (2).

$\therefore \quad \lambda = -u.$

Hence from (5), we have

$$x - u(ax + hy + gz) = 0$$

or $x(1 - au) - huy - guz = 0$

or $$\left(a - \frac{1}{u}\right)x + hy + gz = 0. \qquad ...(8)$$

Similarly from (6) and (7), we have

$$hx + \left(b - \frac{1}{u}\right)y + fz = 0, \qquad ...(9)$$

and $$gx + fy + \left(c - \frac{1}{u}\right)z = 0. \qquad ...(10)$$

Eliminating x, y, z from (8), (9), (10), we get

$$\begin{vmatrix} a - (1/u) & h & g \\ h & b - (1/u) & f \\ g & f & c - (1/u) \end{vmatrix} = 0 \qquad ...(1)$$

Hence the required maximum or minimum values of u are the roots of the equation (11).

Example 27: *Find the minimum value of $x + y + z$, subject to the condition*

$$\left(\frac{a}{x}\right) + \left(\frac{b}{y}\right) + \left(\frac{c}{z}\right) = 1.$$

Solution: Let$u = x + y + z$, ...(1)

where the variables x, y, z are connected by the relation

$$\frac{a}{x} + \frac{b}{y} + \frac{c}{z} = 1. \qquad ...(2)$$

For a maximum or a minimum of u, we have

$$du = 0 \Rightarrow dx + dy + dz = 0. \qquad ...(3)$$

Also differentiating the given equation (2), we get

$$-\frac{a}{x^2}dx - \frac{b}{y^2}dy - \frac{c}{z^2}dz = 0. \qquad ...(4)$$

Multiplying (3) by 1 and (4) by λ, and adding and then equating the coefficients of dx, dy, dz to zero, we get

$$1-\frac{\lambda a}{x^2}=0,\ 1-\frac{\lambda b}{y^2}=0,\ 1-\frac{\lambda c}{z^2}=0.$$

From these, we get $x=\sqrt{(\lambda a)}$, $y=\sqrt{(\lambda b)}$, $z=\sqrt{(\lambda c)}$.

Putting these values of x, y, z in (2), we get

$$\frac{a}{\sqrt{(\lambda a)}}+\frac{b}{\sqrt{(\lambda b)}}+\frac{c}{\sqrt{(\lambda c)}}=1 \text{ or } \frac{1}{\sqrt{\lambda}}\left(\sqrt{a}+\sqrt{b}+\sqrt{c}\right)=1$$

or $\sqrt{l}=\sqrt{a}+\sqrt{b}+\sqrt{c}$.

$\therefore$ u is stationary when

$$\sqrt{l}=\sqrt{a}+\sqrt{b}+\sqrt{c},\ y=\sqrt{b}\left(\sqrt{a}+\sqrt{b}+\sqrt{c}\right),$$

$$z=\sqrt{c}\left(\sqrt{a}+\sqrt{b}+\sqrt{c}\right).$$

Let us regard x and y as independent variables and z as a function of x and y given by (2).

From (1), we have

$$\frac{\partial u}{\partial x}=1+\frac{\partial z}{\partial x}.$$

Differentiating (2) partially w.r.t. x taking y as constant, we get

$$-\frac{a}{x^2}-\frac{c}{z^2}\frac{\partial z}{\partial x}=0 \text{ or } \frac{\partial z}{\partial x}=-\frac{az^2}{cx^2}.$$

$$\therefore \quad \frac{\partial u}{\partial x}=1-\frac{az^2}{cx^2},$$

so that $$\frac{\partial^2 u}{\partial x^2}=\frac{2az^2}{cx^3}-\frac{2az}{cx^2}\frac{\partial z}{\partial x}=\frac{2az^2}{cx^3}+\frac{2az}{cx^2}\cdot\frac{az^2}{cx^2},$$

which is positive for the values of x, y, z found above.

Hence, u is minimum at the stationary point found above and the minimum value of u

$$=\sqrt{a}\left(\sqrt{a}+\sqrt{b}+\sqrt{c}\right)+\sqrt{b}\left(\sqrt{a}+\sqrt{b}+\sqrt{c}\right)+\sqrt{c}\left(\sqrt{a}+\sqrt{b}+\sqrt{c}\right)$$

$$=\left(\sqrt{a}+\sqrt{b}+\sqrt{c}\right)^2.$$

Example 28: *Find the maxima and minima of x^2+y^2 subject to the condition*

$$ax^2+2hxy+by^2=1.$$

Solution: Let $u=x^2+y^2$, ...(1)

where the variables x and y are connected by relation

$$ax^2 + 2hxy + by^2 = 1. \qquad ...(2)$$

For a maximum or a minimum of u, we have du = 0

$\Rightarrow \quad 2xdx + 2ydy = 0$

$\Rightarrow \quad xdx + ydy = 0. \qquad ...(3)$

Also differentiating the given relation (2), get

$$2axdx + 2hxdy + 2hydx + 2bydy = 0$$

or $\quad (ax + hy)\, dx + (hx + by)\, dy = 0 \qquad ....(4)$

Multiplying (3) by 1, (4) by λ and adding, and then equating the coefficients of dx, dy to zero, we have

$$x + \lambda\,(ax + hy) = 0 \qquad ...(5)$$

and $\quad y + \lambda\,(hx + by) = 0 \qquad ...(6)$

Multiplying (5) by x, (6) by y and adding, we get

$$x^2 + y^2 + \lambda\,(ax^2 + 2hxy + by^2) = 0$$

or $\quad u + \lambda\,.\,1 = 0, \qquad$ using (1) and (2).

$\therefore \quad \lambda = -\,u.$

Hence from (5), we have

$$x - u\,(ax + hy) = 0 \text{ or } x\,(1 - au) - huy = 0$$

or $\quad \left(a - \dfrac{1}{u}\right)x + hy + 0 \qquad ...(7)$

Similarly from (6), we have

$$hx + \left(b - \frac{1}{u}\right)y = 0 \qquad ...(8)$$

Eliminating x and y from (7) and (8), we get

$$\begin{vmatrix} a - \frac{1}{u} & h \\ h & b - \frac{1}{u} \end{vmatrix} = 0 \text{ or } \left(a - \frac{1}{u}\right)\left(b - \frac{1}{u}\right) = h^2 \qquad ...(9)$$

Hence the required maximum or minimum values of $u = x^2 + y^2$ are the roots of the equation (9).

Example 29(a): *Find the maximum and minimum values of*

$$\frac{x^2}{a^4} + \frac{y^2}{b^4} + \frac{z^2}{c^4},$$

when $lx + my + nz = 0$ *and* $\frac{x^2}{a^2}+\frac{y^2}{b^2}+\frac{z^2}{c^2}=1$.

Interpret the result geometrically.

Solution: Let $u=\frac{x^2}{a^4}+\frac{y^2}{b^4}+\frac{z^2}{c^4}$. Then for a maximum or a minimum, of u, we have

$$du = 0$$

$$\Rightarrow \qquad \frac{x}{a^4}dx+\frac{y}{b^4}dy+\frac{z}{c^4}dz=0 \qquad ...(1)$$

Also differentiating the two given equations connecting the variables x, y and z, we get

$$l dx + m\,dy + n\,dz = 0, \qquad ...(2)$$

and
$$\frac{x}{a^2}dx+\frac{y}{b^2}dy+\frac{z}{c^2}dz=0 \qquad ...(3)$$

Multiplying (1), (2) and (3) by 1, λ and μ respectively and adding, and then equating to zero the coefficients of dx, dy and dz, we get

$$\frac{x}{a^4}+\lambda l+\mu\frac{x}{a^2}=0, \qquad ...(4)$$

$$\frac{y}{b^4}+\lambda m+\mu\frac{y}{b^2}=0, \qquad ...(5)$$

and
$$\frac{z}{c^4}+\lambda n+\mu\frac{z}{c^2}=0. \qquad ...(6)$$

Multiplying the equations (4), (5) and (6) by x, y and z respectively and adding, we get

$$u + \lambda.0 + \mu.1 = 0 \text{ or } \mu = -u.$$

Putting $\mu = -u$ in (4), we get

$$\frac{x}{a^4}+\lambda l-\frac{xu}{a^2}=0, \text{ or } \frac{x}{a^2}\left\{u-\frac{1}{a^2}\right\}=\lambda l, \text{ or } x=\frac{\lambda u}{a^2u-1}.$$

Similarly from (5) and (6), we get

$$y=\frac{\lambda mb^4}{b^2u-1} \text{ and } z=\frac{\lambda nc^4}{c^2u-1}.$$

Substituting these values of x, y, z in $lx + my + nz = 0$, we get

$$\frac{l^2a^4}{a^2u-1}+\frac{m^2b^4}{b^2u-1}+\frac{n^2c^4}{c^2u-1}=0. \qquad ...(7)$$

The equation (7) gives the required maximum or minimum values of u.

Geometrical Interpretation: The equation of the tangent plane to the ellipsoid $\frac{x^2}{a^2}+\frac{y^2}{b^2}+\frac{z^2}{c^2}=1$ at any point (x, y, z) on it is

$$\frac{Xx}{a^2}+\frac{Yy}{b^2}+\frac{Zz}{c^2}=1. \qquad ...(8)$$

If p be the length of the perpendicular from origin which is also the centre of the ellipsoid to the tangent plane (1), then

$$p^2=\frac{1}{\frac{x^2}{a^4}+\frac{y^2}{b^4}+\frac{z^2}{c^4}}.$$

If the point (x, y, z) on the ellipsoid also lies on the given plane $lx + my + nz = 0$, the problem consists of finding out the maximum or minimum values of the perpendicular distance from the origin to the tangent planes to the ellipsoid at the points common to the plane $lx + my + nz = 0$ and the ellipsoid.

Example 29(b): *Find the maximum value of $x^m y^n z^p$ subject to the condition*

$$x + y + z = a.$$

Solution:

Let $\quad u = x^m y^n z^p, \qquad ...(1)$

where the variables x, y, z are connected by the relation

$$x + y + z = a.$$

From (1), $\log u = m \log x + n \log y + p \log z$.

$$\therefore \quad \frac{1}{u}du=\frac{m}{x}dx+\frac{n}{y}dy+\frac{p}{z}dz.$$

For a maximum or a minimum of u, we have

$$du = 0$$

$$\Rightarrow \quad \frac{m}{x}dx+\frac{n}{y}dy+\frac{p}{y}dz=0. \qquad ...(3)$$

Differentiating the given equation (2) connecting the variables x, y and z, we get

$$dx + dy + dz = 0 \qquad ...(4)$$

Multiplying (3) by 1 and (4) by λ, and adding and then equating the coefficients of dx, dy, dz to zero, we get

$$\frac{m}{x}+\lambda=0,\ \frac{n}{y}+\lambda=0,\ \frac{p}{z}+\lambda=0.$$

From these, we get $x=-\frac{m}{\lambda},\ y=0\frac{n}{\lambda},\ z=-\frac{p}{\lambda}$.

Putting these values of x, y, z in x + y + z = a, we get

$$=\left(\frac{m}{\lambda}+\frac{n}{\lambda}+\frac{p}{\lambda}\right)=a$$

or $$-\frac{1}{\lambda}(m+n+p)=a$$

or $$-\frac{1}{\lambda}=\frac{a}{m+n+p}.$$

$\therefore$ u is stationary when

$$x=\frac{am}{m+n+p},\ y=\frac{an}{m+n+p},\ z=\frac{ap}{m+n+p}.$$

Let us now find the nature of this stationary value of u.

Since the variables x, y and z are connected by the relation (2), only two of them may be regarded as independent.

Let us regard x and y as independent variables and z as function of x and y given by (2).

From (1), we have

$$\log u = m \log x + n \log y + p \log z.$$

$$\therefore \quad \frac{1}{u}\frac{\partial u}{\partial x}=\frac{m}{x}+\frac{p}{z}\frac{\partial z}{\partial x}.$$

Differentiating (2) partially w.r.t. x taking y as constant, we get

$$1=\left(\frac{\partial z}{\partial x}\right)=0 \text{ or } \frac{\partial z}{\partial x}=-1.$$

$$\therefore \quad \frac{1}{u}\frac{\partial u}{\partial x}=\frac{m}{x}-\frac{p}{z},$$

so that $$\frac{1}{u}\frac{\partial^2 u}{\partial x^2}-\frac{1}{u^2}\left(\frac{\partial u}{\partial x}\right)^2=\frac{m}{x^2}+\frac{p}{z^2}\frac{\partial z}{\partial x}=-\frac{m}{x^2}-\frac{p}{z^2}.$$

But at the stationary point, we have $\frac{\partial u}{\partial x}=0$.

$\therefore$ at the stationary point found above, we have

$$\frac{\partial^2 u}{\partial x^2}=-u\left[\frac{m}{x^2}+\frac{p}{z^2}\right]=-x^m y^n z^p\left[\frac{m}{x^2}+\frac{p}{z^2}\right],$$

which is –ive for the values of x, y, z found above.

Hence at the stationary point found above the value of u is maximum and this maximum value

$$=\left(\frac{am}{m+n+p}\right)^m\left(\frac{an}{m+n+p}\right)^n\left(\frac{ap}{m+n+p}\right)^p=\frac{a^{m+n+p}m^m n^n p^p}{(m+n+p)^{m+n+p}}.$$

Example 29(c): *Find the maximum or minimum value of $x^p y^q z^r$ subject to the condition*

$$ax + by + cz = p + q + r.$$

Solution: For complete solution of this question proceed as in solved example.

Let $\quad u = x^p y^q z^r,$ ...(1)

where the variables x, y, z are connected by the relation

$$ax + by + cz = p + q + r. \quad ...(2)$$

From (1), log u = p log x + q log y + r log z.

$$\therefore \quad \frac{1}{u}du=\frac{p}{x}dx+\frac{q}{y}dy+\frac{r}{z}dz.$$

For a maximum or a minimum of u, we have du = 0

$$\Rightarrow \quad \frac{p}{x}dx+\frac{q}{y}dy+\frac{r}{z}dz=0. \quad ...(3)$$

Also differentiating the given equation (2), we get

$$adx + b\,dy + c\,dz = 0. \quad ...(4)$$

Multiplying eqns. (3) by 1 and (4) by λ, and adding and then equating the coefficients of dx, dy, dz to zero, we get

$$\frac{p}{x}+\lambda a=0,\ \frac{q}{y}+\lambda b=0,\ \frac{r}{z}+\lambda c=0.$$

From these, we get $x=-\frac{p}{\lambda a},\ y=-\frac{q}{\lambda b},\ z=-\frac{r}{\lambda c}.$

Putting these values of x, y, z in (2), we get

$$-\left(\frac{p}{\lambda}+\frac{q}{\lambda}+\frac{r}{\lambda}\right)=p+q+r$$

or $$-\frac{1}{\lambda}(p+q+r)=p+q+r$$

or $$\lambda = -1.$$

$\therefore$ u is stationary when $x = \frac{p}{a},\ y = \frac{q}{b},\ z = \frac{r}{c}$.

Now regard x and y as independent variables and z as a function of x and y given by (2).

From (1), we have

$\log u = p \log x + q \log y + r \log z.$

$$\therefore \qquad \frac{1}{u}\frac{\partial u}{\partial x} = \frac{p}{x} + \frac{r}{z}\frac{\partial z}{\partial x}.$$

Differentiating (2) partially w.r.t. x taking y as constant, we get

$$a + c\left(\frac{\partial z}{\partial x}\right) = 0 \text{ or } \frac{\partial z}{\partial x} = -\frac{a}{c}.$$

$$\therefore \qquad \frac{1}{u}\frac{\partial u}{\partial x} = \frac{p}{x} - \frac{r}{z}\cdot\frac{a}{c},$$

so that $\frac{1}{u}\frac{\partial^2 u}{\partial x^2} - \frac{1}{u^2}\left(\frac{\partial u}{\partial x}\right)^2 = -\frac{p}{x^2} + \frac{a}{c}\cdot\frac{r}{z^2}\frac{\partial z}{\partial x} = -\frac{p}{x^2} - \frac{a^2}{c^2}\cdot\frac{r}{z^2}.$

But at the stationary point, we have $\frac{\partial u}{\partial x} = 0$.

$\therefore$ at the stationary point found above, we have

$$\frac{\partial^2 u}{\partial x^2} = -u\left[\frac{p}{x^2} + \frac{z^2}{c^2}\cdot\frac{r}{z^2}\right] = -x^p y^q z^r\left[\frac{p}{x^2} + \frac{a^2}{c^2}\cdot\frac{r}{z^2}\right],$$

which is –ive for the values of x, y, z found above.

Hence u is maximum at the stationary point found above and this maximum value of $u = \left(\frac{p}{a}\right)^p \cdot \left(\frac{q}{b}\right)^q \cdot \left(\frac{r}{c}\right)^r$.

Example 29(d): *Find the maximum or minimum value of $x^p p^q z^r$ subject to the condition*

$$\frac{a}{x} + \frac{b}{y} + \frac{c}{z} = 1.$$

Solution: Let $u = x^p y^q z^r$, ...(1)

where the variables x, y, z are connected by the relation

$$\left(\frac{a}{x}\right) + \left(\frac{b}{y}\right) + \left(\frac{c}{z}\right) = 1. \qquad ...(2)$$

From (1), $\log u = p \log x + q \log y + r \log z.$

$$\therefore \qquad \frac{1}{u}du = \frac{p}{x}dx + \frac{q}{y}dy + \frac{r}{z}dz.$$

For a maximum or a minimum of u, we have du = 0.

$$\therefore \qquad \frac{p}{x}dx + \frac{q}{y}dy + \frac{r}{z}dz = 0. \qquad ...(3)$$

Also differentiating the given equation (2), we get

$$-\frac{a}{x^2}dx - \frac{b}{y^2}dy - \frac{c}{z^2}dz = 0. \qquad ...(4)$$

Multiplying (3) by 1 and (4) λ, and adding and then equating the coefficients of dx, dy, dz to zero, we get

$$\frac{p}{q} - \frac{\lambda a}{x^2} = 0, \frac{q}{y} - \frac{\lambda b}{y^2} = 0, \frac{r}{z} - \frac{\lambda c}{z^2} = 0.$$

From these, we get $x = \frac{a\lambda}{p}, y = \frac{b\lambda}{q}, z = \frac{c\lambda}{r}$.

Putting these values of x, y, z in (2), we get

$$\frac{p}{\lambda} + \frac{q}{\lambda} + \frac{r}{\lambda} = 1 \text{ or } \frac{1}{\lambda}(p + q + r) = 1 \text{ or } \lambda = p + q + r.$$

∴ u is stationary when

$$\frac{px}{a} = \frac{qy}{b} = \frac{rz}{c} = p + q + r.$$

Now, regard x and y as independent variables and z as a function of x and y given by (2).

From (1), we have

$$\log u = p \log x + q \log y + r \log z.$$

$$\therefore \qquad \frac{1}{u}\frac{\partial u}{\partial x} = \frac{p}{x} + \frac{r}{z}\frac{\partial z}{\partial x}.$$

Differentiating (2) partially w.r.t. x taking y as constant, we get

$$-\frac{a}{x^2} - \frac{c}{z^2}\frac{\partial z}{\partial x} = 0 \text{ or } \frac{\partial z}{\partial x} = -\frac{az^2}{cx^2}.$$

$$\therefore \qquad \frac{1}{u}\frac{\partial u}{\partial x} = \frac{p}{x} \cdot \frac{az^2}{cx^2} = \frac{p}{x} - \frac{arz}{cx^2},$$

so that $\frac{1}{u}\frac{\partial^2 u}{\partial x^2} - \frac{1}{u^2}\left(\frac{\partial u}{\partial x}\right)^2 = -\frac{p}{x^2} + \frac{2arz}{cx^3} - \frac{ar}{cx^2}\frac{\partial z}{\partial x}$

$$= -\frac{p}{x^2} + \frac{2arz}{cx^3} + \frac{ar}{cx^2} \cdot \frac{az^2}{cx^2}.$$

But at the stationary point, we have $\frac{\partial y}{\partial x} = 0$.

$\therefore$ at the stationary point found above, we have

$$\frac{\partial^2 u}{\partial x^2} = u\left[-\frac{p}{x^2} - \frac{2arz}{cx^3} + \frac{a^2rz^2}{c^2x^4}\right]$$

$$= x^p y^q z^r \left[-p \cdot \frac{p^2}{a^2(p+q+r)^2} + \frac{2ar}{c} \cdot \frac{c(p+q+r)}{r} \cdot \frac{p^3}{a^2(p+q+r)^2} + \frac{a^2 r}{c^2} \cdot \frac{c^2(p+q+r)^2}{r^2} \cdot \frac{p^4}{a^4(p+q+r)^4}\right]$$

$$= x^p y^q z^r \left[-\frac{p^3}{a^2(p+q+r)^2} + \frac{2p^2}{a^2(p+q+r)^2} + \frac{p^4}{ra^2(p+q+r)^2}\right]$$

$$= x^p y^q z^r \left[\frac{p^3}{a^2(p+q+r)^2} + \frac{p^4}{ra^2(p+q+r)^2}\right],$$

which is +ive for the values of x, y, z found above.

Hence u is minimum at the stationary point given by

$$\frac{px}{a} = \frac{qy}{b} = \frac{rz}{c} = p+q+r.$$

Also the minimum value of u

$$= \left[\frac{a(p+q+r)}{p}\right]^p \left[\frac{b(p+q+r)}{q}\right]^q \left[\frac{c(p+q+r)}{r}\right]^r$$

$$= \frac{a^p b^q c^r}{p^p q^q r^r}(p+q+r)^{p+q+r}$$

Example 30: *A rectangular box open at the top is to have a given capacity. Find the dimensions of the box requiring least material for its construction.*

Solution: Let the given capacity of the box be V, its three edges be x, y, z and its surface be S. Then

$$S = xy + 2yz + 2zx \qquad ...(1)$$

and $\quad xyz = V.$ ...(2)

For a maximum or a minimum of S, we have

$$dS = (y + 2z)\,dx + (x + 2z)\,dy + 2(x + y)\,dz = 0. \quad ...(3)$$

Also from (2), since V is constant, we have

$$yz\,dx + zx\,dy + xy\,dz = 0. \quad ...(4)$$

Multiplying (3) by 1 and (4) by λ, and adding and then equating to zero the coefficients of dx, dy, dz, we get

$$(y + 2z) + \lambda yz = 0, \quad ...(5)$$

and $\quad 2x + 2y + \lambda xy = 0.$ ...(6)

Multiplying (5) by x, (6) by y and subtracting, we get

$$2zx - 2zy = 0 \text{ or } 2z(x - y) = 0, \text{ or } x = y.$$

[The root z = 0 is inadmissible because the depth of the box cannot be zero.]

Similarly, from the equations (6) and (7), we get y = 2z.

Hence the dimensions of the box for a stationary value of S are

$$x = y = 2z = (2V)^{1/3}, \text{ from (2).}$$

Let us now find the nature of this stationary value of S.

Regard x and y as independent variables and z as a function of x and y given by (2).

Then from (1), $\frac{\partial S}{\partial x} = y + 2y\frac{\partial z}{\partial x} + 2z + 2x\frac{\partial z}{\partial x}.$

Differentiating (2) partially w.r.t. x taking y as constant, we get

$$yz + xy\frac{\partial z}{\partial x} = 0 \ \textit{i.e.,} \ \frac{\partial z}{\partial x} = -\frac{z}{x}.$$

$$\therefore \quad \frac{\partial S}{\partial x} = y - \frac{2yz}{x} + 2z - 2z = y - \frac{2yz}{x}$$

and so $\frac{\partial^2 S}{\partial x^2} = \frac{2yz}{x^2} - \frac{2y}{x}\cdot\frac{\partial z}{\partial x} = \frac{2yz}{x^2} + \frac{2yz}{x^2} = \frac{4yz}{x^2} = 2$

at $\quad x = y = 2z.$

Thus, at the stationary point $x = y = 2z = (2V)^{1/3}$, we have

$$r = \frac{\partial^2 S}{\partial x^2} = 2, \text{ which is positive.}$$

Similarly, we can find $s = \frac{\partial^2 S}{\partial x\,\partial y}$ and $s = \frac{\partial^2 S}{\partial x\,\partial y}$

at the stationary point $x = y = 2z = (2V)^{1/3}$ and can show that $rt - s^2$ is positive.

Since at the stationary point $x = y = 2z = (2V)^{1/3}$, $rt - s^2 > 0$ and $r > 0$, therefore the stationary value of S at this point is a minimum.

Hence the dimensions of the box requiring least material for its construction are given by $x = y = 2z = (2V)^{1/3}$.

Example 31: *Prove that the rectangular solid of maximum volume which can be inscribed in a sphere is a cube.*

Solution: Referred to the centre as origin, let the equation of the sphere

$$x^2 + y^2 + z^2 = a^2.$$

Let (x, y, z) denote the coordinates of that vertex of the rectangular parallelopiped inscribed in the sphere which lies in the positive octant and let V denote the volume of the rectangular parallelopiped. Then, we have to find the maximum value of

$$V = 8xyz \qquad ...(1)$$

subject to the condition

$$x^2 + y^2 + z^2 = a^2. \qquad ...(2)$$

From (1), $\log V = \log 8 + \log x + \log y + \log z$.

$$\therefore \quad \frac{1}{V}dV = \frac{1}{x}dx + \frac{1}{y}dy + \frac{1}{z}dz.$$

For a maximum or a minimum of V, we must have $dV = 0$

$$\Rightarrow \quad \frac{1}{x}dx + \frac{1}{y}dy + \frac{1}{z} = 0. \qquad ...(3)$$

Also differentiating (2), we have

$$x\,dx + y\,dy + z\,dz = 0. \qquad ...(4)$$

Multiplying (3) by 1 and (4) by λ, and adding and then equating to zero the coefficients of dx, dy, dz, we get

$$\frac{1}{x} + \lambda x = 0, \ \frac{1}{y} + \lambda y = 0, \frac{1}{z} + \lambda z = 0$$

or $$-\frac{1}{\lambda} = x^2 = y^2 = z^2 \text{ or } x = y = z.$$

Thus V is stationary when $x = y = z = \frac{a}{\sqrt{3}}$, from (2).

The lengths of the edges of the rectangular parallelopiped is a cube.

Now regard x and y as independent variables and z as a function of x and y given by (2).

From (1), log V = log 8 + log x + log y + log z.

$$\therefore \quad \frac{1}{V}\frac{\partial V}{\partial x} = \frac{1}{x} + \frac{1}{z}\cdot\frac{\partial z}{\partial x}.$$

Differentiating (2) partially w.r.t. x taking y as constant, we get

$$2x + 2z\left(\frac{\partial z}{\partial x}\right) = 0 \text{ or } \frac{\partial z}{\partial x} = -\frac{x}{z}.$$

$$\therefore \quad \frac{1}{V}\frac{\partial V}{\partial x} = \frac{1}{x} + \frac{1}{z}\cdot\frac{-x}{z} = \frac{1}{x} - \frac{x}{z^2},$$

$$\text{so that } \frac{1}{V}\frac{\partial^2 V}{\partial x^2} - \frac{1}{\sqrt{2}}\left(\frac{\partial V}{\partial x}\right)^2 = -\frac{1}{x^2} - \frac{1}{z^2} + \frac{2x}{z^3}\frac{\partial z}{\partial x} = -\frac{1}{x^2} - \frac{1}{z^2} - \frac{2x^2}{z^4}$$

But at the stationary point, we have $\frac{\partial V}{\partial x} = 0.$

∴ at the stationary point found above, we have

$$\frac{\partial^2 V}{\partial x^2} = -V\left[\frac{1}{x^2} + \frac{1}{z^2} + \frac{2x^2}{z^4}\right] = -8xyz\left[\frac{1}{x^2} + \frac{1}{z^2} + \frac{2x^2}{z^4}\right],$$

which is –ive when $x = y = z = \frac{a}{\sqrt{3}}$.

Thus V is maximum when $x = y = z = \frac{a}{\sqrt{3}}$.

Hence the rectangular solid of maximum volume inscribed in a sphere is a cube.

Example 32: *Find the maximum value of u, when*

$$u = x^2y^3z^4 \text{ and } 2x + 3y + 4z = a.$$

Solution: Let $u = x^2y^3z^4$, ...(1)

where the variables x, y, z are connected by the relation

$$2x + 3y + 4z = a. \quad ...(2)$$

From (1), log u = 2 log x + 3 log y + 4 log z.

$$\therefore \quad \frac{1}{u}du = \frac{2}{x}dx + \frac{3}{y}dy + \frac{4}{z}dz.$$

For a maximum or a minimum of u, we have

$$du = 0 \Rightarrow \left(\frac{2}{x}\right)dx + \left(\frac{3}{y}\right)dy + \left(\frac{4}{z}\right)dz = 0. \quad ...(3)$$

Differentiating the given equation (2), we have

$$2dx + 3dy + 4dz = 0. \qquad \text{...(4)}$$

Multiplying eqns. (3) by 1 and (4) by λ, and adding and then equating the coefficients of dx, dy, dz to zero, we get

$$\frac{2}{x}+2\lambda=0,\ \frac{3}{y}+3\lambda=0,\ \frac{4}{z}+4\lambda=0.$$

From these, get $x=-\frac{1}{\lambda},\ y=-\frac{1}{\lambda},\ z=-\frac{1}{\lambda}$.

Putting these values of x, y, z in (2), we get

$$-\frac{2}{\lambda}-\frac{3}{\lambda}-\frac{4}{\lambda}=a \text{ or } -\frac{9}{\lambda}=a \text{ or } \lambda=-\frac{9}{a}.$$

∴ u is stationary when $x = y = z = \frac{a}{9}$.

Now regard x and y as independent variables and z as a function of x and y given by (2).

From (1), we have

$$\log u = 2 \log x + 3 \log y + 4 \log z.$$

$$\therefore \qquad \frac{1}{u}\frac{\partial u}{\partial x}=\frac{2}{x}+\frac{4}{z}\frac{\partial z}{\partial x}.$$

Differentiating (2) partially w.r.t. x taking y as constant, we get

$$2+4\left(\frac{\partial z}{\partial x}\right)=0 \text{ or } \frac{\partial z}{\partial x}=-\frac{1}{2}.$$

$$\therefore \qquad \frac{1}{u}\frac{\partial u}{\partial x}=\frac{2}{x}+\frac{4}{2}\cdot\left(-\frac{1}{2}\right)=\frac{2}{x}-\frac{2}{z}.$$

so that $$\frac{1}{u}\frac{\partial^2 u}{\partial x^2}-\frac{1}{u^2}\left(\frac{\partial u}{\partial x}\right)^2=-\frac{2}{x^2}+\frac{2}{z^2}\frac{\partial z}{\partial x}=-\frac{2}{x^2}-\frac{1}{z^2}$$

But at the stationary point, we have $\frac{\partial u}{\partial x}=0$.

∴ at the stationary point found above, we have

$$\frac{\partial^2 u}{\partial x^2}=-u\left(\frac{2}{x^2}+\frac{1}{z^2}\right)=-x^2y^3z^4\left(\frac{2}{x^2}+\frac{1}{z^2}\right),$$

which is –ive for $x = y = z = \frac{a}{9}$.

Hence at the stationary point $x = y = z = \frac{a}{9}$, u is maximum and the maximum value of u

$$=\left(\frac{a}{9}\right)^2\left(\frac{a}{9}\right)^3\left(\frac{a}{9}\right)^4=\left(\frac{a}{9}\right)^9.$$

Example 33(a): *Show that the volume of the greatest rectangular parallelopiped that can be inscribed in the ellipsoid*

$$\frac{x^2}{a^2}+\frac{y^2}{b^2}+\frac{z^2}{c^2}=1$$

is $$\frac{8abc}{3\sqrt{3}}.$$

or

Find the maximum value of u:

$$u = 8xyz, \text{ given } \frac{x^2}{a^2}+\frac{y^2}{b^2}+\frac{z^2}{c^2}=1.$$

Solution: Let (x, y, z) denote the coordinates of the vertex of the rectangular parallelopiped which lies in the positive octant and let V denote its volume. Then, we have to find the maximum value of

$$V = 8xyz \qquad ...(1)$$

subject the condition

$$\frac{x^2}{a^2}+\frac{y^2}{b^2}+\frac{z^2}{c^2}=1. \qquad ...(2)$$

For a maximum or a minimum of V, we have

$$dV = 8yz\,dx + 8zx\,dy + 8xy\,dz = 0$$

i.e., $$yz\,dx + zx\,dy + xy\,dz = 0. \qquad ...(3)$$

Also differentiating (2), we get

$$\frac{2x}{a^2}dx+\frac{2y}{b^2}dy+\frac{2z}{c^2}dz=0$$

i.e., $$\frac{x}{a^2}dx+\frac{y}{b^2}dy+\frac{z}{c^2}dz=0. \qquad ...(4)$$

Multiplying (3) by 1 and (4) by λ, and adding and then equating the coefficients of dx, dy, dz to zero, we get

$$yz+\frac{\lambda x}{a^2}=0,\ zx+\frac{\lambda y}{b^2}=0 \text{ and } xy+\frac{\lambda z}{c^2}=0.$$

From these, we get

$$\frac{x}{a^2}=-\frac{yz}{\lambda},\ \frac{y}{b^2}=-\frac{zx}{\lambda},\ \frac{z}{c^2}=-\frac{xy}{\lambda}$$

or $$\frac{x^2}{a^2}=-\frac{xyz}{\lambda},\ \frac{y^2}{b^2}=-\frac{xyz}{\lambda},\ \frac{z^2}{c^2}=-\frac{xyz}{\lambda}$$

or $$\frac{x^2}{a^2}=\frac{y^2}{b^2}=\frac{z^2}{c^2}=\frac{x^2/a^2+y^2/b^2+z^2/c^2}{3}=\frac{1}{3}, \text{ using (2)}$$

or $$x=\frac{a}{\sqrt{3}}, y=\frac{b}{\sqrt{3}}, z=\frac{c}{\sqrt{3}}.$$

Thus, V is stationary when $x=\frac{a}{\sqrt{3}}, y=\frac{b}{\sqrt{3}}, z=\frac{c}{\sqrt{3}}$.

Now regard x and y as independent variables and z as a function of x and y given by (2).

Then from (1), $\frac{\partial V}{\partial x}=8yz+8xy\frac{\partial z}{\partial x}$.

Differentiating (2) partially w.r.t. x taking y as constant, we get

$$\frac{2x}{a^2}+\frac{2z}{c^2}\frac{\partial z}{\partial x}=0 \text{ or } \frac{\partial z}{\partial x}=-\frac{c^2x}{a^2z}.$$

$$\therefore \quad \frac{\partial V}{\partial x}=8yz+8xy\cdot\left(-\frac{c^2x}{a^2z}\right)=8yz-\frac{8c^2x^2y}{a^2z}$$

and so $$\frac{\partial^2 V}{\partial x^2}=8y\frac{\partial z}{\partial x}-\frac{16c^2xy}{a^2z}+\frac{8c^2x^2y}{a^2z^2}\cdot\frac{\partial z}{\partial x}$$

$$=8y\cdot\left(\frac{-c^2x}{a^2z}\right)-\frac{16c^2xy}{a^2z}-\frac{8c^2x^2y}{a^2z}\cdot\frac{c^2x}{a^2z},$$

which is –ive when $x=\frac{a}{\sqrt{3}}, y=\frac{b}{\sqrt{3}}, z=\frac{c}{\sqrt{3}}$.

Hence V is maximum when $x=\frac{a}{\sqrt{3}}, y=\frac{b}{\sqrt{3}}, z=\frac{c}{\sqrt{3}}$ and the maximum value of $V=\frac{8abc}{3\sqrt{3}}$.

Note: In complicated problems to show that whether the stationary value of a function is maximum or minimum, it will be sufficient to see whether the second partial differential coefficient of the function w.r.t. any of the independent variables is negative or positive.

Example 33(b): *Find the triangle of maximum area inscribed in a circle.*

Solution: Let x, y, z denote the angles of a triangle inscribed in a given circle of radius k. If u be the area of the triangle, then

$$u=\frac{1}{2}k^2\ (\sin 2x+\sin 2y+\sin 2z), \qquad \text{...(1)}$$

where the variables x, y, z are connected by the relation

$$x + y + z = \pi. \quad ...(2)$$

From (1), $du = k^2 (\cos 2x \, dx + \cos 2y \, dy + \cos 2z \, dz)$.

For a maximum or a minimum of u, we must have $du = 0$

$$\Rightarrow \cos 2x \, dx + \cos 2y \, dy + \cos 2z \, dz = 0. \quad ...(3)$$

Also differentiating (2), we have

$$dx + dy + dz = 0 \quad ...(4)$$

Multiplying (3) by 1 and (4) by λ, and adding and then equating to zero the coefficients of dx, dy and dz, we get

$$\cos 2x + \lambda = 0, \ \cos 2y + \lambda = 0, \ \cos 2z + \lambda = 0$$

or $-\lambda = \cos 2x = \cos 2y = \cos 2z$

or $1 - 2\sin^2 x = 1 - 2\sin^2 y = 1 - 2\sin^2 z$

or $\sin^2 x = \sin^2 y = \sin^2 z$

or $\sin x = \sin y = \sin^2 z$

or $x = y = z = \frac{\pi}{3}$, from (2).

Thus, u is stationary when $x = y = z = \frac{\pi}{3}$ *i.e.*, the triangle is equilateral.

Now to show that the stationary value of u given by $x = y = z = \frac{\pi}{3}$ is maximum.

Let us regard x and y as independent variables and z as a function of x and y given by (2).

From eqn. (1), $\frac{\partial u}{\partial x} = k^2\left(\cos 2x + \cos 2z \frac{\partial z}{\partial x}\right)$.

Also differentiating (2) partially w.r.t. x taking y as a constant, we get

$$1 + \left(\frac{\partial z}{\partial x}\right) = 0$$

or $\frac{\partial z}{\partial x} = -1$.

$\therefore \quad \frac{\partial u}{\partial x} = k^2(\cos 2x - \cos 2z)$,

so that $\frac{\partial^2 u}{\partial x^2} = k^2\left[-2\sin 2x + 2\sin 2z \cdot \frac{\partial z}{\partial x}\right]$

$$= -2k^2 (\sin 2x + \sin 2z).$$

Also $$\frac{\partial^2 u}{\partial x\, \partial y} = k^2\left(2\sin 2z \frac{\partial z}{\partial y}\right) = -2k^2 \sin 2z.$$

Hence putting $x = y = z = \frac{\pi}{3}$, we get

$$r = \frac{\partial^2 u}{\partial x^2} = -2k^2\left(\frac{\sqrt{3}}{2} + \frac{\sqrt{3}}{2}\right) = -2k^2\sqrt{3}.$$

$$s = \frac{\partial^2 u}{\partial x\, \partial y} = -2k^2 \cdot \frac{\sqrt{3}}{2} = -k^2\sqrt{3}.$$

Also by symmetry, $t = \frac{\partial^2 u}{\partial y^2} = -2k^2\sqrt{3}$.

$\therefore$ at the stationary point $x = y = z = \frac{\pi}{3}$, we have

$$rt - s^2 = 12k^4 - 3k^4 = 9k^4 \text{ which is } > 0$$

and $r = -2k^2\sqrt{3}$ which is –ive.

$\therefore$ at the stationary point $x = y = z = \frac{\pi}{3}$, u is maximum.

Hence the triangle of maximum area inscribed in a circle is equilateral.

Example 33(c): *Show that if the perimeter of a triangle is constant, its area is a maximum when it is equilateral.*

Solution: Let a, b, c denote the sides of a triangle, 2s is constant perimeter and u its area.

Then $$u^2 = s(s - a)(s - b)(s - c), \qquad ...(1)$$

where the variables a, b, c are connected by the relation

$$a + b + c = 2s. \qquad ...(2)$$

From (1), $2 \log u = \log s + \log (s - a) + \log (s - b) + \log (s - c)$.

$$\therefore \quad \frac{2}{u} du = -\frac{1}{s-a} da - \frac{1}{s-b} db - \frac{1}{s-c} dc.$$

For a maximum or a minimum of u, we must have $du = 0$

$$\Rightarrow \quad \frac{da}{s-a} + \frac{db}{s-a} + \frac{dc}{s-c} = 0. \qquad ...(3)$$

Also differentiating (2), we have

$$da + db + dc = 0. \qquad ...(4)$$

Multiplying (3) by 1 and (4) by λ, and adding and when equating to zero the coefficients of da.db and dc, we get

$$\frac{1}{s-a}+\lambda=0,\ \frac{1}{s-b}+\lambda=0,\ \frac{1}{s-c}+\lambda=0.$$

$\therefore$ $$-\lambda=\frac{1}{s-a}=\frac{1}{s-b}=\frac{1}{s-c}$$

or $s - a = s - b = s - c$ or $a = b = c$.

Hence, u is stationary when $a = b = c$ *i.e.*, the triangle is equilateral.

Now, to show that the stationary value of u given by $a = b = c$ is maximum.

Let us regard a and b as independent variables and c as a function of a and b given by (2).

From (1), differentiating logarithmically, we have

$$\frac{2}{u}\frac{\partial u}{\partial a}=-\frac{1}{s-a}-\frac{1}{s-c}\frac{\partial c}{\partial a}.$$

But from (2), $1+\left(\frac{\partial c}{\partial a}\right)=0$ or $\frac{\partial c}{\partial a}=-1$.

$\therefore$ $$\frac{2}{u}\frac{\partial^2 u}{\partial a^2}-\frac{2}{u^2}\left(\frac{\partial u}{\partial a}\right)^2=-\frac{1}{(s-a)^2}+\frac{1}{(s-c)^2}\cdot\frac{\partial c}{\partial a}$$

$$=-\left[\frac{1}{(s-a)^2}+\frac{1}{(s-c)^2}\right].$$

But at the stationary point, we have $\frac{\partial u}{\partial a}=0$.

$\therefore$ at the stationary point found above, we have

$$\frac{2}{u}\frac{\partial^2 u}{\partial a^2}=-\left[\frac{1}{(s-a)^2}+\frac{1}{(s-c)^2}\right]$$

or $$\frac{\partial^2 u}{\partial a^2}=-\frac{u}{2}\left[\frac{1}{(s-a)^2}+\frac{1}{(s-c)^2}\right],$$ which is –ive.

Hence u is maximum when $a = b = c$ *i.e.*, the area of the triangle is maximum when it is equilateral.

Example 34: *Given* $u=\frac{5xyz}{(x+2y+4z)}$. *Find the values of x, y, z for which u is maximum subject to the condition* $xyz = 8$.

Solution: We have $u=\frac{5xyz}{(x+2y+4z)}$ ...(1)

where the variables x, y, z are connected by the relation

$$xyz = 8. \qquad ...(2)$$

From (1) and (2), we have $u = \dfrac{40}{(x+2y+4z)}$.

$$\therefore \qquad du = \frac{-40}{(x+2y+4z)^2}(dx+2dy+4dz).$$

For a maximum of a minimum of u, we have du = 0

$$\Rightarrow \qquad dx + 2dy + 4dz = 0. \qquad ...(3)$$

From (2), log x + log y + log z = log 8.

Differentiating this, we get

$$\left(\frac{1}{x}\right)dx + \left(\frac{1}{y}\right)dy + \left(\frac{1}{z}\right)dz = 0. \qquad ...(4)$$

Multiplying eqns. (3) by 1 and (4) by λ, and adding and then equating to zero the coefficients of dx, dy and dz, we get

$$1 + \left(\frac{\lambda}{x}\right) = 0,\ 2 + \left(\frac{\lambda}{y}\right) = 0,\ 4 + \left(\frac{\lambda}{z}\right) = 0.$$

From these, we get $x = -\lambda$, $y = -\dfrac{\lambda}{2}$, $z = -\dfrac{\lambda}{4}$.

Putting these values of x, y, z in (2), we get

$$-\frac{\lambda^3}{8} = 8 \text{ or } \lambda^3 = -64 \text{ or } \lambda = -4.$$

$\therefore$ u is stationary at the given by x = 4, y = 2, z = 1.

Now regard x and y as independent variables and z as a function of x and y given by (2).

We have $u = \dfrac{40}{(x+2y+4z)}$.

$$\therefore \qquad \frac{\partial u}{\partial x} = -\frac{40}{(x+2y+4z)^2}\left[1 + 4\frac{\partial z}{\partial x}\right].$$

From (2), log x + log y + log z = log 8.

$$\therefore \qquad \left(\frac{1}{x}\right) + \left(\frac{1}{z}\right)\left(\frac{\partial z}{\partial x}\right) = 0$$

$$\text{or} \qquad \frac{\partial z}{\partial x} = -\frac{z}{x}.$$

$$\therefore \quad \frac{\partial u}{\partial x} = -\frac{40}{(x+2y+4z)^2}\left[1-4\frac{z}{x}\right],$$

so that $$\frac{\partial^2 u}{\partial x^2} = \frac{80}{(x+2y+4z)^3}\left[1+4\frac{\partial z}{\partial x}\right]\left[1-4\frac{z}{x}\right]$$

$$-\frac{40}{(x+2y+4z)^2}\left[\frac{4z}{x^2}-\frac{4}{x}\frac{\partial z}{\partial x}\right]$$

$$= \frac{80}{(x+2y+4z)^3}\left[1-\frac{4z}{x}\right]^2 - \frac{40}{(x+2y+4z)^2}\left[\frac{4z}{x^2}+\frac{4z}{x^2}\right].$$

$\therefore$ at the stationary point (4, 2, 1) found above, we have

$$\frac{\partial^2 u}{\partial x^2} = \frac{80}{12^3}[1-1]^2 - \frac{40}{144}\left[\frac{1}{4}+\frac{1}{4}\right],$$ which is –ive.

$\therefore$ u is maximum at the point given by x = 4, y = 2, z = 1.

Example 35(a): *Find the plane triangle ABC such that*

$u = \sin^m A \sin^n B \sin^p C$ has maximum value.

Solution: We have $u = \sin^m A \sin^n B \sin^p C$, ...(1)

where the variables A, B and C are connected by the relation

$$A + B + C = \pi. \quad ...(2)$$

From (1), log u = m log sin A + n log sin B + p log sin C.

$$\therefore \quad \frac{1}{u}du = m \cot A\, dA + n \cot B\, dB + p \cot C\, dC.$$

For a maximum or a minimum of u we must have du = 0

$$\Rightarrow \quad m \cot A\, dA + n \cot B\, dB + p \cot C\, dC = 0. \quad ...(3)$$

Also differentiating (2), we get

$$dA + dB + dC = 0. \quad ...(4)$$

Multiplying eqns. (3) by 1 and (4) by λ, and adding and then equating to zero the coefficients of dA.dB and dC, we get

$$m \cot A + \lambda = 0,\ n \cot B + \lambda = 0,\ p \cot C + \lambda = 0.$$

$$\therefore \quad -\lambda = m \cot A = n \cot B = p \cot C.$$

Hence u is stationary when A, B, C are given by

$$m \cot A = n \cot B = p \cot C$$

or $$\frac{\tan A}{m} = \frac{\tan B}{n} = \frac{\tan C}{p}.$$

Now to show that the above stationary value of u is maximum.

Let us regard A and B as independent variables and C as a function of A and B given by (2).

From (1), log u = m log sin A + n log sin B + p log sin C.

$$\therefore \quad \frac{1}{u}\frac{\partial u}{\partial A} = m \cot A + p \cot C \cdot \frac{\partial C}{\partial A}.$$

But from (2), $1 + \left(\frac{\partial C}{\partial A}\right) = 0$ or $\frac{\partial C}{\partial A} = -1.$

$$\therefore \quad \frac{1}{u}\frac{\partial u}{\partial A} = m \cot A - p \cot C,$$

so that $$\frac{1}{u}\frac{\partial^2 u}{\partial A^2} - \frac{1}{u^2}\left(\frac{\partial u}{\partial A}\right)^2 = -m \operatorname{cosec}^2 A + p \operatorname{cosec}^2 C \cdot \frac{\partial C}{\partial A}$$

$$= -(m \operatorname{cosec}^2 A + p \operatorname{cosec}^2 C).$$

But at the stationary point found above, we have

$$\frac{\partial^2 u}{\partial A^2} = -u\left(m \operatorname{cosec}^2 A + p \operatorname{cosec}^2 C\right)$$

$$= -\sin^m A \sin^n B \sin^p C\,(m \operatorname{cosec}^2 A + p \operatorname{cosec}^2 C),$$

which is obviously –ive if A, B, C are the angles of a triangle.

Hence u is maximum when A, B, C are given by

$$\frac{\tan A}{m} = \frac{\tan B}{n} = \frac{\tan C}{p}.$$

Example 35(b): *Divide a number a into three parts such that their product will be maximum.*

Solution: Let u = xyz, ...(1)

where the variables x, y and z are connected by the relation

$$x + y + z = a. \qquad ...(2)$$

From (1), log u + log x + log y + log z.

$$\therefore \quad \left(\frac{1}{u}\right)du = \left(\frac{1}{x}\right)dx + \left(\frac{1}{y}\right)dy + \left(\frac{1}{z}\right)dz.$$

For a maximum or a minimum of u, we have du = 0

$$\Rightarrow \qquad \left(\frac{1}{x}\right)dx + \left(\frac{1}{y}\right)dy + \left(\frac{1}{z}\right)dz. \qquad ...(3)$$

Also differentiating the equation (2), we have

$$dx + dy + dz = 0. \qquad ...(4)$$

Multiplying (3) by 1 and (4) by λ, and adding and then equating the coefficients of dx, dy, dz to zero, we get

$$\frac{1}{x} + \lambda = 0, \frac{1}{y} + \lambda = 0, \frac{1}{z} + \lambda = 0.$$

From these, we get $x = y = z = -\frac{1}{\lambda}$.

Putting these values of x, y, z in (2), we get

$$-\frac{3}{\lambda} = a \text{ or } \lambda = -\frac{3}{a}.$$

$\therefore$ u is stationary at the point given by

$$x = y = z = \frac{a}{3}.$$

Now regard x and y as independent variables and z as a function of x and y given by (2).

From (1), log u = log x + log y + log z.

$$\therefore \qquad \frac{1}{u}\frac{\partial u}{\partial x} = \frac{1}{x} + \frac{1}{z}\frac{\partial z}{\partial x}$$

But from eqn. (2) 1, + $\left(\frac{\partial z}{\partial x}\right) = 0$

$$\text{or} \qquad \frac{\partial z}{\partial x} = -1$$

$$\frac{1}{u}\frac{\partial u}{\partial x} = \frac{1}{x} - \frac{1}{z}$$

so that $\frac{1}{u}\frac{\partial^2 u}{\partial x^2} - \frac{1}{u^2}\left(\frac{\partial u}{\partial x}\right)^2 = -\frac{1}{x^2} + \frac{1}{z^2}\frac{\partial z}{\partial x} = -\frac{1}{x^2} - \frac{1}{z^2}.$

But at the stationary point, we have $\frac{\partial u}{\partial x} = 0.$

$\therefore$ at the stationary point $\left(\frac{a}{3}, \frac{a}{3}, \frac{a}{3}\right)$, we have

$$\frac{\partial^2 u}{\partial x^2} = -u\left[\frac{1}{x^2} + \frac{1}{z^2}\right] = -xyz\left[\frac{1}{x^2} + \frac{1}{z^2}\right]$$

which is negative $x = y = z = \dfrac{a}{3}$.

Hence u is maximum when $x = y = z = \dfrac{a}{3}$ and the maximum value of $u = \left(\dfrac{a}{3}\right)^3$.

Ans. The required three parts of a are $\dfrac{a}{3}, \dfrac{a}{3}, \dfrac{a}{3}$ and the maximum value of the product is $\left(\dfrac{a}{3}\right)^3$.

Example 35(c): *In a plane triangle ABC, find the maximum value of $u = \cos A \cos B \cos C$.*

Solution: We have $u = \cos A \cos B \cos C$, ...(1)

where the variables A, B and C are connected by the relation

$$A + B + C = \pi. \qquad ...(2)$$

From eqn. (1), $\log u = \log \cos A + \log \cos B = \log \cos C$.

$$\therefore \quad \frac{1}{u}du = -\tan A\, dA - \tan B\, dB - \tan C\, dC.$$

For a maximum or a minimum of u, we must have $du = 0$

$$\Rightarrow \quad \tan A\, dA + \tan B\, dB + \tan C\, dC = 0. \qquad ...(3)$$

Also differentiating (2), we get

$$dA + dB + dC = 0. \qquad ...(4)$$

Multiplying eqns. (3) by 1 and (4) by λ, and adding and then equating to zero the coefficients of dA, dB and dC, we get

$$\tan A + \lambda = 0, \tan B + \lambda = 0, \tan C + \lambda = 0$$

or $$-\lambda = \tan A = \tan B = \tan C$$

or $$A = B = C = \frac{\pi}{3}, \text{ from (2).}$$

Thus u is stationary when $A = B = C = \dfrac{\pi}{3}$ *i.e.*, the triangle is equilateral.

Now to show that the stationary value of u given by

$$A = B = C = \frac{\pi}{3} \text{ is maximum.}$$

Let us regard A and B is independent variables and C as a function of A and B given by (2).

From (1), $\log u = \log \cos A + \log \cos B + \log \cos C$.

$$\therefore \quad \frac{1}{u}\frac{\partial u}{\partial A} = -\tan A - \tan C \cdot \frac{\partial C}{\partial A}.$$

Also differentiating (2) partially w.r.t. A taking B as constant, we get

$$1+\left(\frac{\partial C}{\partial A}\right)=0 \text{ or } \frac{\partial C}{\partial A}=-1.$$

$$\therefore \qquad \frac{1}{u}\frac{\partial u}{\partial A}=-\tan A+\tan C,$$

so that $\frac{1}{u}\frac{\partial^2 u}{\partial^2 A}-\frac{1}{u^2}\left(\frac{\partial u}{\partial A}\right)^2=-\sec^2 A+\sec^2 C\cdot\frac{\partial C}{\partial A}$

$$= -(\sec^2 A + \sec^2 C).$$

But at the stationary point $\frac{\partial u}{\partial A}=0.$

$\therefore$ at the stationary point found above, we have

$$\frac{\partial^2 u}{\partial A^2}=-u\left(\sec^2 A+\sec^2 C\right)$$

$$= -\cos A \cos B \cos C (\sec^2 A + \sec^2 A),$$

which is –ive for $A = B = C = \frac{\pi}{3}$.

Hence u is maximum when $A = B = C = \frac{\pi}{3}$ and the maximum value of

$$u=\left(\cos\frac{\pi}{3}\right)^3=\left(\frac{1}{2}\right)^3=\frac{1}{8}.$$

Example 35(d): *Show that the maximum and minimum of the radii vectors of the sections of the surface*

$$(x^2 + y^2 + z^2) = \frac{x^2}{a^2}+\frac{y^2}{b^2}+\frac{z^2}{c^2}$$

by the plane $\lambda x + \mu y + yz = 0$

are given by the equation

$$\frac{a^2\lambda^2}{1-a^2r^2}+\frac{b^2\mu^2}{1-b^2r^2}+\frac{c^2v^2}{1-c^2r^2}=0.$$

Solution: We have to find the maximum and minimum values of r, where

$$r^2 = x^2 + y^2 + z^2. \qquad ...(1)$$

Also the variables x, y, z are connected by the relations

$$\frac{x^2}{a^2}+\frac{y^2}{b^2}+\frac{z^2}{c^2}=\left(x^2+y^2+z^2\right)^2=r^4 \qquad ...(2)$$

and $\qquad \lambda x + \mu y + vz = 0. \qquad ...(3)$

From (1), $2r\,dr = 2x\,dx + 2y\,dy + 2z\,dz$.

For a maximum or a minimum of r, we have

$$dr = 0 \Rightarrow x\,dx + y\,dy + z\,dz = 0. \quad ...(4)$$

Differentiating (2), we get

$$\frac{2x}{a^2}dx + \frac{2y}{b^2}dy + \frac{2z}{c^2}dz = 4r^3 dr.$$

But for a maximum or a minimum of r, we have $dr = 0$.

$$\therefore \quad \frac{x}{a^2}dx + \frac{y}{b^2}dy + \frac{z}{c^2}dz = 0. \quad ...(5)$$

Also differentiating (3), we get

$$\lambda dx + \mu dy + v dz = 0. \quad ...(6)$$

Multiplying (4) by 1, (5) by λ_1 and (6) by λ_2, and adding and then equating to zero the coefficients of dx, dy, dz, we get

$$x + \frac{x}{a^2}\lambda_1 + \lambda\lambda_2 = 0 \quad ...(7)$$

$$y + \frac{y}{b^2}\lambda_1 + \mu\lambda_2 = 0 \quad ...(8)$$

$$z + \frac{z}{c^2}\lambda_1 + v\lambda_2 = 0. \quad ...(9)$$

$\therefore$ from eqn. (7), we have $x - \frac{a}{a^2} \cdot \frac{1}{r^2} + \lambda\lambda_2 = 0$

$$\text{or} \quad x = \frac{a^2 r^2 \lambda\lambda_2}{1 - a^2 r^2}.$$

Similarly from (8) and (9), we have

$$y = \frac{b^2 r^2 \mu\lambda_2}{1 - b^2 r^2} \text{ and } z = \frac{c^2 r^2 v\lambda_2}{1 - c^2 r^2}.$$

Substituting these values of x, y, z in $\lambda x + \mu y + vz = 0$, we get

$$\frac{a^2 r^2 \lambda^2 \lambda_2}{1 - a^2 r^2} + \frac{b^2 r^2 \mu^2 \lambda_2}{1 - b^2 r^2} + \frac{c^2 r^2 v^2 \lambda_2}{1 - c^2 r^2} = 0$$

$$\text{or} \quad \frac{a^2\lambda^2}{1 - a^2 r^2} + \frac{b^2\mu^2}{1 - b^2 r^2} + \frac{c^2 v^2}{1 - c^2 r^2} = 0. \quad ...(10)$$

The equation (10) gives the maximum and minimum values of r.

Example 36: *Discuss the maxima and minima of the function*

$$u = \sin x \sin y \sin z,$$

where x, y, z are the angles of a triangle.

Solution: We have $u = \sin x \sin y \sin z$, ...(1)

where $x + y + z = \pi$. ...(2)

For a maximum or a minimum of u, we must have

$$du = \cos x \sin y \sin z \, dx + \sin x \cos y \sin z \, dy + \sin x \sin y \cos z \, dz = 0 \quad ...(3)$$

Also form (2), we have

$$dx + dy + dz = 0. \quad ...(4)$$

Multiplying (3) by 1 and (4) by λ and adding and then equating to zero the coefficients of dx, dy, dz, we get

$$\cos x \sin y \sin z + \lambda = 0,$$

$$\sin x \cos y \sin z + \lambda = 0,$$

and $\sin x \sin y \cos z + \lambda = 0.$

From these, we get

$$-\lambda = \cos x \sin y \sin z = \sin x \cos y \sin z = \sin x \sin y \cos z$$

or $\cot x = \cot y = \cot z$ (dividing by sin x sin y sin z)

i.e., $x = y = z = \frac{\pi}{3}$, from (2).

Thus, u is stationary when $x = y = z = \frac{\pi}{3}$.

Let us now find the nature of this stationary value of u.

Since variables x, y and z are connected by the relation (2), only two of them may be regarded as independent.

Let us regard x and y as independent and z to be dependent on them by the relation (2).

Then from (1),

$$\frac{\partial u}{\partial x}\frac{\partial u}{\partial x} = \sin y \sin z \cos x + \sin x \sin y \cos z \frac{\partial z}{\partial x}.$$

Also form (2),

$$1 + \frac{\partial z}{\partial x} = 0 \text{ or } \frac{\partial z}{\partial x} = -1.$$

$$\therefore \quad \frac{\partial u}{\partial x} = \sin y \sin z \cos x - \sin x \sin y \cos z$$

and
$$\frac{\partial^2 u}{\partial x^2} = -\sin y \sin z \sin x - \sin y \cos x \cos z \frac{\partial z}{\partial x} - \cos x \sin y \cos z + \sin x \sin y \sin z \frac{\partial z}{\partial x}$$

$$= -2 \sin x \sin y \sin z - 2 \sin y \cos x \cos z.$$

Also $\dfrac{\partial^2 u}{\partial x\, \partial y} = \cos y \sin z \cos x + \sin y \cos x \cos z \dfrac{\partial z}{\partial y}$

$$- \sin x \cos y \cos z + \sin x \sin y \sin z \frac{\partial z}{\partial y}$$

$$= \cos y \sin z \cos x - \sin y \cos x \cos z - \sin x \cos y \cos z + \sin x \sin y \sin z.$$

Hence putting $x = y = z = \dfrac{\pi}{3}$, we get

$$r = \frac{\partial^2 u}{\partial x^2} = -2 \cdot \frac{\sqrt{3}}{2} \cdot \frac{\sqrt{3}}{2} \cdot \frac{\sqrt{3}}{2} - 2 \cdot \frac{\sqrt{3}}{2} \cdot \frac{1}{2} \cdot \frac{1}{2} = -\frac{3\sqrt{3}}{4} - \frac{\sqrt{3}}{4} = -\sqrt{3},$$

$$s = \frac{\partial^2 u}{\partial x\, \partial y} = \frac{\sqrt{3}}{8} - \frac{\sqrt{3}}{8} - \frac{\sqrt{3}}{8} - \frac{3\sqrt{3}}{8} = -\frac{\sqrt{3}}{2}.$$

Also by symmetry, $t = \dfrac{\partial^2 u}{\partial y^2} = -\sqrt{3}$.

$\therefore$ at the stationary point $x = y = z = \dfrac{\pi}{3}$, we have

$$rt - s^2 = 3 - \left(\frac{3}{4}\right) = \frac{9}{4} \text{ which is } > 0$$

and $\quad r = -\sqrt{3}$ which is < 0.

$\therefore$ at the stationary point $x = y = z = \dfrac{\pi}{3}$, u is maximum.

Hence u is maximum when $x = y = z = \dfrac{\pi}{3}$ and the maximum value of

$$u = \left(\frac{\sqrt{3}}{2}\right)^3 = \frac{3\sqrt{3}}{8}.$$

EXERCISES

1. A monopoly manufactures two goods with demand functions and; and the combined cost function is. Find the maximum profit achievable and quantities produced of the two commodities in order to achieve the maximum profit.
2. For a firm under perfect competition, it is given that total cost function is: $C = 1/3x^3 - 5x^2 + 30x + 10$. For price p = Rs. 10, find for what value of x, the profit will be maximum?
3. For a firm under perfect competition, it is given that total cost function is: $C = 1/3q^3 - 5q^2 + 15q + 10$, where q denotes units of

output. If the price has been determined as Rs 6 per unit, find quantity produced at which profit will be maximum and the amount of maximum profit.

4. If the demand function is given by $p = \sqrt{9 - x}$, find the level of output for which the total revenue will be maximum. What will be the maximum revenue?

5. If total revenue of a firm is $4000x - 33x^2$, and total cost is $2x^3 - 3x^2 + 400x + 5000$, find quantity produced at which profit will be maximum and the amount of maximum profit.

6. A monopolist sells two products x and y for which the demand functions are and ; and the combined cost function is . Find :
   (i) profit maximizing level of output for each product
   (ii) the profit maximizing price for each product; and
   (iii) the maximum profit.

7. Show that the curve $y = x + 1/x$ has one maximum at $x = -1$ and one minimum at $x = +1$

8. Find the maxima and (or) minima of the following functions subject to the given constraints:
   (i) $x^2 - 10y^2$; subject to $x - y = 18$
   (ii) $3x^2 + 4y^2 - xy$; subject to $2x + y = 21$
   (iii) $160x - 3x^2 + 120y - 2y^2 - 2xy - 18$; subject to $x + y = 34$
   (iv) $10x + 20y - x^2 - y^2 - 2xy - 18$; subject to $3x + 4y = 5$
   (v) $x^2 - 5y^2 + 3xy$; subject to $2x + 3y = 6$

9. For a firm under perfect competition, it is given that total cost function is: $C = 1/3q^3 - 5q^2 + 28q + 27$, where q denotes units of output. For price p = Rs. 19, find quantity should be produced for attaining maximum profit and the amount of maximum profit.

4

Integral Calculus

INTRODUCTION

The integration and differentiation are inverse operations, as are addition and subtraction or multiplication and division. But, the techniques of integration are most difficult and less intuitive than those of differentiation. The process of integration may be divided into two, viz., definite integration and indefinite integration. We have seen that if $f(x)$ is an integral of $\varphi(x)$, then $f(x)+c$, for any constant c, is also an integral of $\varphi(x)$. Thus, a function whose derivative is given cannot be completely determined, since it always contains an arbitrary additive constant.

$$\int \varphi(x)\, dx = f(x) + c$$

as *indefinite integrals.*

The idea of finding the limiting value of the quotient obtained by dividing the increment of a function by the corresponding increment of its independent variable. But, in many problems, especially of Economics the derivative is known and the function itself must be determined. Integral Calculus is the means by which such determination may be made. Consider the function $\varphi(x) = x^2$. Let its derivative be denoted by $f(x)$. Then $f(x) = 2x$. We call x^2 the *antiderivative* or *primitive* of $2x$. That is, $2x$ is the derivative of x^2 and x^2 is the 'antiderivative' of $2x$. Now, if we start with $f(x) = 2x$, how can we arrive at its antiderivative, $\varphi(x) = x^2$? This can be done by the inverse process of differentiation, which is called *Integration*. The antiderivative x^2 of $2x$ is called the *integral* of $2x$.

$$\int \varphi(x)\, dx = f(x) + c$$

where the symbol '$\int$' is the *integral sign*, $\varphi(x)$ is the *integrand*, $f(x)$ is the *integral* and c is a constant known as the *constant of integration*. [The constant of integration arise since the derivative of c is zero, for any constant c. Hence, for example, the derivative of x^2 and $x^2 + c$, with respect to x is $2x$]. The implication of dx is that the integration is done with respect to x.

Geometrically, $y = f(x) + c$ represents a family of curves, for different value of c. The slope of the tangent to any one of them at the point having abscissa x is $\varphi(x)$. Thus this family of curves has the property that, given any point (x_0, y_0), there is one and only one curve of the family which passes through that point. This will uniquely specify the value of c. In other words, the constant of integration is uniquely determined if one point through which the curve representing the integral passes is specified. This specification is usually refereed to as an 'initial condition'.

DEFINITE INTEGRATION

In indefinite integration we are concerned only with finding a function with given derivative. But, definite integration concerns with evaluating the integral on a particular interval. This evaluation may be interpreted as finding the area under a curve in a particular interval. The concepts of indefinite and definite integrations are connected by the following theorem, known as the *Fundamental Theorem of Integral Calculus*. The theorem states that:

If f(x) is continuous on the interval (a, b) and if F(x) $\int f(x)dx$ = is any indefinite integral of f(x), then, the value of the definite integral $\int_a^b f(x)dx$ is given by F(b) – F(a).

This is interpreted as the area bounded by the curve $y = f(x)$, the x–axis and the lines $x = a$ and $x = b$ (see fig.12.2). This process of integration is known as definite integration, since; its value is independent of the constant of integration.

APPLICATIONS IN ECONOMICS

Integration is used to solve a number of problems in Economics. For example, integration is used to find the total revenue (TR) function from marginal revenue (MR) function, total cost (TC) function from marginal cost (MC) function, or total profit (TP) function from marginal profit function. While integrating the marginal revenue function, we use the fact that the TR is zero for zero output, in order to find the constant of integration. i.e., TR = 0, when $x = 0$. While finding the total cost, the constant of integration may be determined using the fact that, the cost of producing zero units is the fixed cost. Also note that when a total cost function is represented in polynomial form, the constant of integration will be the fixed cost of production. The equilibrium output or the profit maximizing output is given by the solving the equation MR = MC. The value of x thus obtained, when substituted in the profit function will yield the maximum profit. Also, if the consumption function is given by $c = f(y)$, where c is total national consumption and y is total national income, then the marginal propensity to consume (MPC) is the derivative with respect to y of the consumption function. Hence if the marginal propensity to consume is given, then its integral with respect to y will give the consumption function.

i. TR = $\int (\text{MR})dx$

ii. TC = $\int (\text{MC})dx$

iii. TP = $\int (\text{MP})dx$

Domar's Model of Public Debt and National Income

Let Y denotes national income where $Y(t)$ be its value at any time t. Let the national income at beginning be a. i.e., $Y(0) = a$. If $Y(t)$ grows continuously at a constant rate r, then from equation (A) above, we have

$$Y_t = Y_0 e^{rt} = ae^{rt}$$

Let D denotes the public debt, where D(t) be its value at any time t. Then the debt at time t will be

$$\text{D}(t) = \text{D}_0 + \int_0^t \alpha Y_t \, dt$$

$$= \text{D}_0 + \alpha \int_0^t ae^{rt} dt = \text{D}_0 + \alpha a \frac{1}{r}(e^{rt} - 1)$$

The ratio of debt to income becomes

$$\frac{D_t}{Y_t} = \frac{D_0}{Y_t} + \frac{\alpha a}{rY_t}(e^{rt} - 1)$$

$$= \frac{D_0}{ae^{rt}} + \frac{\alpha a}{rae^{rt}}(e^{rt} - 1) = \frac{D_0}{ae^{rt}} + \frac{\alpha}{r}(1 - e^{-rt})$$

Thus, if we let $t \to \infty$, then

$$\lim_{t\to\infty} \frac{D_t}{Y_t} = \lim_{t\to\infty} \frac{D_0}{ae^{rt}} + \lim_{t\to\infty} \frac{\alpha}{r}(1 - e^{-rt})$$

$$= 0 + \frac{\alpha}{r} = \frac{\alpha}{r}$$

INTEGRATION OF RATIONAL ALGEBRAIC FUNCTIONS

1. To find $\int \frac{dx}{ax^2+bx+c}$; where the denominator is non–resolvable.

   This can be integrated by completing the square of the denominator.

2. To find $\int \frac{px+q}{ax^2+bx+c}\,dx$; where the denominator is non–resolvable

   This can be integrated by expressing the numerator as:

   Numerator = A×(Derivative of the Denominator) + B

3. To find $\int \frac{dx}{ax^2+bx+c}$ or $\int \frac{px+q}{ax^2+bx+c}\,dx$; where the denominator is resolvable.

   This can be integrated by resolving the denominator and then putting the function in to partial fractions.

4. If the integrand is a rational function with the degree of the numerator greater than or equal to the degree of the denominator, then divide the numerator by the denominator and express the integrand as the sum of a polynomial and a proper fraction. Then use appropriate method of integration.

STANDARD RULES OF INTEGRATION

i. If $f(x)$ is an integrable function and k is a constant, then

$$\int k\,f(x)dx = k\int f(x)dx$$

ii. If $f(x)$ and $g(x)$ are integrable functions, then

$$\int (f(x)\pm g(x))\,dx = \int f(x)dx \pm \int g(x)dx$$

Example: *Integrate the following with respect to x:*

i. x^8 *ii.* $\sqrt{x}$

iii. $\frac{1}{\sqrt{x}}$ *iv.* $\frac{1}{x^2}$

Solution:

i. $\int x^8 dx = \frac{x^{8+1}}{8+1} + c = \frac{x^{8+1}}{8+1} + c$

ii. $\int\sqrt{x}\,dx = \int x^{1/2}\,dx = \frac{x^{1/2+1}}{1/2+1} + c = \frac{x^{3/2}}{3/2} + c = \frac{2x^{3/2}}{3} + c$

iii. $\int\frac{1}{\sqrt{x}}\,dx = \int x^{-1/2}\,dx = \frac{x^{-1/2+1}}{-1/2+1} + c = \frac{x^{1/2}}{1/2} + c = 2\sqrt{x} + c$

iv. $\int\frac{1}{x^2}\,dx = \int x^{-2}\,dx = \frac{x^{-2+1}}{-2+1} + c = \frac{x^{-1}}{-1} + c = \frac{-1}{x} + c$

INTEGRATION OF FUNCTIONS INVOLVING $a^2 \pm x^2$

Example 1: *To find* $\int\frac{dx}{a^2+x^2}$ *put* $x = at$ *; then* $dx = a\,dt$ *and* $t = \frac{x}{a}$

$$\therefore \quad \int\frac{dx}{a^2+x^2} = \int\frac{adt}{a^2+(at)^2} = \int\frac{adt}{a^2(1+t^2)}$$

$$= \frac{1}{a}\int\frac{dt}{1+t^2} = \frac{1}{a}\tan^{-1}t = \frac{1}{a}\tan^{-1}\left(\frac{x}{a}\right)$$

Example 2: *To find* $\int\frac{dx}{a^2-x^2}$

First put $\frac{1}{a^2-x^2}$ *into partial fractions (Refer Chapter–1 for the method of partial fractions)*

$$\frac{1}{a^2-x^2} = \frac{1}{(a+x)(a-x)} = \frac{A}{(a+x)} + \frac{B}{(a-x)}$$

$$= \frac{1}{2a}\left[\frac{1}{a+x} + \frac{1}{a-x}\right] \text{ (since } A = B = \frac{1}{2a}\text{)}$$

$$\therefore \quad \int\frac{dx}{a^2-x^2} = \int\frac{1}{2a}\left[\frac{1}{a+x} + \frac{1}{a-x}\right]dx = \frac{1}{2a}\left[\int\frac{dx}{a+x} + \int\frac{dx}{a-x}\right]$$

$$= \frac{1}{2a}\left[\log(a+x) - \log(a+x)\right] = \frac{1}{2a}\log\frac{a+x}{a-x}$$

INTEGRATION BY PARTS

If $u(x)$ and $v(x)$ are two functions of x, then:

$$\int u(x)\,v(x)dx = u(x)\int v(x)dx - \int\left[\frac{du}{dx}\int v(x)dx\right]dx$$

Notes:

i. When we employ the method of integration by parts, the second function in the product should have known integral.

ii. As far as possible x or its powers should be chosen as the first function, provided the integral of the other function is known.

iii. Integrals of some functions may be determined using the method of integration by parts by taking 1 as the second function.

iv. In certain cases, the rule of integration by parts may have to be repeated more than once.

DEFINITE INTEGRAL AS THE LIMIT OF A SUM

Though we had defined integration as the inverse process of differentiation, it is funny to see that integration was invented much before the invention of differentiation. Integral Calculus was developed for the purpose of calculating areas by supposing them to be divided into an infinite number of infinitesimally small parts whose sum gives the required area. In fact the integral sign '$\int$' is the elongated version of S, represents 'sum'.

Let be a single valued function continuous on the closed interval $[a, b]$. Divide the interval into n sub–intervals, not necessarily of equal length; say by points $x_1, x_2, \ldots, x_{n-1}$ as in figure 12.3(a). Let ξ_i be any point in the interval $[x_{i-1}, x_i)$. Let the length of the interval $[x_{i-1}, xi)$ is Δx_i For convenience, we take $a = x_0$ and $b = x_n$.

Consider the sum,

$$S_n = f(\xi_1)\Delta x_1 + f(\xi_2)\Delta x_2 + \ldots + f(\xi_n)\Delta x_n = \sum_{i=1}^{n} f(\xi_i)\Delta x_i$$

If the limit of S_n exists as $n \to \infty$ and *max* $\Delta x_i \to 0$, and if this limit is independent of the choice of and of the method of subdivision, then this limit is called the definite integral of from a to b, which is denoted by $\int_a^b f(x)dx.$

i.e., $$\lim_{n\to\infty,\ \max \Delta x_i \to 0} \sum_{i=1}^{n} f(\xi_i)\Delta x_i = \int_a^b f(x)dx$$

Here a and b are called the *lower* and *upper limits* of integration, respectively. is said to be 'integrable' over $[a, b]$, if the above limit exists. It can be shown that a single valued function, which is continuous on an interval, is integrable. Also, from the definition, it follows that,

i. $\int_a^b f(x)dx = -\int_b^a f(x)dx$

ii. $\int_a^a f(x)dx = 0$

iii. $\int_a^b f(x)dx = \int_b^c f(x)dx + \int_c^b f(x)dx$; where $a \le c \le b$

INTEGRATION BY SUBSTITUTION

A. Functions of the form $f(ax+b)$

Method

To find $\int f(ax+b)dx$

Put $ax+b=t$; where = $\int f(t)dt = \varphi(t)+c..$

Taking differentials,

$$a\,dx = dt\ ,\ dx = \frac{1}{a}dt$$

Hence we get,

$$\int f(ax+b)dx = \int f(t)\frac{1}{a}dt$$

$$= \frac{1}{a}\int f(t)dt = \frac{1}{a}\varphi(t)+c$$

B. Functions of the form $f(x^n)\,x^{n-1}$

Method

To find $\int f(x^n)x^{n-1}dx$

Put $x^n = t$; where = $\int f(t)dt = \varphi(t)+c$

Taking differentials,

$$nx^{n-1}\,dx = dt$$

$$\therefore \quad x^{n-1}\,dx = \frac{1}{n}dt$$

Hence we get, $\int f(ax+b)dx = \int f(t)\frac{1}{a}dt$

$$= \frac{1}{a}\int f(t)dt = \frac{1}{a}\int f(t)dt$$

C. Functions of the form $f(x)^n f¢(x)$

Method

To find $\int [f(x)]^n f'(x)dx$; $n \neq -1$

Put $f(x) = t$; then $f'(x)\, dx = dt$

Hence we get,

$$\int [f(x)]^n f'(x)dx = \int t^n dt = \frac{t^{n+1}}{n+1} + c = \frac{[f(x)]^{n+1}}{n+1} + c$$

D. Functions of the form $f'(x)/f(x)$

Method

To find $\int [f'(x)/f(x)]dx$

Put $f(x) = t$; then $f'(x)\, dx = dt$.

Hence we get,

$$\int [f'(x)/f(x)]dx = \int (1/t)dt = \log t + c = \log f(x) + c$$

E. Functions of the form f[φ(x)]φ′(x)

Method

To find $\int f[\varphi(x)]\varphi'(x)dx$

Put $\varphi(x) = t$; where $\int f(t)dt = F(t)$

then $\varphi'(x)dx = dt$

Hence we get,

$$\int f[\varphi(x)]\varphi'(x)dx = \int f(t)dt = F(t) = F[\varphi(x)]$$

Example 1: *To find* $\int \frac{dx}{x^2 - a^2}$

First put $\frac{1}{x^2 - a^2}$ into partial fractions.

$$\frac{1}{x^2 - a^2} = \frac{1}{(x-a)(x+a)} = \frac{1}{2a}\left[\frac{1}{x-a} - \frac{1}{x+a}\right]$$

$$\therefore \int \frac{dx}{x^2 - a^2} = \int \frac{1}{2a}\left[\frac{1}{x-a} - \frac{1}{x+a}\right] dx = \frac{1}{2a}\left[\int \frac{dx}{x-a} - \int \frac{dx}{x+a}\right]$$

$$= \frac{1}{2a}\left[\log(x-a) - \log(x+a)\right] = \frac{1}{2a}\log\frac{x-a}{x+a}$$

Example 2: *Integrate the following with respect to x:*

i. $x^2(2x-1)(x+1)$

ii. $\frac{x^3-2x^2+5x-1}{x^2}$

iii. $\left(\sqrt{x}+\frac{1}{\sqrt{x}}\right)^2$

iv. 5^{x+5}

Solution:

i. $\int\left[x^2(2x-1)(x+1)\right]dx = \int\left(2x^4+x^3-x^2\right)dx = \frac{2x^5}{5}+\frac{x^4}{4}-\frac{x^3}{3}+c$

ii. $\int\frac{x^3-2x^2+5x-1}{x^2}dx = \int x-2+\frac{5}{x} - \frac{1}{x^2}dx$

$$= \frac{x^2}{2}-2x+5\log x+\frac{1}{x}+c$$

iii. $\int\left(\sqrt{x}+\frac{1}{\sqrt{x}}\right)^2 dx = \int\left(x+2+\frac{1}{x}\right)dx = \frac{x^2}{2}+2x+\log x+c$

iv. $\int 5^{x+2}dx = \int 5^2.5^x dx = 25\int 5^x dx = 25.\left(\frac{5^x}{\log 5}\right)+c = \frac{5^{x+2}}{\log 5}+c$

2. Integrate the following with respect to x:

i. $\sin x + e^x - \sqrt{x}$

ii. $\frac{1}{1+x^2}+\sec^2 x$

iii. $\tan^2 x$

iv. $\frac{1}{\sin^2 x\ \cos^2 x}$

Solution:

i. $\int\left(\sin x+e^x-\sqrt{x}\right)dx = -\cos x+e^x-\frac{x^{3/2}}{3/2}+c$

$$= -\cos x+e^x-\frac{2x^{3/2}}{3}+c$$

ii. $\int\left(\frac{1}{1+x^2}+\sec^2 x\right)dx = \tan^{-1}x+\tan x+c$

iii. $\int \tan^2 x \, dx = \int (\sec^2 x - 1) dx = \tan x - x + c$

iv. $\int \frac{1}{\sin^2 x \cos^2 x} dx = \int \frac{\sin^2 x + \cos^2 x}{\sin^2 x \cos^2 x} dx$

$$= \int \left[\frac{\sin^2 x}{\sin^2 x \cos^2 x} + \frac{\cos^2 x}{\sin^2 x \cos^2 x} \right] dx$$

$$= \int \left[\frac{1}{\cos^2 x} + \frac{1}{\sin^2 x} \right] dx$$

$$= \int \left[\sec^2 x + \operatorname{cosec}^2 \right] dx$$

$$= \tan x - \cot x + c$$

3. Integrate the following with respect to x:

i. $x (a^2 - x^2)^{7/2}$

ii. $\frac{2x+3}{(x^2+3x-1)^5}$

iii. $\sin^2 x \cos x$

iv. $\frac{e^{2x}}{(e^{2x}+1)^4}$

v. $\frac{x^9}{1+x^{10}}$

vi. $\frac{1}{x \log x}$

vii. $\frac{\tan x}{\sec x + \tan x}$

viii. $\cos^9 x$

Solution:

i. put $a^2 - x^2 = t$; then $x \, dx = dt$

$$\int x(a^2 - x^2)^{7/2} dx = \frac{-1}{2} \int (t)^{7/2} dt = \frac{-1}{2} \frac{t^{9/2}}{9/2} + c = \frac{-1}{9}(a^2 - x^2)^{9/2} + c$$

ii. put $x^2 + 3x - 5 = t$; then $(2x + 3) \, dx = dt$

$$\int \frac{2x+3}{(x^2+3x-1)^5} dx = \int \frac{1}{t^5} dt = \frac{t^{-5+1}}{-5+1} + c$$

$$= -4t^{-4} + c = \frac{-4}{(x^2+3x-1)^4} + c$$

iii. put $\sin x = t$; then $\cos x\, dx = dt$

$$\int \sin^2 x \cos x\, dx = \int t^2 dt = \frac{t^3}{3} + c = \frac{\sin^3 x}{3} + c$$

iv. put $e^{2x} + 1 = t$; then $e^{2x}\, dx = \frac{1}{2} dt$

$$\int \frac{e^{2x}}{\left(e^{2x}+1\right)^4} dx = \frac{1}{2}\int \frac{dt}{t^4} = \frac{1}{2}\ \frac{t^{-4+1}}{-4+1}$$

$$= \frac{-1}{6t^3} + c = \frac{-1}{6(e^{2x}+1)^3} + c$$

v. put $1 + x^{10} = t$; then $x^9\, dx = \frac{1}{10} dt$

$$\int \frac{x^9}{1+x^{10}} dx = \frac{1}{10}\int \frac{dt}{t} = \frac{1}{10}\log t + c$$

$$= \frac{1}{10}\log(1+x^{10}) + c$$

vi. put $\log x = t$; then $\frac{1}{x} dx = dt$

$$\int \frac{1}{x \log x} dx = \int \frac{(1/x)}{\log x} dx = \int \frac{dt}{t}$$

$$= \log t + c = \log(\log x) + c$$

vii. $\dfrac{\tan x}{\sec x + \tan x} = \dfrac{\tan x(\sec x - \tan x)}{(\sec x + \tan x)(\sec x - \tan x)}$;

by multiplying and dividing by $(\sec x - \tan x)$

$$= \frac{\tan x \sec x - \tan^2 x}{\sec^2 x - \tan^2 x} = \frac{\tan x \sec x - (\sec^2 x - 1)}{1}$$

$$= \tan x \sec x - \sec^2 x + 1$$

$$\int \frac{\tan x}{\sec x + \tan x} dx = \int (\tan x \sec x - \sec^2 x + 1) dx$$

$$= \sec x - \tan x + x + c \cos^9 x$$

viii. $\cos^9 x = \cos^8 x \cos x = (\cos^2 x)^4 \cos x$

$$= (1 - \sin^2 x)^4 \cos x$$

$$\therefore \quad \int \cos^9 x\, dx = \int (1-\sin^2 x)^4 \cos x\, dx$$

put $\sin x = t$; then $= dt$

$$\therefore \quad \int \cos^9 x\, dx = \int (1-\sin^2 x)^4 \cos x\, dx$$

$$= \int (1-t^2)^4 dt = \int (1-4t^2+6t^4-4t^6+t^8) dt$$

$$= t - 4.\frac{t^3}{3} + 6.\frac{t^5}{5} - 4.\frac{t^7}{7} + \frac{t^9}{9} + c$$

$$= \sin x - \frac{4}{3}\sin^3 x + \frac{6}{5}\sin^5 x - \frac{4}{7}\sin^7 x + \frac{1}{9}\sin^9 x + c$$

Example 3: *Integrate the following functions with respect to x*

i. $\sqrt{2x-6}$

ii. $\dfrac{3}{2\sqrt{2-3x}}$

iii. $\dfrac{3}{5(2x+1)^7}$

iv. $\dfrac{1}{\sqrt{ax+b}}$

v. $\sec(5x)\tan(5x)$

vi. $\sin(3x)\sin(6x)$

vii. $\sin^2 x$

viii. $\cos^3 x$

ix. $\sqrt{1+\sin x}$

x. $\dfrac{1}{\sqrt{1-4x^2}}$

xi. $\dfrac{\log x}{x}$

xii. $\dfrac{x}{x^4+2}$

xiii. $\dfrac{x+3}{x+5}$

xiv. $\dfrac{1}{e^x-1}$

xv. $\dfrac{x}{x^4+2}$

xvi. $\dfrac{\tan(\log x+1)}{x}$

xvii. $\sqrt{1+\cos 2x}$

xviii. $\dfrac{(1+\log x)^n}{x}$

Solution:

i. put $2x - 6 = t$; then $2dx = dt$ $\therefore$ $dx = \frac{1}{2}dt$

$$\int \sqrt{2x-6}\, dx = \frac{1}{2}\int \sqrt{t}\, \mathrm{dt} = \frac{1}{2}\,\frac{t^{3/2}}{3/2} = \frac{t^{3/2}}{3} + c$$

$$= \frac{1}{3}(2x-6)^{3/2}$$

ii. put $2 - 3x = t$; then $dx = \frac{-1}{3} dt$

$$\int \frac{3}{2\sqrt{2-3x}} dx = \int \frac{3}{2} \frac{1}{\sqrt{t}} \frac{-1}{3} dt$$

$$= \frac{-1}{2} \int \frac{1}{\sqrt{t}} dt = \frac{-1}{2} \int t^{-1/2} dt$$

$$= -\sqrt{t} + c = -\sqrt{2-3x}$$

iii. put $2x + 1 = t$; then $dx = \frac{1}{2} dt$

$$\int \frac{3}{5(2x+1)^7} dx = \frac{3}{5} . \frac{1}{2} \int \frac{1}{t^7} dt = \frac{3}{10} \int t^{-7} dt = \frac{3}{10} \frac{t^{-6}}{-6} + c$$

$$= \frac{1}{20\, t^6} + c = \frac{1}{20\,(2x+1)^6} + c$$

iv. put $ax + b = t$; then $dx = \frac{-1}{3} dt$

$$\int \frac{3}{2\sqrt{2-3x}} dx = \int \frac{3}{2} \frac{1}{\sqrt{t}} \frac{-1}{3} dt = \frac{-1}{2} \int \frac{1}{\sqrt{t}} dt = \frac{-1}{2} \int t^{-1/2} dt$$

$$= -\sqrt{t} + c = -\sqrt{2-3x} + c$$

v. put $5x = t$; then $dx = \frac{1}{5} dt$

$$\int \sec(5x) \tan(5x) dx = \frac{1}{5} \int \sec t \tan t \; dt = \frac{1}{5}(\sec t) + c = \frac{1}{5} \sec(5x) + c$$

vi. We know that:

$$\sin 3x \sin 6x = \frac{1}{2}(\cos 3x - \cos 9x)$$

Put $3x = t$; then $dx = \frac{1}{5} dt.$

Also put $3x = t$; then $dx = \frac{1}{5} dt.$

$$\int \sec(5x) \tan(5x) dx = \frac{1}{5} \int \sec t \tan t \; dt = \frac{1}{5}(\sec t) + c = \frac{1}{5} \sec(5x) + c$$

vii. We know the formula: $\sin^2 x = \frac{1}{2}(1-\cos 2x)$

$$\int \sin^2 x\,dx = \int \frac{1}{2}(1-\cos 2x)\,dx$$

$$= \frac{1}{2}\int dx - \frac{1}{2}\int \cos 2x dx = \frac{1}{2}x - \frac{1}{2}\frac{\sin 2x}{2}$$

(by the substitution $2x = t$ in the second integral)

$$= \frac{x}{2} - \frac{\sin 2x}{4}$$

viii. We have $\cos^3 x = \frac{1}{4}(\cos 3x - 3\cos x)$

$$\therefore \quad \int \cos^3 x\,dx = \int \frac{1}{4}(\cos 3x - 3\cos x)\,dx = \frac{1}{4}\left(\frac{\sin 3x}{3} - 3\sin x\right)$$

ix. We have $\sin x = 2\sin(x/2)\cos(x/2)$, and

$\sin^2(x/2) + \cos^2(x/2) = 1$

$$\therefore \quad \sqrt{1+\sin x} = \sqrt{\sin^2(x/2) + \cos^2(x/2) + 2\sin(x/2)\cos(x/2)}$$

$$= \sqrt{[\sin(x/2) + \cos(x/2)]^2} = \sin(x/2) + \cos(x/2)$$

$$\int \sqrt{1+\sin x}\,dx = \int [\sin(x/2) + \cos(x/2)]dx$$

$$= \frac{-\cos(x/2)}{1/2} + \frac{\sin(x/2)}{1/2}$$

$$= 2[\sin(x/2) - \cos(x/2)]$$

x. $$\int \frac{dx}{\sqrt{1-4x^2}} = \int \frac{dx}{\sqrt{1-(2x)^2}}$$

$$= \frac{\sin^{-1} 2x}{2}, \text{ by substituting } t = 2x$$

xi. put $\log x = t$; then $\frac{1}{x}dx = dt$

$$\int \frac{\log x}{x}\,dx = \int \log x \,\frac{1}{x}\,dx = \int t dt = \frac{t^2}{2} = \frac{(\log x)^2}{2}$$

xii. $$\int \frac{x}{x^4+2}\,dx = \int \frac{x}{(x^2)^2 + (\sqrt{2})^2}\,dx$$

put $x^2 = t$; $2x\ dx = dt$

$$\therefore \quad \int \frac{x}{x^4+2}dx = \frac{1}{2}\int \frac{dt}{t^2+(\sqrt{2})^2}$$

$$= \frac{1}{2}\frac{1}{\sqrt{2}}\tan^{-1}\left(t/\sqrt{2}\right) = \frac{1}{2\sqrt{2}}\ \tan^{-1}\left(x^2/\sqrt{2}\right)$$

xiii. $$\int \frac{x+3}{x+5}dx = \int \frac{x+5-2}{x+5}dx$$

$$= \int \left[\frac{x+5}{x+5} - \frac{2}{x+5}\right]dx = \int \left[1 - \frac{2}{x+5}\right]dx$$

$$= x - 2\log(x+5)$$

xiv. $$\int \frac{dx}{e^x-1} = \int \frac{e^{-x}dx}{1-e^{-x}}$$; by multiplying Nr. and Dr. by e^{-x}

put $t = 1 - e^{-x}$; then $dt = e^{-x}dx$

$$\therefore \quad \int \frac{dx}{e^x-1} = \int \frac{e^{-x}dx}{1-e^{-x}} = \int \frac{dt}{t} = \log t = \log(1 - e^{-x})$$

xv. $$\int \frac{xdx}{x^4+2}$$

put $t = x^2$; then $dt = 2xdx$

$$\therefore \quad \int \frac{xdx}{x^4+2} = \frac{1}{2}\int \frac{dt}{t^2+2}$$

$$= \frac{1}{2}\tan^{-1}\frac{t}{\sqrt{2}} + c = \frac{1}{2}\tan^{-1}\frac{x^2}{\sqrt{2}} + c$$

xvi. $$\frac{\tan(\log x+1)}{x}$$

put $t = \log x + 1$; then $dt = \frac{1}{x}dx$

$$\therefore \int \frac{\tan(\log x+1)}{x}dx = \int \tan t\ dt$$

$$= \log\sec t + c = \log\sec(\log x+1) + c$$

xvii. $$\sqrt{1+\cos 2x} = \sqrt{2\cos^2 x} = \sqrt{2}\cos x$$

$$\therefore \int \sqrt{1+\cos 2x}\ dx = \int \sqrt{2}\ \cos x\ dx = \sqrt{2}\sin x + c$$

xviii. $\int \frac{(1+\log x)^n}{x} dx$

put $t = 1 + \log x$; then $dt = \frac{1}{x} dx$

$$\therefore \int \frac{(1+\log x)^n}{x} dx = \int t^n \, dt$$

$$= \frac{t^{n+1}}{n+1} + c = \frac{(1+\log x)^{n+1}}{n+1} + c$$

SOLVED EXAMPLES

Example 1: *Integrate the following with respect to x:*

i. $\frac{1}{9-4x^2}$ *ii.* $\frac{1}{4+25x^2}$

iii. $\frac{1}{49x^2-16}$ *iv.* $\frac{1}{\sqrt{9-16x^2}}$

Solution:

i. $\int \frac{1}{9-4x^2} dx = \int \frac{dx}{4[(9/4)-x^2]}$

$= \frac{1}{4} \int \frac{dx}{(3/2)^2 - x^2}$ (put a = 3/2 in Result–2)

$= \frac{1}{4}\left[\frac{1}{2(3/2)} \log \frac{(3/2)+x}{(3/2)-x}\right] = \frac{1}{12} \log \frac{3+2x}{3-2x}$

ii. $\int \frac{1}{4+25x^2} dx = \int \frac{dx}{25[(4/25)+x^2]}$

$= \frac{1}{25} \int \frac{dx}{[(2/5)^2 + x^2]}$ (put a = 2/5 in Result–1)

$= \frac{1}{25} \; \frac{1}{2/5} \tan^{-1} \frac{x}{2/5} = \frac{1}{10} \tan^{-1} \frac{5x}{2}$

iii. $\int \frac{1}{49x^2-16} dx = \frac{1}{49} \int \frac{dx}{x^2 - (16/49)}$

$= \frac{1}{49} \int \frac{dx}{x^2 - (4/7)^2}$ (put a = 4/7 in Result–3)

$= \frac{1}{49} \; \frac{1}{2(4/7)} \log \frac{x-(4/7)}{x+(4/7)} = \frac{1}{56} \; \log \frac{7x-4}{7x+4}$

iv. $\int \frac{1}{\sqrt{9-16x^2}}dx = \int \frac{dx}{4\sqrt{(9/16)-x^2}}$

$= \int \frac{dx}{4\sqrt{(3/4)^2-x^2}}$ (put a = 3/4 in Result–4)

$= \frac{1}{4}\sin^{-1}\left(\frac{x}{3/4}\right) = \frac{1}{4}\sin^{-1}\left(\frac{4x}{3}\right)$

Example 2: *Integrate the following with respect to x:*

i. $\frac{1}{x^2+6x+25}$ *ii.* $\frac{2x+1}{x^2+2x+3}$

Solution:

i. $x^2 +6x +25$ is not resolvable.

Completing the square, we get:

$$x^2 + 6x + 25 = x^2 + 6x + 9 + 16 = (x + 3)^2 + 4^2$$

$$\int \frac{dx}{x^2+6x+25} = \int \frac{dx}{(x+3)^2+4^2} = \frac{1}{4}\tan^{-1}\left(\frac{x+3}{4}\right)$$

ii. $x^2 + 2x +3$ is not resolvable.

Put $2x + 1$ = A × (derivative of $x^2 + 2x +3$) + B

= A($2x$ +2) + B

Equating coefficients of similar terms, we get,

A = 1, B = –1

Hence $2x + 1 = (2x +2) - 1$

(in this case, this step may also be obtained by simple inspection)

$$\therefore \int \frac{(2x+1)dx}{x^2+2x+3} = \int \frac{(2x+2-1)dx}{x^2+2x+3} = \int \frac{(2x+2)dx}{x^2+2x+3} - \int \frac{dx}{x^2+2x+3}$$

$$= \int \frac{(2x+2)dx}{x^2+2x+3} - \int \frac{dx}{(x+1)^2+(\sqrt{2})^2}$$

$$= \log (x^2 + 2x +3) - \frac{1}{\sqrt{2}}\tan^{-1}\frac{x+1}{\sqrt{2}}$$

Example 3(a): *Sketch the curve* $y = \frac{1}{2}x^2$ *and find the area bounded by the curve, the x–axis and the* $x = 0$ *lines and* $x = 2$.

Solution: The required area = $\int_a^b y\, dx$; where a = 0 and b= 2

$$= \int_0^2 \frac{1}{2}x^2\,dx = \frac{1}{2}\left[\frac{x^3}{3}\right]_0^2$$

$$= \frac{1}{2}\left[\frac{2^3}{3} - 0\right] = \frac{1}{2}\left[\frac{2^3}{3} - 0\right] = \frac{4}{3} \text{ square units}$$

Example 3(b): *Integrate the following with respect to x:*

i. $x(1+x^2)^3$ *ii.* $\dfrac{\sin^{-1} x}{\sqrt{1-x^2}}$

Solution:

i. put $1+x^2 = t$; then $xdx = \frac{1}{2}dt$

$$\int x(1+x^2)^3\,dx = \frac{1}{2}\int t^3 dt = \frac{t^4}{8} + c = \frac{(1+x^2)^4}{8} + c$$

ii. put $\sin^{-1} x = t$; then $dx = \frac{1}{\sqrt{1-x^2}}\,dt$

$$\int \frac{\sin^{-1} x}{\sqrt{1-x^2}}dx = \int t\,dt = \frac{t^2}{2} + c = \frac{(\sin^{-1} x)^2}{2} + c$$

Example 3(c): *Integrate the following with respect to x:*

i. $\dfrac{\sin(\log x)}{x}$ *ii.* $\dfrac{1}{e^x+1}$

Solution:

i. put $\log x = t$; then $\frac{1}{x}dx = dt$

$$\int \frac{\sin(\log x)}{x}dx = \int \sin(\log x)(1/x)dx$$

$$= \int \sin t\,dt = -\cos t + c = -\cos(\log x) + c$$

ii. $\dfrac{1}{e^x+1} = \dfrac{e^{-x}}{e^{-x}(e^x+1)}$; by multiplying and dividing by e^{-x}

$$= \frac{e^{-x}}{(1+e^{-x})}$$

Put $1+e^{-x} = t$; then $e^{-x}dx = -dt$

$$\int \frac{1}{e^x+1}dx = \int \frac{e^{-x}}{(1+e^{-x})}dx$$

$$= -\int \frac{dt}{t} = -\log t + c = -\log(1+e^{-x}) + c$$

Example 4: *Integrate the following with respect to x:*

i. $x \sin x^2$ ii. $\dfrac{2x^4}{1+x^{10}}$

Solution:

i. put $x^2 = t$; then $x dx = \dfrac{1}{2} dt$

$$\int x \sin x^2 dx = \frac{1}{2}\int \sin t\, dt = \frac{1}{2}(-\cos t) + c = \frac{-\cos x^2}{2} + c$$

ii. put $x^5 = t$; then $x^4\ dx = \dfrac{1}{5} dt$

$$\int \frac{2x^4 dx}{1+x^{10}} = \frac{2}{5}\int \frac{dt}{1+t^2} = \frac{2}{5}\tan^{-1} t = \frac{2}{5}\tan^{-1}(x^5) + c$$

Example 5: *Integrate the following with respect to x:*

i. $\tan x$ ii. $\dfrac{2x+3}{x^2+3x+2}$

Solution:

i. $\tan x = \dfrac{\sin x}{\cos x}$

put $\cos x = t$; then $\sin x\ dx = -dt$

$$\int \tan x\ dx = \int \frac{\sin x}{\cos x} dx = -\int \frac{dt}{t}$$

$$= -\log t + c = -\log(\cos x) + c$$

Aliter: $\tan x = \dfrac{\sec x \tan x}{\sec x}$;

put $\sec x = t$;

Then $\sec x \tan x\ dx = dt$

$$\therefore \int \tan x\ dx = \int \frac{\sec x \tan x}{\sec x} dx = \int \frac{dt}{t}$$

$$= \log t + c = \log(\sec x) + c$$

ii. put $x^2 + 3x + 2 = t$; then $(2x + 3)dx = dt$

$$\int \frac{2x+3}{x^2+3x+2}\,dx = \int \frac{dt}{t} = \log t + c = \log(x^2+3x+2)+c$$

Example 6: *Integrate the following with respect to x:*

i. $(2x + 3)^8$ ii. $\frac{1}{3+5x}$

iii. $\sec^2 (2 - x)$ iv. $e^{2(3-x)}$

Solution:

i. Put $2x + 3 = t$;

Then, $2\,dx = dt$, $dx = \frac{1}{2}dt$

$$\int (2x+3)^8\,dx = \int (t)^8 \frac{1}{2}dt = \frac{1}{2}\int (t)^8\,dt = \frac{1}{2}\frac{t^9}{9}$$

$$= \frac{t^9}{18} + c = \frac{(2x+3)^9}{18} + c$$

ii. Put $3 + 5x = t$

Then $5\,dx = dt$, $dx = \frac{1}{2}dt$

$$\int \frac{1}{3+5x}\,dx = \int \frac{1}{t}\,\frac{1}{5}dt$$

$$= \frac{1}{5}\int \frac{1}{t}\,dt = \frac{1}{5}(\log t) + c = \frac{1}{5}\log(3+5x) + c$$

iii. Put $2 - x = t$

Then $-dx = dt$, $dx = -dt$

$$\int \sec^2(2-x)dx = \int \sec^2 t\,(-dt)$$

$$= -\int \sec^2 t\,(-dt) = -\tan t + c = -\tan(2-x) + c$$

iv. Put $2(3 - x) = t$

i.e., $6 - 2x = t$

Then $(-2)dx = dt$, $dx = \frac{-1}{2}dt$

$$\int e^{2(3-x)}dx = \int e^t\left(\frac{-1}{2}dt\right)$$

$$= \frac{-1}{2}\int e^t dt = -\frac{e^t}{2} + c = -\frac{e^{2(3-x)}}{2} + c$$

Example 7(a): *Integrate the following functions with respect to x*

i. $\dfrac{\cos x}{16+9\sin^2 x}$ ii. $\dfrac{e^x}{e^{2x}-4}$

iii. $\dfrac{1}{x^2-6x-16}$ iv. $\dfrac{1}{3x^2-2x+4}$

v. $\int\dfrac{dx}{1-x-x^2}$ vi. $\int\dfrac{dx}{1+\cos^2 x}$

vii. $\dfrac{2x+1}{x^2+4x+6}$ viii. $\dfrac{5x+7}{6x^2+4x-1}$

ix. $\dfrac{4x}{(x-2)(x-1)}$ x. $\dfrac{1}{(x-1)(x^2+1)}$

xi. $\dfrac{1}{x^3+1}$ xii. $\dfrac{5}{(1+2x)(1-3x)}$

xiii. $\dfrac{x^2}{x^4+x^2-2}$ xiv. $\dfrac{1}{x(x^3+1)}$

xv. $\dfrac{3x+5}{2x^2+5x+3}$ xvi. $\dfrac{x^2}{x^2+7x+10}$

Solution:

i. $\int\dfrac{\cos x}{16+9\sin^2 x}dx$ put $\sin x = t$; then $\cos x\ dx = dt$

$$\therefore \int\frac{\cos x}{16+9\sin^2 x}dx = \int\frac{dt}{16+9t^2}$$

$$= \frac{1}{9}\int\frac{dt}{(4/3)^2+t^2} = \frac{1}{9}\frac{1}{(4/3)}\tan^{-1}\left(\frac{t}{4/3}\right)$$

$$= \frac{1}{12}\tan^{-1}\left(\frac{3t}{4}\right) = \frac{1}{12}\tan^{-1}\left(\frac{3\sin x}{4}\right)$$

ii. $\int\dfrac{e^x dx}{e^{2x}-4}$ put $e^x = t$; then $e^x\ dx = dt$

$$\int\frac{e^x dx}{e^{2x}-4} = \int\frac{dt}{t^2-2^2} = \frac{1}{2.2}\log\frac{t-2}{t+2} = \frac{1}{4}\log\frac{e^x-2}{e^x+2}$$

iii. $\int\dfrac{dx}{x^2-6x-16} = \int\dfrac{dx}{(x-3)^2-5^2} = \dfrac{1}{10}\log\dfrac{x-8}{x-2}$

iv. $\int\dfrac{dx}{3x^2-2x+4}$

$$3x^2-2x+4 = 3\,[x^2-(2/3)x+(4/3)]$$

$$= 3\left[\left(x-\frac{1}{3}\right)^2 - \frac{1}{9}+\frac{4}{3}\right] = 3\left[\left(x-\frac{1}{3}\right)^2 - \left(\frac{\sqrt{11}}{3}\right)^2\right]$$

$$\int\frac{dx}{3x^2-2x+4} = \frac{1}{3}\int\frac{dx}{(x-(1/3))^2-\left(\sqrt{11}/3\right)^2}$$

$$= \frac{1}{3}\frac{1}{\sqrt{11}/3}\tan^{-1}\frac{x-(1/3)}{\sqrt{11}/3} = \frac{1}{\sqrt{11}}\tan^{-1}\frac{3x-1}{\sqrt{11}}$$

v. $\int\frac{dx}{1-x-x^2}$

$$1-x-x^2 = 1+\frac{1}{4}-\left(\frac{1}{4}+x+x^2\right) = \frac{5}{4}-\left(x+\frac{1}{2}\right)^2$$

$$\therefore \int\frac{1}{1-x-x^2}\,dx = \int\frac{dx}{\left(\sqrt{5}/2\right)^2-[x+(1/2)]^2}$$

$$= \frac{1}{2(\sqrt{5}/2)}\log\left(\frac{(\sqrt{5}/2)+[x+(1/2)]}{(\sqrt{5}/2)-[x+(1/2)]}\right)$$

$$= \frac{1}{\sqrt{5}}\log\frac{\sqrt{5}+1+2x}{\sqrt{5}-1-2x}$$

vi. $\int\frac{dx}{1+\cos^2 x} = \int\frac{\sec^2 x\,dx}{\sec^2 x+1}$; by multiplying Nr. and Dr. by $\sec^2 x$

$$= \int\frac{\sec^2 x\,dx}{1+\tan^2 x+1} = \int\frac{\sec^2 x\,dx}{2+\tan^2 x}$$

Put $\tan x = t$; then $\sec^2 x\,dx = dt$

$$\therefore \int\frac{dx}{1+\cos^2 x} = \int\frac{\sec^2 x\,dx}{2+\tan^2 x} = \int\frac{dt}{2+t^2}$$

$$= \frac{1}{\sqrt{2}}\tan^{-1}\frac{t}{\sqrt{2}} = \frac{1}{\sqrt{2}}\tan^{-1}\left(\frac{\tan x}{\sqrt{2}}\right)$$

vii. $\int\frac{2x+1}{x^2+4x+6}\,dx$

Put $2x+1=(2x+4)-3$

$$\int\frac{2x+1}{x^2+4x+6}\,dx = \int\frac{(2x+4-3)dx}{x^2+4x+6}$$

$$= \int \frac{(2x+4)dx}{x^2+4x+6} - \int \frac{3dx}{x^2+4x+6}$$

$$= \int \frac{(2x+4)dx}{x^2+4x+6} - \int \frac{3dx}{x^2+4x+6}$$

$$= \int \frac{(2x+4)dx}{x^2+4x+6} - 3\int \frac{dx}{(x+2)^2+(\sqrt{2})^2}$$

$$= \log(x^2+4x+6) - \frac{3}{\sqrt{2}} \tan^{-1} \frac{x+2}{\sqrt{2}}$$

viii. $\int \frac{(5x+7)dx}{6x^2+4x-1}$

Put $5x+7 = A(12x+4)+B$

Equating similar terms on both sides we get,

A = 5/12 , B = 16/3

$$5x+7 = (5/12)(12x+4)+(16/3)$$

$$\int \frac{(5x+7)dx}{6x^2+4x-1} = \frac{5}{12}\int \frac{(12x+4)dx}{6x^2+4x-1} + \frac{16}{3}\int \frac{dx}{6x^2+4x-1}$$

$$= \frac{5}{12}\int \frac{(12x+4)dx}{6x^2+4x-1} + \frac{16}{3}.\frac{1}{6}\int \frac{dx}{(x+(1/3))^2-(\sqrt{10}/6)^2}$$

$$= \frac{5}{12}\log(6x^2+4x-1) + \frac{8}{9}.\frac{1}{2.(\sqrt{10}/6)}\log\frac{x+(1/3)-(\sqrt{10}/6)}{x+(1/3)+(\sqrt{10}/6)}$$

$$= \frac{5}{12}\log(6x^2+4x-1) + \frac{8}{3\sqrt{10}}\log\frac{6x+2-\sqrt{10}}{6x+2+\sqrt{10}}$$

ix. $\int \frac{4xdx}{(x-2)(x-1)}$

Using the method of partial fractions,

$$\frac{1}{(x-2)(x-1)} = \frac{8}{(x-2)} - \frac{4}{(x-1)}$$

$$\therefore \int \frac{4xdx}{(x-2)(x-1)} = \int \frac{8dx}{(x-2)} - \int \frac{4dx}{(x-1)}$$

$$= 8\log(x-2) - 4\log(x-1)$$

$$= 4[2\ \log(x-2) - \log(x-1)]$$

$$= 4[\log(x-2)^2 - \log(x-1)]$$

$$= 4\log\frac{(x-2)^2}{x-1}$$

x. Using the method of partial fractions,

$$\frac{1}{(x-1)(x^2+1)} = \frac{1/2}{(x-1)} - \frac{(1/2)(x+1)}{(x^2+1)}$$

$$\therefore \int\frac{dx}{(x-1)(x^2+1)} = \frac{1}{2}\log(x-1) - \frac{1}{2}\int\frac{(x+1)dx}{(x^2+1)}$$

Put $(x + 1) = (1/2)2x + 1$

$$\int\frac{(x+1)dx}{(x^2+1)} = \int\frac{(1/2)2xdx}{(x^2+1)} + \int\frac{dx}{(x^2+1)}$$

$$= \frac{1}{2}\int\frac{2xdx}{(x^2+1)} + \int\frac{dx}{(x^2+1)} = \frac{1}{2}\log(x^2+1) + \tan^{-1}x$$

$$\therefore \int\frac{dx}{(x-1)(x^2+1)} = \frac{1}{2}\log(x-1) - \frac{1}{2}\left[\frac{1}{2}\log(x^2+1) + \tan^{-1}x\right]$$

$$= \frac{1}{2}\log(x-1) - \frac{1}{4}\log(x^2+1) - \frac{1}{2}\tan^{-1}x$$

xi. $x^3 + 1 = (x+1)(x^2 - x + 1)$

Using the method of partial fractions,

$$\frac{1}{x^3+1} = \frac{1}{3(x-1)} - \frac{x-2}{3(x^2-x+1)}$$

$$\therefore \int\frac{dx}{x^3+1} = \frac{1}{3}\int\frac{dx}{(x-1)} - \frac{1}{3}\int\frac{(x-2)dx}{(x^2-x+1)}$$

$$= \frac{1}{3}\log(x-1) - \frac{1}{3}\int\frac{(x-2)dx}{(x^2-x+1)}$$

Put $x - 2 = (1/2)(2x-1) - (3/2)$

$$\int\frac{(x-2)dx}{(x^2-x+1)} = \frac{1}{2}\int\frac{(2x-1)dx}{(x^2-x+1)} - \frac{3}{2}\int\frac{dx}{(x^2-x+1)}$$

$$= \frac{1}{2}\int\frac{(2x-1)dx}{(x^2-x+1)} - \frac{3}{2}\int\frac{dx}{(x-1/2)^2 + (\sqrt{3}/2)^2}$$

$$= \frac{1}{2}\log(x^2 - x + 1) - \frac{3}{2}.\frac{1}{\sqrt{3}/2}\tan^{-1}\frac{x - 1/2}{\sqrt{3}/2}$$

$$\therefore \quad \int\frac{dx}{x^3 + 1} = \frac{1}{3}\log(x-1) - \frac{1}{3}\left[\frac{1}{2}\log(x^2 - x + 1) - \frac{3}{2}.\frac{1}{\sqrt{3}/2}\tan^{-1}\frac{x - 1/2}{\sqrt{3}/2}\right]$$

$$= \frac{1}{3}\log(x-1) - \frac{1}{6}\log(x^2 - x + 1) + \frac{1}{\sqrt{3}}\tan^{-1}\frac{2x-1}{\sqrt{3}}$$

xii. Using the method of partial fractions,

$$\frac{5}{(1+2x)(1-3x)} = \frac{2}{(1+2x)} + \frac{3}{(1-3x)}$$

$$\therefore \int\frac{5dx}{(1+2x)(1-3x)} = \int\frac{2dx}{(1+2x)} + \int\frac{3dx}{(1-3x)}$$

$$= \log(1+2x) - \log(1-3x) = \log\frac{1+2x}{1-3x}$$

xiii. $\int\frac{x^2dx}{x^4 + x^2 - 2}$

$$x^4 + x^2 - 2 = (x^2+2)(x^2-1)$$

$$\frac{x^2}{x^4 + x^2 - 2} = \frac{x^2}{(x^2+2)(x^2-1)} = \frac{2/3}{(x^2+2)} + \frac{1/3}{(x^2-1)};$$

by the method of partial fractions

$$\therefore \quad \int\frac{x^2dx}{x^4 + x^2 - 2} = \frac{2}{3}\int\frac{dx}{(x^2+2)} + \frac{1}{3}\int\frac{dx}{(x^2-1)}$$

$$= \frac{2}{3}\frac{1}{\sqrt{2}}\tan^{-1}\frac{x}{\sqrt{2}} + \frac{1}{3}.\frac{1}{2}\log\frac{x-1}{x+1}$$

$$= \frac{\sqrt{2}}{3}\tan^{-1}\frac{x}{\sqrt{2}} + \frac{1}{6}\log\frac{x-1}{x+1}$$

xiv. $\frac{1}{x(x^3+1)} = \frac{x^2}{x^3(x^3+1)}$; by multiplying Nr. and Dr. by x^2

put $t = x^3$; then $dt = 3x^2$ dx

$$\therefore \quad \int\frac{1}{x(x^3+1)}dx = \int\frac{x^2dx}{x^3(x^3+1)} = \frac{1}{3}\int\frac{dt}{t(t+1)}$$

$$= \frac{1}{3}\int\left[\frac{1}{t}-\frac{1}{(t+1)}\right]dt;$$

by the method of partial fractions

$$= \frac{1}{3}[\log t - \log(t+1)] = \frac{1}{3}\log\frac{t}{t+1} = \frac{1}{3}\log\frac{x^3}{x^3+1}$$

xv. $2x^2 + 5x + 3$ is resolvable

$$\frac{3x+5}{2x^2+5x+3} = \frac{3x+5}{(x+1)(2x+3)}$$

Putting into partial fractions:

$$\frac{3x+5}{(x+1)(2x+3)} = \frac{A}{x+1}+\frac{B}{2x+3} = \frac{2}{x+1}-\frac{1}{2x+3}$$

$$\therefore \int\frac{3x+5}{2x^2+5x+3}\,dx = \int\frac{2dx}{x+1}-\int\frac{dx}{2x+3}$$

$$= 2\log(x+1) - \frac{1}{2}\log(2x+3)$$

xvi. Since the integrand is an improper algebraic fraction, first divide the numerator by the denominator.

By dividing we get:

$$\frac{x^2}{x^2+7x+10} = 1 - \frac{7x+10}{x^2+7x+10}$$

Also, $x^2 + 7x + 10 = (x+5)(x+2)$

Putting into partial fractions:

$$\frac{7x+10}{x^2+7x+10} = \frac{7x+10}{(x+5)(x+2)} = \frac{(25/3)}{(x+5)}-\frac{(4/3)}{(x+2)}$$

$$\therefore \int\frac{x^2dx}{x^2+7x+10} = \int\left[1-\frac{(25/3)}{(x+5)}+\frac{(4/3)}{(x+2)}\right]dx$$

$$= x-\frac{25\log(x+5)}{3}+\frac{4\log(x+2)}{3}$$

Example 7(b): *Integrate the following with respect to x:*

i. $x\sin x$

ii. $x^2\log x$

iii. $\log x$

iv. $e^{2x}x$

Solution:

i. $\int x \sin x \, dx = x.(-\cos x) - \int 1.(-\cos x)dx = -x\cos x + \int \cos x \, dx$

ii. $\int x^2 \log x \, dx = \int (\log x)\, x^2 \, dx = \log x.\frac{x^3}{3} - \int \frac{1}{x}\,\frac{x^3}{3}\, dx$

$= \frac{x^3}{3}\log x - \frac{1}{3}\int x^2 \, dx = \frac{x^3}{3}\log x - \frac{1}{3}\frac{x^3}{3} = \frac{x^3 \log x}{3} - \frac{x^3}{9}$

iii. $\int \log x \, dx = \int \log x \,.1 \, dx = \log x.\, x - \int \frac{1}{x}\,.x \, dx$

$= x\log x - \int dx = x\log x - x = x(\log x - 1)$

iv. $\int e^{2x} x \, dx = \int xe^{2x} \, dx = x.\frac{e^{2x}}{2} - \int 1.\frac{e^{2x}}{2}\, dx$

$= \frac{xe^{2x}}{2} - \frac{1}{2}\int e^{2x} \, dx = \frac{xe^{2x}}{2} - \frac{1}{2}\frac{e^{2x}}{2} = \frac{xe^{2x}}{2} - \frac{e^{2x}}{4}$

Example 7(c): *Evaluate the following definite integrals:*

i. $\int_1^3 (1+2x)dx$ ii. $\int_{-1}^{-1} \frac{1}{\sqrt{x-1}}\, dx$

Solution:

i. $\int_1^3 (1+2x)dx = \left[x + 2\frac{x^2}{2}\right]_1^3$

$= \left[x + x^2\right]_1^3 = \left[(3+3^2) - (1+1^2)\right]$

$= \left[(3+9) - (1+1)\right] = 12 - 2 = 10$

ii. $\int_{-1}^{-1} \frac{1}{\sqrt{x-1}}\, dx = \left[2\sqrt{x-1}\right]_{-1}^{1}$

$= \left[2(1-1) - 2(-1-1)\right] = \left[0+4\right] = 4.$

Example 8: *Integrate the following functions with respect to x (use the method of integration by parts wherever necessary)*

i. $x^2 \cos 3x$ ii. xe^{9x}

iii. $e^x(x^2+1)$ iv. $xe^{\sqrt{x}}$

v. $x\tan^2 x$ vi. $x\log x$

vii. $x^n \log x$ viii. $x^2 \log(1+x)$

ix. $x\tan^{-1}x$

x. $\tan^{-1}x$

xi. $(\log x)^2$

xii. $\dfrac{xe^x}{(1+x)^2}$

xiii. $x\sec^2 x$

xiv. $\sin^{-1}x$

Solution:

i. $$\int x^2\cos 3x dx = x^2\frac{\sin 3x}{3} - \int 2x.\frac{\sin 3x}{3}dx$$

$$= \frac{x^2\sin 3x}{3} + \frac{2}{3}\int x\sin 3x\, dx$$

$$= \frac{x^2\sin 3x}{3} + \frac{2}{3}\left[x.\frac{-\cos 3x}{3} - \int 1.\frac{-\cos 3x}{3}dx\right]$$

$$= \frac{x^2\sin 3x}{3} + \frac{2}{3}\left[\frac{-x\cos 3x}{3} + \frac{\sin 3x}{9}\right]$$

$$= \frac{x^2\sin 3x}{3} - \frac{2x\cos 3x}{9} + \frac{2\sin 3x}{27}$$

ii. $$\int xe^{9x}dx = x.\frac{e^{9x}}{9} - \int 1.\frac{e^{9x}}{9}dx = \frac{xe^{9x}}{9} - \frac{e^{9x}}{81}$$

iii. $$\int e^x(x^2+1)dx = \int (x^2+1)e^x dx = (x^2+1)e^x - \int 2xe^x dx$$

$$= (x^2+1)e^x - 2\left[xe^x - \int 1.e^x dx\right]$$

$$= (x^2+1)e^x - 2xe^x + 2e^x = (x^2-2x+3)e^x$$

iv. $\int x\, e^{\sqrt{x}}dx$

Put $\sqrt{x}$ = t ; x = t^2 ; then, $dx = 2\ t\ dt$

$$\therefore \quad \int x\, e^{\sqrt{x}}dx = \int e^t t^2 2t\ dt = 2\int t^3 e^t dt$$

$$= 2\left[t^3e^t - \int 3t^2e^t dt\right] = 2t^3e^t - 6\int t^2e^t dt$$

$$= 2t^3e^t - 6\left[t^2e^t - \int 2te^t dt\right]$$

$$= 2t^3e^t - 6t^2e^t + 12\int te^t dt$$

$$= 2t^3e^t - 6t^2e^t + 12\left[te^t - \int 1.e^t dt\right]$$

$$= 2t^3e^t - 6t^2e^t + 12te^t - 12e^t$$

$$= \left[2t^3 - 6t^2 + 12t - 12\right]e^t$$

$$= \left[2x\sqrt{x} - 6x + 12\sqrt{x} - 12\right]e^{\sqrt{x}}$$

v. $\int x\tan^2 x\,dx = \int x(\sec^2 x - 1)\,dx$

$$= \int (x\sec^2 x - x)\,dx = \int x\sec^2 x\,dx - \int x dx$$

$$= x\tan x - \int 1.\tan x\,dx - \frac{x^2}{2}$$

$$= x\tan x - \log(\sec x) - \frac{x^2}{2}$$

vi. $\int x\log x\,dx = \int \log x\; x\,dx$

$$= \log x\frac{x^2}{2} - \int \frac{1}{x}\frac{x^2}{2}\,dx = \frac{x^2\log x}{2} - \frac{1}{2}\int x\,dx$$

$$= \frac{x^2\log x}{2} - \frac{1}{2}\frac{x^2}{2} = \frac{x^2\log x}{2} - \frac{x^2}{4}$$

vii. $\int x^n\log x\,dx = \int \log x\; x^n dx$

$$= \log x\frac{x^{n+1}}{n+1} - \int \frac{1}{x}\frac{x^{n+1}}{n+1}dx = \frac{x^{n+1}\log x}{n+1} - \frac{1}{n+1}\int x^n dx$$

$$= \frac{x^{n+1}\log x}{n+1} - \frac{1}{n+1}\frac{x^{n+1}}{n+1} = \frac{x^{n+1}\log x}{n+1} - \frac{x^{n+1}}{(n+1)^2}$$

viii. $\int x^2\log(1+x)dx = \int \log(1+x)x^2 dx$

$$= \log(1+x)\frac{x^3}{3} - \int \frac{1}{1+x}\frac{x^3}{3}dx$$

$$= \frac{x^3\log(1+x)}{3} - \frac{1}{3}\int \frac{x^3}{1+x}dx$$

$$= \frac{x^3\log(1+x)}{3} - \frac{1}{3}\int \left[x^2 - x + 1 - \frac{1}{x+1}\right]dx$$

(by dividing x^3 by $1+x$)

$$= \frac{x^3\log(1+x)}{3} - \frac{1}{3}\left[\frac{x^3}{3} - \frac{x^2}{2} + x - \log(x+1)\right]$$

$$= \frac{1}{18}\left[6(x^3+1)\log(1+x) - 2x^3 + 3x^2 - 6x\right]$$

ix. $\int x \tan^{-1} x \, dx = \int \tan^{-1} x . x \, dx$

$$= \tan^{-1} x . \frac{x^2}{2} - \int \frac{1}{1+x^2} . \frac{x^2}{2} \, dx$$

$$= \frac{x^2 \tan^{-1} x}{2} - \frac{1}{2} \int \frac{x^2}{1+x^2} \, dx$$

$$= \frac{x^2 \tan^{-1} x}{2} - \frac{1}{2} \int \left[1 - \frac{1}{1+x^2} \right] dx$$

$$= \frac{x^2 \tan^{-1} x}{2} - \frac{1}{2} \left[x - \tan^{-1} x \right] = \frac{(x^2+1) \tan^{-1} x - x}{2}$$

x. $\int \tan^{-1} x \, dx = \int \tan^{-1} x . 1 \, dx$

$$= \tan^{-1} x . x - \int \frac{1}{1+x^2} . x \, dx$$

$$= x \tan^{-1} x - \int \frac{x}{1+x^2} \, dx$$

$$= x \tan^{-1} x - \frac{1}{2} \log(1+x^2)$$

xi. $\int (\log x)^2 dx = \int (\log x)^2 .1 \, dx$

$$= (\log x)^2 .x - \int \left(2 \log x . \frac{1}{x} \right) x \, dx$$

$$= x(\log x)^2 - 2 \int \log x \, dx = x(\log x)^2 - 2 \int \log x \, .1 \, dx$$

$$= x(\log x)^2 - 2 \left[\log x . x - \int \frac{1}{x} x dx \right]$$

$$= x(\log x)^2 - 2 \left[x \log x - \int dx \right]$$

$$= x(\log x)^2 - 2 \left[x \log x - x \right]$$

$$= x(\log x)^2 - 2x \log x + 2x$$

xii. $\int \frac{xe^x}{(1+x)^2} dx = \int \frac{(1+x-1)e^x}{(1+x)^2} dx = \int \frac{(1+x)e^x}{(1+x)^2} dx - \int \frac{e^x}{(1+x)^2} dx$

$$= \int \frac{e^x}{1+x} dx - \int \frac{e^x}{(1+x)^2} dx$$

$$= \int \frac{1}{1+x} .e^x \, dx - \int \frac{e^x}{(1+x)^2} dx$$

(Integrate the first integral using integration by parts and keep the second integral as such)

$$= \frac{1}{1+x}.e^x - \int \frac{-1}{(1+x)^2}.e^x\, dx - \int \frac{e^x}{(1+x)^2} dx$$

$$= \frac{e^x}{1+x}$$

xiii. $\int x \sec^2 x\, dx = x.\tan x - \int 1.\tan x\, dx$

$$= x.\tan x - \log(\sec x) + c$$

xiv. $\int \sin^{-1} x\, dx = \int \sin^{-1} x.\, 1\, dx = \sin^{-1} x.\, x - \int \frac{1}{\sqrt{1-x^2}}.\, x\, dx$

$$= x\sin^{-1} x - \int \frac{x}{\sqrt{1-x^2}}\, dx = x\sin^{-1} x + \sqrt{1-x^2}$$

Example 9: *Evaluate the following definite integrals:*

i. $\int_0^3 x^2 dx$

ii. $\int_a^b x dx$

iii. $\int_0^2 \sqrt{1+x}\, dx$

iv. $\int_0^{\pi/2} \cos x\, dx$

v. $\int_{-1}^1 dx$

vi. $\int_0^3 \frac{1}{1+x} dx$

vii. $\int_0^{\pi/2} \sin x\, dx$

viii. $\int_0^{\sqrt{2}} x\, dx$

ix. $\int_3^{-2} (1-x)\, dx$

x. $\int_{-1}^1 x(1-x^2)\, dx$

Solution:

i. $\int_0^3 x^2 dx = \left[\frac{x^3}{3}\right]_0^3 = \left[\frac{3^3}{3} - 0\right] = \frac{9}{3} = 3$

ii. $\int_a^b x dx = \left[\frac{x^2}{2}\right]_a^b = \left[\frac{b^2}{2} - \frac{a^2}{2}\right] = \frac{1}{2}(b^2 - a^2)$

iii. $\int_0^2 \sqrt{1+x}\, dx = \left[\frac{2(1+x)^{3/2}}{3}\right]_0^2 = \frac{2}{3}\left[3^{3/2} - 0\right] = 2\sqrt{3}$

iv. $\int_0^{\pi/2} \cos x\, dx = [\sin x]_0^{\pi/2} = [\sin(\pi/2) - \sin 0] = 1 - 0 = 1$

v. $\int_{-1}^{1} dx = [x]_{-1}^{1} = 1 - (-1) = 2$

vi. $\int_{0}^{3} \frac{1}{1+x} dx = [\log(1+x)]_0^3 = [\log 4 - \log 1]$

$= \log 2^2 - 0 = 2\log 2$

vii. $\int_{0}^{\pi/2} \sin x \, dx = [-\cos x]_0^{\pi/2}$

$= [-\cos(\pi/2) + \cos 0] = 0 + 1 = 1$

viii. $\int_{0}^{\sqrt{2}} x \, dx = \left[\frac{x^2}{2}\right]_0^{\sqrt{2}} = \frac{2}{2} - 0 = 1$

ix. $\int_{3}^{-2} (1-x) \, dx = \left[x - \frac{x^2}{2}\right]_3^{-2}$

$= \left[(-2) - \frac{4}{2}\right] - \left[3 - \frac{9}{2}\right] = [-2-2] - \left[3 - \frac{9}{2}\right]$

$= -4 + 3 - \frac{9}{2} = -1 - \frac{9}{2} = -\frac{11}{2}$

x. $\int_{-1}^{1} x(1-x^2) \, dx = \int_{-1}^{1} (x - x^3) \, dx$

$= \left[\frac{x^2}{2} - \frac{x^4}{4}\right]_{-1}^{1} = \left[\left(\frac{1}{2} - \frac{1}{4}\right) - \left(\frac{1}{2} - \frac{1}{4}\right)\right] = 0$

Example 10: *For each of the following curves, find the area bounded by the curve, the x–axis, and the given ordinates:*

i. $y = \sqrt{x}$; $x = 1$, $x = 16$

ii. $y = 2x + 1$; $x = 0$, $x = 4$

iii. $y = 3x^2$; $x = 1$, $x = 3$

iv. $y = x^2 - 3x$; $x = -1$, $x = 4$

Solution:

i. The required area $= \int_{a}^{b} y dx$

$= \int_{1}^{16} \sqrt{x} \, dx = \left[\frac{2x^{3/2}}{3}\right]_1^{16} = \frac{2}{3}\left[16^{3/2} - 1\right]$

$$= \frac{2}{3}[64-1] = \frac{2}{3}[63] = 42 \text{ square units}$$

ii. The required area $= \int_0^4 (2x+1)\,dx = \left[\frac{2x^2}{2} + x\right]_0^4$

$$= \left[x^2 + x\right]_0^4 = 4^2 + 4 - 0 = 20 \text{ square units}$$

iii. The required area $= \int_1^3 3x^2\,dx = \left[\frac{3x^3}{3}\right]_1^3$

$$= 27 - 1 = 26 \text{ square units}$$

iv. The required area $= \int_{-1}^4 (x^2 - 3x)\,dx = \left[\frac{x^3}{3} - \frac{3x^2}{2}\right]_{-1}^4$

$$= \left[\left(\frac{4^3}{3} - 3\frac{4^2}{2}\right) - \left(\frac{-1}{3} - \frac{3}{2}\right)\right] = \left[\left(\frac{64}{3} - 24\right) + \frac{11}{6}\right]$$

$$= 26 \text{ square units}$$

Example 11: *Find the total revenue (TR) and deduce the demand function if the marginal revenue (MR) function is given as:* $\frac{20}{(x+2)^2} - 2$

Solution: Given that MR $= \frac{20}{(x+2)^2} - 2$

$$\text{TR} = \int (\text{MR})dx = \int \left(\frac{20}{(x+2)^2} - 2\right)dx = \frac{-20}{(x+2)} - 2x + c$$

We know that, TR is zero when $x = 0$.

i.e., $\frac{-20}{(x+2)} - 2x + c = 0$, when $x = 0$

i.e., $\frac{-20}{2} - 0 + c = 0$

i.e., $c = 10$

$\therefore$ $\text{TR} = \frac{-20}{(x+2)} - 2x + 10 = \frac{10x}{(x+2)} - 2x$

Demand function, $p = \frac{\text{TR}}{x}$

i.e., $p = \frac{1}{x}\left[\frac{10x}{(x+2)} - 2x\right] = \frac{10}{(x+2)} - 2$

Example 12: *Find the total revenue at 18 units of output of a firm if its marginal revenue is* $10-2x+x^2$

Solution: TR (at 10 units of output) = $\int_0^{18}(\text{MR})dx$

$$= \int_0^{18}\left(10-2x+x^2\right)dx = \left[10x-2\frac{x^2}{2}+\frac{x^3}{3}\right]_0^{18}$$

$$= \left(180-324+\frac{5832}{3}\right)-(0)$$

$$= \frac{1000}{3} = 1800$$

$\therefore$ TR at 18 units of output = Rs. 1800

Example 13: *Marginal cost function of a firm is given as:* $2+4x+30x^2$. *If the fixed cost is Rs. 100, determine the total cost function of the firm.*

Solution: Given that MC = $2+4x+30x^2$

$$\text{TC} = \int(\text{MC})dx = \int\left(2+4x+30x^2\right)dx$$

$$= 2x+4\frac{x^2}{2}+30\frac{x^3}{3}+c = 2x+2x^2+10x^3+c$$

We know that, at x = 0, TC = fixed cost.

Hence c = 100

$\therefore$ TC = $100+2x+2x^2+10x^3$

Example 14: *The marginal cost function of a firm is given as: MC = 1 + 5* e^x. *If fixed cost is Rs. 1000, find the total cost function. Find average cost and marginal cost for x = 5 units.*

Solution: Given that MC = 1 + 5 e^x

$$\text{TC} = \int(\text{MC})dx$$

$$= \int\left(1+5e^x\right)dx = x+5e^x+c$$

Given that fixed cost = 1000.

i.e., $0+5e^0+c$ = 1000 (note that e^0 =1)

i.e., $5+c$ = 1000; or c = 995

$\therefore$ TC = $x+5e^x+995$

Also, Average Cost (AC) $= \dfrac{\text{TC}}{x}$

$$= \frac{x+5e^x+995}{x} = \frac{x+5e^x+995}{x}$$

When $x = 5$, AC $= \dfrac{5+5e^5+995}{5} = 1+e^5+199 = e^5+200$

When $x = 5$, MC $= 5e^5+1$

Example 15: *If the marginal cost function is given as* $\dfrac{a}{\sqrt{ax+b}}$ *and if the cost of zero output is zero, find the total cost as a function of x.*

Solution: Given that MC $= \dfrac{a}{\sqrt{ax+b}}$

$$\text{TC} = \int(\text{MC})dx$$

$$= \int \frac{a}{\sqrt{ax+b}}\, dx = 2\sqrt{ax+b}+c$$

We have, TC $= 0$ for $x = 0$

i.e., $2\sqrt{0+b}+c = 0$ or, $2\sqrt{b}+c = 0$

Hence, $c = -2\sqrt{b}$

$\therefore$ $\text{TC} = 2\sqrt{ax+b}-2\sqrt{b} = 2\left[\sqrt{ax+b}-\sqrt{b}\right]$

Example 16: *If the marginal revenue function is given by* $\dfrac{2}{(x+2)^2}+3$, *find the total revenue function and demand function in terms of x.*

Solution:

Given that $\text{MR} = \dfrac{2}{(x+2)^2}+3$

$$\text{TR} = \int(\text{MR})dx$$

$$= \int\left[\frac{2}{(x+2)^2}+3\right]dx = \frac{-2}{x+2}+3x+c$$

Since TR $= 0$ when $x = 0$,

$$\frac{-2}{x+2}+3x+c = 0$$

i.e., $-1+c = 0$; or $c = 1$

$\therefore$ $$\text{TR} = \frac{-2}{x+2}+3x+1 = 3x+1-\frac{2}{x+2}$$

$$= 3x+\frac{x+2-2}{x+2} = 3x+\frac{x}{x+2}$$

Demand function, $p = \text{AR} = \dfrac{\text{TR}}{x} = 3+\dfrac{1}{x+2}$

Example 17: *If the marginal cost function of a firm is given as* $20+\dfrac{x}{20}$ *and its marginal revenue is Rs. 30, determine the maximum profit and the number of items produced for this profit. The fixed cost is given as Rs. 200.*

Solution:

Given that $\text{MC} = 20+\dfrac{x}{20}$

$$\text{TC} = \int\left(20+\frac{x}{20}\right)dx = 20x+\frac{x^2}{40}+c_1$$

We have $c_1 = 200$

$\therefore$ $$\text{TC} = 20x+\frac{x^2}{40}+200$$

Also, $\text{MR} = 30$

$$\text{TR} = \int(30)\,dx = 30x+c_2$$

Since $\text{TR} = 0$, for $x = 0$, we get $c_2 = 0$

$\therefore$ $\text{TR} = 30x$

Profit $$\pi = \text{TR} - \text{TC} = 30x-\left(20x+\frac{x^2}{40}+200\right)$$

$$= 10x-\frac{x^2}{40}-200$$

For profit maximizing output, MC = MR

i.e., $20+\dfrac{x}{20} = 30$;

or $\dfrac{x}{20} = 10$

$\therefore$ $x = 200$

The maximum profit $= 10\times 200 - \dfrac{200^2}{40} - 200 =$ Rs. 800

Example 18: *If the marginal revenue function is given as* $R'(x) = 12 - 8x + x^2$. *Determine the revenue and demand functions. Is there any restriction on the value of x?*

Solution:

Revenue, $R = \int R'(x)dx = \int \left(12 - 8x + x^2\right) dx$

$$= 12x - 8\frac{x^2}{2} + \frac{x^3}{3} + c = 12x - 4x^2 + \frac{x^3}{3} + c$$

When $x = 0$, R = 0, which implies, $c = 0$

$$\therefore \quad R = 12x - 4x^2 + \frac{x^3}{3}$$

Also, demand $p = AR = \dfrac{R}{x}$

$$= 12 - 4x + \frac{x^2}{3} = \frac{(6-x)^2}{3}$$

Hence x should not be greater than 6.

Example 19: *The marginal propensity to consume is given as:* $100 + \dfrac{1}{50\sqrt{y}}$. *When income is zero, consumption is 150. Find the consumption function.*

Solution: Marginal propensity to consume, MPC $= 100 + \dfrac{1}{50\sqrt{y}}$

$\therefore$ Consumption, $c = \int \text{MPC}\, dy$

$$= \int \left[100 + \frac{1}{50\sqrt{y}}\right] dy = 100y + \frac{2\sqrt{y}}{50} + c_1;$$

where c_1 is the constant of integration

$$= 100y + \frac{\sqrt{y}}{25} + c_1$$

When $y = 0$, $c = 150$, which gives $c_1 = 150$.

Hence, consumption $= 100y + \dfrac{\sqrt{y}}{25} + 150.$

Example 20: *The marginal propensity to consume out of income for the economy as a whole is given as 4/5. It is known that when income is zero, consumption equals Rs. 12 billion. Find the function relating to aggregate consumption to national income.*

Solution: Marginal propensity to consume, MPC = 4/5

$\therefore$ Consumption, $c = \int \text{MPC}\, dy$

$$= \int \frac{4}{5}\, dy = \frac{4}{5}y + c_1$$

Given that when $y = 0$, $c = 12$.

Hence $c_1 = 12$ billion

$$\therefore \quad c = \frac{4}{5}y + 12.$$

Example 21: *A company suffers a loss of Rs.110if its product does not sell at all. Marginal revenue and marginal cost functions for the product is given by MR = 50 – 4x and MC = –10 + x. Find the profit function and the equilibrium level of output. Also find the maximum profit and the break even point.*

Solution:

Profit function, $\pi = \int [50 - 4x - (-10 + x)]\, dx$

$$= \int [60 - 5x]\, dx = 60x - \frac{5x^2}{2} + c$$

It is given that $\pi = -110$, when $x = 0$.

[The minus sign is due to loss]

Hence $c = -110$

$\therefore$ Profit function, $\pi = 60x - \frac{5x^2}{2} - 110$

We know that for equilibrium, MR = MC.

i.e., $50 - 4x = -10 + x$; which gives $x = 12$.

$\therefore$ The equilibrium level of output = 12

When $x = 12$, $\pi = 60 \times 12 - \frac{5 \times 12^2}{2} - 110 = 250$

For break–even point, $\pi = 0$

i.e., $60x - \frac{5x^2}{2} - 110 = 0$

or, $\quad x^2 - 24x + 44 = 0$

or, $\quad (x-2)(x-22) = 0$

This gives $x = 2$ or $x = 22$.

Hence the break even point is at $x = 2$.

Example 22: *If the marginal cost function is* $f'(q) = 2 + 3\sqrt{q} + \frac{5}{\sqrt{q}}$, *find the total cost function when* $f(1) = 21$.

Solution: $f'(q) = 2 + 3\sqrt{q} + \frac{5}{\sqrt{q}}$

$$\Rightarrow \quad f(q) = \int\left(2 + 3\sqrt{q} + \frac{5}{\sqrt{q}}\right)dq$$

$$= 2q + 3\frac{q^{3/2}}{3/2} + 5.2\sqrt{q} + c$$

$$= 2q + 2q^{3/2} + 10\sqrt{q} + c$$

$$\Rightarrow \quad f(1) = 2 + 2 + 10 + c = 21$$

$$\Rightarrow \quad c = 7.$$

EXERCISES

1. Find the area bounded by the curve $y = 6x - x^2$ and the x–axis, between $x = 0$ and $x = 6$
2. Find the area bounded by the curve $y = (x-2)^3$, the x–axis, and the ordinates at $x = 1$ and $x = 3$
3. The marginal cost of production is found to be MC = 1000 ▮ $20x + x^2$, where x is the number of units produced. The fixed cost of production is Rs.9000. Find the cost function.
4. A company has determined that the marginal cost function for a product of a particular commodity is given by MC = $125 + 10x - \frac{x^2}{9}$. If the fixed cost is Rs. 250, what is the cost of producing 15 units?
5. A firm has the marginal revenue function given as MR = $\frac{a}{(x+b)^2} - c$; where a, b and c are constants. Show that the demand function is given by $x = \frac{a}{b(p+c)} - b$

[hint: first obtain the demand function in the form $p = \frac{a}{b(x+b)} - c$. Then find x from the equation]

6. A firm has the marginal revenue function for x units of outputs is MR = $\frac{a}{(x+b)} - c$; where a, b and c are constants. Show that the demand law is given by $p = \frac{a}{x}\log\frac{x+b}{b} - c$

7. Marginal cost as a function of units produced x is given by $1.064 - 0.005x$. If the fixed cost is 16.3, show that the total cost function is $16.3 + 1.064x - 0.0025x^2$.

8. If the marginal revenue function is given as: $8 - 6x - 2x^2$, find the revenue and the demand law.

9. If the marginal revenue function is given by MR = $\frac{60}{(x+3)^2} - 2$, find total revenue. Also, deduce the demand function.

10. The marginal revenue function of a firm is given by MR = $240 - 4x$. Find total revenue function and the demand function. At what level of output is total revenue maximum? Find the maximum total revenue.

11. If the marginal revenue function of a firm is given by MR = $30 - 10x + x^2$, find total revenue of the firm at 12 units of output.

12. If the marginal cost function of a firm is given by MC = $4 + 6x + 30x^2$, where x is the number of units of output. Find total cost function of the firm, given that fixed cost is Rs. 500.

13. If the marginal cost function of a firm is given by MC = $5 + 3e^x$, where x denotes thousand units of output. Find
    i. Total cost, if total cost at zero is Rs. 250
    ii. Average cost
    iii. Evaluate TC for 500 units of output

14. The marginal propensity to consume (in billion dollars) is $\frac{9}{10} + \frac{3}{10\sqrt{x}}$. When income is zero, consumption is 10 billion dollars. Find the consumption function.

15. If the marginal revenue function is $\frac{6}{(x+3)^2} - 10$, show that the demand law is: $p = \frac{2}{(x+3)} - 10$

5

Differentiation

THE MEANING OF DIFFERENTIABILITY

Let y = f(x) be a continuous function. Let P and Q are two points on the curve whose co–ordinates are (x, y) and $(x+\Delta x, y+\Delta y)$ respectively. From P and Q draw perpendiculars PM and QN to x–axis and PR to NQ. Draw the tangent of the curve at P, and let it makes an angle θ with the positive direction of x–axis.

$PR = \Delta x$ and $RQ = \Delta y$.

Hence we have,

$$\tan \angle QPR = \frac{\Delta y}{\Delta x}$$

Obviously, as $\Delta x \to 0, Q \to P$. In the limiting position, the secant PQ becomes the tangent at P and $\angle QPR$ will tend to angle θ, which the tangent at P makes with the positive direction of the x–axis.

Thus $\lim\limits_{\Delta x \to 0} \frac{\Delta y}{\Delta x} = \lim\limits_{Q \to P} \tan \angle QPR = \tan \theta$

i.e., $\frac{dy}{dx} = \tan \theta$ = slope of the tangent at $P(x, y)$

Thus, $\frac{dy}{dx}$ at any point gives the slope of the tangent of at that point.

DIFFERENTIATION FROM FIRST PRINCIPLES

The method of finding the differential coefficient of a function from the very definition of differentiation is called differentiation from first principles or by *ab–initio* method or by delta method. Differentiation is a part of *Calculus*, which has widespread applications in Science, both pure and applied, and in Engineering. Theory of Integration stands as the counter part of differentiation. Calculus is used to solve complex and expansive problems for which Algebra alone is not sufficient. Mathematicians all over the world contributed to the development but, two most recognised discoverers are Issac Newton and Gottfried Wilhelm Leibnitz. Consider the production function $y = f(x)$, where y denotes the level of output corresponding to the use of x units of labour. When $x = 0$, we have $y = 0$, since if no labour is employed the output is zero. When we employ a labour of x_1 units, the level of output becomes $f(x_1)$. Let it be y_1.

When the labour employed is x_2 units, the level of output becomes $f(x_2)$, say y_2. Thus whenever a change in the value of x occurs, it corresponds a change in the value of y. Then we can find the average rate of change of output, by taking the ratio $\dfrac{\text{change in output}}{\text{change in labour}}$.

Let the labour is increased by a small amount Δx. Let the corresponding change in y be Δy. Then the average rate of change of output is given by the ratio $\dfrac{\Delta y}{\Delta x}$. For calculating the marginal product of labour we are interested in determining the rate of change of output at a 'point' of labour. This is possible only for the limiting case when = 0.

i.e., when $\lim\limits_{\Delta x \to 0} \dfrac{\Delta y}{\Delta x}$

Value of this limit, if it exists, is called the *derivative* or *differential coefficient* of the function $y = f(x)$.

For another example, consider a particle moving in a straight line. Let the distance y it covers at any time x is defined by the rule, $y = f(x)$. As x changes, y also will change. We know that the average speed is the ratio of its change in distance to the change in time. i.e., $\dfrac{\Delta y}{\Delta x}$ gives its average speed. For getting instantaneous speed, we take as in the previous example, $\lim\limits_{\Delta x \to 0} \dfrac{\Delta y}{\Delta x}$. Thus, the derivative of distance with respect to time gives speed.

In Economics we need rate of change of quantities occasionally; and it is termed as 'marginal' change. For example, we know that the 'marginal propensity to consume' is the amount of change in consumption for a small unit change in income.

Consider a consumption function,

$$C = a + bY$$

where C is the consumption expenditures and Y the income. We know that a change in Y will result in a change in C. i.e., if ΔY is a small increment in Y which results in a corresponding change ΔC in C, then, the consumption function takes the form:

$$C + \Delta C = a + b\,(Y + \Delta Y)$$

Hence the change in consumption is:

$$\begin{aligned} \Delta D &= a + b\,(Y + \Delta Y) - C \\ &= a + b\,(Y + \Delta Y) - (a + bY) \\ &= b\,\Delta Y \end{aligned} \qquad \ldots(1)$$

Now, the amount of change in consumption for a small unit change in income is obtained by dividing the change in consumption ΔC by change in income ΔY.

$\therefore$ Marginal propensity to consume $= \dfrac{\Delta C}{\Delta Y} = b$, from equation (1)

Differentiation is the process of finding the differential coefficient or derivative of a function. By derivative we mean the rate at which a variable quantity is changing. The instantaneous (marginal) rate of change of a function is its derivative. Derivative of a function $y = f(x)$ with respect to x is denoted by $\dfrac{dy}{dx}$ [note that the symbols Dy or $f^{1}(x)$ or $f¢(x)$ or y^{1} or $y¢$ or y_1 may be used to denote derivative of $y = f(x)$].

$$\therefore \quad \frac{dy}{dx} = \lim_{\Delta x \to 0} \frac{\Delta y}{\Delta x}$$

DIFFERENTIATION OF A FUNCTION WITH RESPECT TO ANOTHER FUNCTION

So far we have discussed the methods of finding derivative of one variable with respect to another variable. It is also possible to find derivative of one function with respect to another function. If u and v are two functions of x, then, derivative of u with respect to v,

$$\text{i.e.,} \quad \frac{du}{dv} = \frac{du/dx}{dv/dx}.$$

That is, to find the derivative of $u(x)$ with respect to $v(x)$, first find the derivatives of $u(x)$ with respect to x and derivative of $v(x)$ with respect x, and then divide derivative of $u(x)$ with respect to x by that of $v(x)$ with respect to x.

Example: *Differentiate sin x with respect to cos x*

Solution: Let $u = \sin x$ and $v = \cos x$

$$\frac{du}{dx} = \cos x, \quad \frac{dv}{dx} = \sin x$$

$$\therefore \quad \frac{du}{dv} = \frac{du/dx}{dv/dx} = \frac{\cos x}{\sin x} = \cot x$$

DIFFERENTIATION OF PARAMETRIC FUNCTIONS

When x and y are given as functions of a single variable say t, then, to find the derivative of y with respect to x, first find derivatives of both x and y with respect to t and then divide the derivative of y by that of x. That is, if $x = \varphi(t)$, and $y = \psi(t)$, then,

$$\frac{dy}{dx} = \frac{dy/dt}{dx/dt}$$

This method of differentiation is called parametric differentiation and the single variable by which x and y are expressed is called the parameter.

HIGHER ORDER DERIVATIVES

We have seen that if y is a function of x then its derivative with respect to x is again, in general, a function of x, so the derivative can also be differentiated. The derivative $\frac{dy}{dx}$ is called the *first order derivative* of y with respect to x and the 'derivative of' $\frac{dy}{dx}$ is called the *second order derivative* of y with respect to x, and is denoted as $\frac{d^2y}{dx^2}$.

i.e., $$\frac{d^2y}{dx^2} = \frac{d}{dx}\left(\frac{dy}{dx}\right)$$

Generally, $\frac{d^n y}{dx^n}$ denotes the nth order derivative of y with respect to x. The notations y_n, $y^{(n)}$ and $D^n(y)$ are also used for denoting n[th] order derivative of y.

DIFFERENTIATION BY TRIGONOMETRICAL TRANSFORMATIONS

It may be sometimes easy to differentiate trigonometrical functions by using transformations. It is more useful for functions of inverse trigonometric ratios. Trigonometrical identities and results are given in Chapter–1. Some important results on trigonometric ratios are given below for quick reference:

I i. $\sin 2x = 2 \sin x \cos x = \dfrac{2\tan x}{1+\tan^2 x}$

ii. $\cos 2x = \cos^2 x - \sin^2 x$

$= 2\cos^2 x - 1$

$= 1 - 2\sin^2 x = \dfrac{1-\tan^2 x}{1+\tan^2 x}$

iii. $\tan 2x = \dfrac{2\tan x}{1-\tan^2 x}$

II i. $\sin 3x = 3\sin x - 4\sin^3 x$

ii. $\cos 3x = 4\cos^3 x - 3\cos x$

iii. $\tan 3x = \dfrac{3\tan x - \tan^3 x}{1-3\tan^2 x}$

III i. $\sin^{-1} x \pm \sin^{-1} y = \sin^{-1}\left(x\sqrt{1-y^2} \pm y\sqrt{1-x^2}\right)$

ii. $\cos^{-1} x \pm \cos^{-1} y = \cos^{-1}\left(xy \mp \sqrt{1-x^2}\sqrt{1-y^2}\right)$

iii. $\tan^{-1} x \pm \tan^{-1} y = \tan^{-1}\left(\dfrac{x \pm y}{1 \mp xy}\right)$

IV i. $\sin^{-1} x + \cos^{-1} x = \dfrac{\pi}{2}$

ii. $\tan^{-1} x + \cot^{-1} x = \dfrac{\pi}{2}$

iii. $\sec^{-1} x + \text{cosec}^{-1} x = \dfrac{\pi}{2}$

V i. $\sin^{-1}(-x) = -\sin^{-1} x$

ii. $\cos^{-1}(-x) = \pi - \cos^{-1} x$

iii. $\tan^{-1}(-x) = -\tan^{-1} x$

VI i. $\text{cosec}^{-1}(1/x) = \sin^{-1} x$

ii. $\sec^{-1}(1/x) = \cos^{-1} x$

iii. $\cot^{-1}(1/x) = \tan^{-1} x$

LOGARITHMIC DIFFERENTIATION

This method of differentiation is mainly used when a function contains variable powers like: $x^{\cos x}$ or $(\sin x)^{\log x}$. This method is also convenient if the function is given as product or quotient of too many functions or as their roots.

DIFFERENTIATION OF IMPLICIT FUNCTIONS

If the variables x and y are connected by a relation of the form, $f(x, y) = 0$ and if y cannot be conveniently expressed in terms of x, then y is said to be an implicit function of x. In order to find $\frac{dy}{dx}$ in such cases, we differentiate each term of the relation with respect to x, keeping in mind that derivative of x with respect to x is 1 but derivative of y with respect to x is $\frac{dy}{dx}$.

For example,

$$\frac{d}{dx}(\log x) = \frac{1}{x}\frac{dx}{dx} = \frac{1}{x}.1 = \frac{1}{x};$$

but, $$\frac{d}{dx}(\log y) = \frac{1}{y}\frac{dy}{dx}$$

SOME FUNDAMENTAL RULES OF DIFFERENTIATION

Rule–1

If c is a constant, $=\frac{d(c)}{dx}\ 0$

Proof:

Let $y = c$.

Consider a small increment in Δx and let Δy be the corresponding change in y. Then,

$$y + \Delta y = c \text{ and}$$

$$\Delta y = c - y = y - y = 0$$

Hence $$\frac{dy}{dx} = \lim_{\Delta x \to 0} \frac{\Delta y}{\Delta x} = 0.$$

Rule–2

If u is a differentiable function of x and c is a constant,

$$\frac{d(cu)}{dx} = c\frac{du}{dx}$$

Proof:

Let $y = cu$.

Consider a small increment Δx in x. Let Δy be the corresponding change in y and Δu be the corresponding change in u.

Then,

$$y + \Delta y = c(u + \Delta u) \text{ and}$$

$$\Delta y = c(u + \Delta u) - cu = c\Delta u.$$

Hence $$\frac{dy}{dx} = \lim_{\Delta x \to 0} \frac{\Delta y}{\Delta x} = \lim_{\Delta x \to 0} \frac{c\Delta u}{\Delta x}$$

$$= c.\lim_{\Delta x \to 0} \frac{\Delta u}{\Delta x} = c\frac{du}{dx}$$

RELATION BETWEEN $\frac{dy}{dx}$ AND $\frac{dx}{dy}$

By definition, $$\frac{dy}{dx} = \lim_{\Delta x \to 0} \frac{\Delta y}{\Delta x} = \text{ and}$$

$$\frac{dy}{dx} = \lim_{\Delta y \to 0} \frac{\Delta x}{\Delta y}$$

We have, $$\frac{\Delta y}{\Delta x}\frac{\Delta x}{\Delta y} = 1$$

i.e., $$\lim_{\Delta x \to 0} \frac{\Delta y}{\Delta x}\frac{\Delta x}{\Delta y} = 1$$

i.e., $$\lim_{\Delta x \to 0} \frac{\Delta y}{\Delta x} \lim_{\Delta x \to 0} \frac{\Delta x}{\Delta y} = 1$$

i.e., $$\lim_{\Delta x \to 0} \frac{\Delta y}{\Delta x} \lim_{\Delta y \to 0} \frac{\Delta x}{\Delta y} = 1 \quad [\because \Delta x \to 0 \Rightarrow \Delta y \to 0]$$

i.e., $$\frac{dy}{dx}\frac{dy}{dx} = 1$$

$$\therefore \quad \frac{dy}{dx} = \frac{1}{\left(\frac{dx}{dy}\right)}$$

DIFFERENTIABILITY OF A FUNCTION

Let $f(x)$ be a real valued function defined on an open interval (a, b). Let $c \in (a, b)$. Then $f(x)$ is said to be *differentiable* at $x = c$ iff,

$$\lim_{x \to c} \frac{f(x) - f(c)}{x - c}$$

exists. The limit is called the derivative or differential coefficient of the function at $x = c$.

$f(x)$ is said to be differentiable in an open interval (a, b) if it is differentiable at each point of (a, b). If $f(x)$ is differentiable in the open interval (a, b) and at the endpoints a and b, then it is said to be differentiable in the closed interval $[a, b]$. *A function is said to be a differentiable function if it is differentiable at every point of its domain.* A function, which is differentiable in every point of the real line, is said to be an everywhere differentiable function.

DERIVATIVES OF INVERSE TRIGONOMETRIC FUNCTIONS

Inverse trigonometric functions are the corresponding inverses of trigonometric functions. Using the derivatives of trigonometric functions, we can find the derivatives of inverse trigonometric functions as well.

SOLVED EXAMPLES

Example 1: *Find the derivatives of the following functions with respect to x from first principles.*

(i) $2x - 1$

(ii) $5x^2 - 3x + 7$

Solution: (i) Let $y = 2x - 1$

Consider a small increment in Δx and let be the corresponding change in Δy. Then,

$$y + \Delta y = 2(x + \Delta x) - 1$$

$$\Delta y = 2(x + \Delta x) - 1 - y$$

$$= 2(x + \Delta x) - 1 - (2x - 1)$$

$$= 2x + 2\Delta x - 1 - 2x + 1 = 2\Delta x$$

$$\frac{\Delta y}{\Delta x} = \frac{2\Delta x}{\Delta x} = 2$$

$$\frac{dy}{dx} = \lim_{\Delta x \to 0} \frac{\Delta y}{\Delta x} = \lim_{\Delta x \to 0} (2) = 2$$

ii. Let $y = 5x^2 - 3x + 7$

Consider a small increment in x and let be the corresponding change in y. Then,

$$y + \Delta y = 5(x + \Delta x)^2 - 3(x + \Delta x) + 7$$

$$\Delta y = 5(x + \Delta x)^2 - 3(x + \Delta x) + 7 - y$$
$$= 5(x + \Delta x)^2 - 3(x + \Delta x) + 7 - (5x^2 - 3x + 7)$$
$$= 5(\Delta x)^2 + 10\, x\, \Delta x - 3\Delta x$$
$$\frac{\Delta y}{\Delta x} = 5\Delta x + 10\, x - 3$$
$$\frac{dy}{dx} = \lim_{\Delta x \to 0} \frac{\Delta y}{\Delta x} = \lim_{\Delta x \to 0} (5\Delta x + 10\, x - 3) = 10\, x - 3$$

Example 2: *Find the derivatives of the following functions with respect to x from first principles.*

i. $\sqrt{x}$

ii. x^n *, n is a rational number*

Solution:

i. Let $y = \sqrt{x}$

Consider a small increment in Δx and let Δy be the corresponding change in y. Then,

$$y + \Delta y = \sqrt{x + \Delta x}$$
$$\Delta y = \sqrt{x + \Delta x} - \sqrt{x}$$
$$= \frac{\left(\sqrt{x + \Delta x} - \sqrt{x}\right)\left(\sqrt{x + \Delta x} + \sqrt{x}\right)}{\left(\sqrt{x + \Delta x} + \sqrt{x}\right)}$$
$$= \frac{(x + \Delta x) - x}{\left(\sqrt{x + \Delta x} + \sqrt{x}\right)} = \frac{\Delta x}{\left(\sqrt{x + \Delta x} + \sqrt{x}\right)}$$
$$\frac{\Delta y}{\Delta x} = \frac{1}{\left(\sqrt{x + \Delta x} + \sqrt{x}\right)}$$
$$\frac{dy}{dx} = \lim_{\Delta x \to 0} \frac{\Delta y}{\Delta x} = \lim_{\Delta x \to 0} \frac{1}{\left(\sqrt{x + \Delta x} + \sqrt{x}\right)} = \frac{1}{2\sqrt{x}}$$
$$\frac{d}{dx}\left(\sqrt{x}\right) = \frac{1}{2\sqrt{x}}$$

ii. Let $y = x^n$

Consider a small increment in Δx and let Δy be the corresponding change in y. Then,

$$y + \Delta y = (x + \Delta x)^n$$

$$\Delta y = (x + \Delta x)^n - x^n$$

$$\frac{\Delta y}{\Delta x} = \frac{(x+\Delta x)^n - x^n}{\Delta x}$$

$$\frac{dy}{dx} = \lim_{\Delta x \to 0} \frac{\Delta y}{\Delta x} = \lim_{\Delta x \to 0} \frac{(x+\Delta x)^n - x^n}{\Delta x}$$

$$= \lim_{x+\Delta x \to x} \frac{(x+\Delta x)^n - x^n}{(x+\Delta x) - x}$$

$$= n\, x^{n-1} \left[\because \lim_{x \to a} \frac{x^n - a^n}{x - a} = n\, a^{n-1}\right]$$

$$\frac{d}{dx}\left(x^n\right) = nx^{n-1}$$

Remark: the following illustration shows how this result may be used to find differential coefficients of functions.

Example 3(a): *Find the derivatives of the following functions with respect to x.*

i. x^{10}	*ii.* $\frac{1}{x}$
iii. $\frac{1}{x^2}$	*iv.* $\sqrt{x}$
v. $\frac{1}{\sqrt{x}}$	*vi.* $\sqrt[3]{x^2}$

Solution:

i. Let $y = x^{10}$

$$\frac{dy}{dx} = 10\, x^{10-1} = 10\, x^9$$

ii. Let $y = x$

$$\frac{dy}{dx} = 1\, x^{1-1} = x^0 = 1$$

iii. Let $y = \frac{1}{x} = x^{-1}$

$$\frac{dy}{dx} = -1\, x^{-1-1} = -1\, x^{-2} = \frac{-1}{x^2}$$

iv. Let $y = \frac{1}{x^2} = x^{-2}$

$$\frac{dy}{dx} = -2\ x^{-2-1} = -2\ x^{-3} = \frac{-2}{x^3}$$

v. Let $y = \sqrt{x} = x^{1/2}$

$$\frac{dy}{dx} = \frac{1}{2}\ x^{1/2-1} = \frac{1}{2}x^{-1/2} = \frac{1}{2\sqrt{x}}$$

vi. Let $y = \frac{1}{\sqrt{x}} = x^{-1/2}$

$$\frac{dy}{dx} = -\ \frac{1}{2}x^{-1/2-1} = \frac{1}{2}x^{-3/2} = \frac{1}{2x\sqrt{x}}$$

vii. Let $y = \sqrt[3]{x^2} = x^{3/2}$

$$\frac{dy}{dx} = -\ \frac{1}{2}x^{3/2-1} = \frac{1}{2}x^{1/2} = \frac{\sqrt{x}}{2}.$$

Example 3(b): *Find the derivatives of the following functions with respect to x from first principles.*

i. e^x

ii. a^x, $a > 0$, $a \neq 1$

iii. $\log x$

Solution:

i. Let $y = e^x$

Consider a small increment in Δx and let Δy be the corresponding change in y. Then,

$$y + \Delta y = e^{x+\Delta x}$$

$$\Delta y = e^{x+\Delta x} - e^x = e^x e^{\Delta x} - e^x$$

$$= e^x\left(e^{\Delta x} - 1\right)$$

$$\frac{\Delta y}{\Delta x} = \frac{e^x(e^{\Delta x} - 1)}{\Delta x}$$

$$\frac{dy}{dx} = \lim_{\Delta x \to 0} \frac{\Delta y}{\Delta x}$$

$$= \lim_{\Delta x \to 0} \frac{e^x (e^{\Delta x} - 1)}{\Delta x}$$

$$= e^x \lim_{\Delta x \to 0} \frac{(e^{\Delta x} - 1)}{\Delta x}$$

$$= e^x \times 1 = e^x \quad [\because \lim_{x \to 0} \frac{(e^x - 1)}{x} = 1]$$

$$\frac{d}{dx}\left(e^x\right) = e^x$$

ii. Let $y = a^x$

Consider a small increment in Δx and let Δy be the corresponding change in y. Then,

$$y + \Delta y = a^{x+\Delta x}$$

$$\Delta y = a^{x+\Delta x} - a^x = a^x a^{\Delta x} - a^x$$

$$= a^x \left(a^{\Delta x} - 1\right)$$

$$\frac{\Delta y}{\Delta x} = \frac{a^x (a^{\Delta x} - 1)}{\Delta x}$$

$$\frac{dy}{dx} = \lim_{\Delta x \to 0} \frac{\Delta y}{\Delta x} = \lim_{\Delta x \to 0} \frac{a^x (a^{\Delta x} - 1)}{\Delta x}$$

$$= a^x \lim_{\Delta x \to 0} \frac{(a^{\Delta x} - 1)}{\Delta x} = a^x \log_e a$$

$$\left[\because \lim_{x \to 0} \frac{(a^x - 1)}{x} = \log_e a\right]$$

iii. Let $y = \log x$

Consider a small increment in Δx and let Δy be the corresponding change in y. Then,

$$y + \Delta y = \log (x + \Delta x)$$

$$\Delta y = \log (x + \Delta yx) - \log x$$

$$= \log\left(\frac{x + \Delta x}{x}\right) = \log\left(1 + \frac{\Delta x}{x}\right)$$

$$\frac{\Delta y}{\Delta x} = \frac{1}{\Delta x} \log \left(1 + \frac{\Delta x}{x}\right)$$

$$\frac{dy}{dx} = \lim_{\Delta x \to 0} \frac{\Delta y}{\Delta x}$$

$$= \lim_{\Delta x \to 0} \frac{1}{\Delta x} \log\left(1 + \frac{\Delta x}{x}\right)$$

$$= \lim_{\Delta x \to 0} \frac{1}{x} \frac{\log\left(1 + \frac{\Delta x}{x}\right)}{\left(\frac{\Delta x}{x}\right)}$$

$$= \frac{1}{x} \times 1 = \frac{1}{x} \qquad \left[\because \lim_{x \to 0} \frac{\log(1+x)}{x} = 1\right]$$

$$\boxed{\frac{d}{dx}(\log x) = \frac{1}{x}}$$

Example 3(c): *Find the derivatives of the following functions with respect to x from first principles.*

i. sin x ii. cos x

iii. tan x iv. cot x

v. sec x vi. cosec x

Solution:

i. Let $y = \sin x$

Consider a small increment in Δx and let Δy be the corresponding change in y. Then,

$$y + \Delta y = \sin(x + \Delta x)$$

$$\Delta y = \sin(x + \Delta x) - \sin x$$

$$= 2\cos\frac{x + \Delta x + x}{2} \sin\frac{x + \Delta x - x}{2}$$

(refer Chapter–1 for the formula)

$$= 2\cos\left(x + \frac{\Delta x}{2}\right) \sin\frac{\Delta x}{2}$$

$$\frac{\Delta y}{\Delta x} = \frac{2}{\Delta x} \cos\left(x + \frac{\Delta x}{2}\right) \sin\frac{\Delta x}{2}$$

$$= \cos\left(x + \frac{\Delta x}{2}\right) \frac{\sin\left(\frac{\Delta x}{2}\right)}{\left(\frac{\Delta x}{2}\right)}$$

$$\frac{dy}{dx} = \lim_{\Delta x \to 0} \frac{\Delta y}{\Delta x}$$

$$= \lim_{\Delta x \to 0} \cos\left(x + \frac{\Delta x}{2}\right) \frac{\sin\left(\frac{\Delta x}{2}\right)}{\left(\frac{\Delta x}{2}\right)}$$

$$= \lim_{\Delta x \to 0} \cos\left(x + \frac{\Delta x}{2}\right) \lim_{\frac{\Delta x}{2} \to 0} \frac{\sin\left(\frac{\Delta x}{2}\right)}{\left(\frac{\Delta x}{2}\right)}$$

$$= \cos x \times 1 = \cos x \quad [\because \lim_{x \to 0} \frac{\sin x}{x} = 1]$$

$$\boxed{\frac{d}{dx}(\sin x) = \cos x}$$

ii. Let $y = \cos x$

Consider a small increment in Δx and let Δy be the corresponding change in y. Then,

$$y + \Delta y = \cos (x + \Delta x)$$

$$\Delta y = \cos (x + \Delta x) - \cos x$$

$$= -2 \sin \frac{x + \Delta x + x}{2} \sin \frac{x + \Delta x - x}{2}$$

$$= -2 \sin \left(x + \frac{\Delta x}{2}\right) \sin \frac{\Delta x}{2}$$

$$\frac{\Delta y}{\Delta x} = \frac{-2}{\Delta x} \sin \left(x + \frac{\Delta x}{2}\right) \sin \frac{\Delta x}{2}$$

$$= - \sin \left(x + \frac{\Delta x}{2}\right) \frac{\sin\left(\frac{\Delta x}{2}\right)}{\left(\frac{\Delta x}{2}\right)}$$

$$\frac{dy}{dx} = \lim_{\Delta x \to 0} \frac{\Delta y}{\Delta x}$$

$$= - \lim_{\Delta x \to 0} \sin \left(x + \frac{\Delta x}{2}\right) \frac{\sin\left(\frac{\Delta x}{2}\right)}{\left(\frac{\Delta x}{2}\right)}$$

$$= - \lim_{\Delta x \to 0} \sin\left(x+\frac{\Delta x}{2}\right) \lim_{\frac{\Delta x}{2} \to 0} \frac{\sin\left(\frac{\Delta x}{2}\right)}{\left(\frac{\Delta x}{2}\right)}$$

$$= -\sin \times 1 = -\sin x \quad [\because \lim_{x \to 0} \frac{\sin x}{x} = 1]$$

$$\boxed{\frac{d}{dx}(\cos x) = -\sin x}$$

iii. Let $y = \tan x$

Consider a small increment in Δx and let Δy be the corresponding change in y. Then,

$$y + \Delta y = \tan(x + \Delta x)$$

$$\Delta y = \tan(x + \Delta x) - \tan x$$

$$= \frac{\sin(x+\Delta x)}{\cos(x+\Delta x)} - \frac{\sin x}{\cos x}$$

$$= \frac{\sin(x+\Delta x)\cos x - \cos(x+\Delta x)\sin x}{\cos(x+\Delta x)\cos x}$$

$$= \frac{\sin(x+\Delta x - x)}{\cos(x+\Delta x)\cos x} = \frac{\sin(\Delta x)}{\cos(x+\Delta x)\cos x}$$

$$\frac{\Delta y}{\Delta x} = \frac{1}{\Delta x}\frac{\sin(\Delta x)}{\cos(x+\Delta x)\cos x}$$

$$= \frac{1}{\cos(x+\Delta x)\cos x}\frac{\sin(\Delta x)}{\Delta x}$$

$$\frac{dy}{dx} = \lim_{\Delta x \to 0}\frac{\Delta y}{\Delta x}$$

$$= \lim_{\Delta x \to 0}\frac{1}{\cos(x+\Delta x)\cos x}\frac{\sin(\Delta x)}{\Delta x}$$

$$= \lim_{\Delta x \to 0}\frac{1}{\cos(x+\Delta x)\cos x}\lim_{\Delta x \to 0}\frac{\sin \Delta x}{\Delta x}$$

$$= \frac{1}{\cos(x+0)\cos x} \times 1$$

$$= \frac{1}{\cos^2 x} = \sec^2 x$$

$$\boxed{\frac{d}{dx}(\tan x) = \sec^2 x}$$

iv. Let $y = \cot x$

Consider a small increment in Δx and let Δy be the corresponding change in y. Then,

$$y + \Delta y = \cot (x + \Delta x)$$

$$\Delta y = \cot (x + \Delta x) - \cot x$$

$$= \frac{\cos(x+\Delta x)}{\sin(x+\Delta x)} - \frac{\cos x}{\sin x}$$

$$= \frac{\sin x \cos(x+\Delta x) - \cos x \sin(x+\Delta x)}{\sin(x+\Delta x)\sin x}$$

$$= \frac{\sin[x-(x+\Delta x)]}{\sin(x+\Delta x)\sin x} = \frac{\sin(-\Delta x)}{\sin(x+\Delta x)\sin x}$$

$$= \frac{-\sin(\Delta x)}{\sin(x+\Delta x)\sin x}$$

$$\frac{\Delta y}{\Delta x} = \frac{-1}{\Delta x}\frac{\sin(\Delta x)}{\sin(x+\Delta x)\sin x}$$

$$= \frac{-1}{\sin(x+\Delta x)\sin x}\frac{\sin(\Delta x)}{\Delta x}$$

$$\frac{dy}{dx} = \lim_{\Delta x \to 0}\frac{\Delta y}{\Delta x}$$

$$= \lim_{\Delta x \to 0}\frac{-1}{\sin(x+\Delta x)\sin x}\frac{\sin(\Delta x)}{\Delta x}$$

$$= -\lim_{\Delta x \to 0}\frac{1}{\sin(x+\Delta x)\sin x}\lim_{\Delta x \to 0}\frac{\sin \Delta x}{\Delta x}$$

$$= \frac{-1}{\sin(x+0)\sin x}\times 1$$

$$= \frac{-1}{\sin^2 x} = -\operatorname{cosec}^2 x$$

$$\boxed{\frac{d}{dx}(\cot x) = -\operatorname{cosec}^2 x}$$

v. Let $y = \sec x$

Consider a small increment in Δx and let Δy be the corresponding change in y. Then,

$$y + \Delta y = \sec(x + \Delta x)$$

$$\Delta y = \sec(x + \Delta x) - \sec x$$

$$= \frac{1}{\cos(x+\Delta x)} - \frac{1}{\cos x}$$

$$= \frac{\cos x - \cos(x+\Delta x)}{\cos(x+\Delta x)\cos x}$$

$$= \frac{-2\sin\left(\frac{x+x+\Delta x}{2}\right)\sin\left(\frac{x-(x+\Delta x)}{2}\right)}{\cos(x+\Delta x)\cos x}$$

$$= \frac{-2\sin\left(x+\frac{\Delta x}{2}\right)\sin\left(\frac{-\Delta x}{2}\right)}{\cos(x+\Delta x)\cos x}$$

$$= \frac{2\sin\left(x+\frac{\Delta x}{2}\right)\sin\left(\frac{\Delta x}{2}\right)}{\cos(x+\Delta x)\cos x}$$

$$\frac{\Delta y}{\Delta x} = \frac{1}{\Delta x}\frac{2\sin\left(x+\frac{\Delta x}{2}\right)\sin\left(\frac{\Delta x}{2}\right)}{\cos(x+\Delta x)\cos x}$$

$$= \frac{\sin\left(x+\frac{\Delta x}{2}\right)}{\cos(x+\Delta x)\cos x}\cdot\frac{\sin\left(\frac{\Delta x}{2}\right)}{\left(\frac{\Delta x}{2}\right)}$$

$$\frac{dy}{dx} = \lim_{\Delta x\to 0}\frac{\Delta y}{\Delta x}$$

$$= \lim_{\Delta x\to 0}\frac{\sin\left(x+\frac{\Delta x}{2}\right)}{\cos(x+\Delta x)\cos x}\cdot\frac{\sin\left(\frac{\Delta x}{2}\right)}{\left(\frac{\Delta x}{2}\right)}$$

$$= \lim_{\Delta x\to 0}\frac{\sin\left(x+\frac{\Delta x}{2}\right)}{\cos(x+\Delta x)\cos x}\lim_{\frac{\Delta x}{2}\to 0}\frac{\sin\left(\frac{\Delta x}{2}\right)}{\left(\frac{\Delta x}{2}\right)}$$

$$= \frac{\sin(x+0)}{\cos(x+0)\cos x} \times 1$$

$$= \frac{\sin x}{\cos^2 x} = \sec x \tan x$$

$$\boxed{\frac{d}{dx}(\sec x) = \sec x \tan x}$$

vi. Let $y = \operatorname{cosec} x$

Consider a small increment in Δx and let Δy be the corresponding change in y. Then,

$$y + \Delta y = \operatorname{cosec}(x + \Delta x)$$

$$\Delta y = \operatorname{cosec}(x + \Delta x) - \operatorname{cosec} x$$

$$= \frac{1}{\sin(x+\Delta x)} - \frac{1}{\sin x}$$

$$= \frac{\sin x - \sin(x+\Delta x)}{\sin(x+\Delta x)\sin x}$$

$$= \frac{2\cos\left(\frac{x+x+\Delta x}{2}\right)\sin\left(\frac{x-(x+\Delta x)}{2}\right)}{\sin(x+\Delta x)\sin x}$$

$$= \frac{2\cos\left(x+\frac{\Delta x}{2}\right)\sin\left(\frac{-\Delta x}{2}\right)}{\sin(x+\Delta x)\sin x}$$

$$= \frac{-2\cos\left(x+\frac{\Delta x}{2}\right)\sin\left(\frac{\Delta x}{2}\right)}{\sin(x+\Delta x)\sin x}$$

$$\frac{\Delta y}{\Delta x} = \frac{1}{\Delta x}\frac{-2\cos\left(x+\frac{\Delta x}{2}\right)\sin\left(\frac{\Delta x}{2}\right)}{\sin(x+\Delta x)\sin x}$$

$$= \frac{-\cos\left(x+\frac{\Delta x}{2}\right)}{\sin(x+\Delta x)\sin x} \cdot \frac{\sin\left(\frac{\Delta x}{2}\right)}{\left(\frac{\Delta x}{2}\right)}$$

$$\frac{dy}{dx} = \lim_{\Delta x \to 0}\frac{\Delta y}{\Delta x}$$

$$= \lim_{\Delta x \to 0} \frac{-\cos\left(x+\frac{\Delta x}{2}\right)}{\sin(x+\Delta x)\sin x} \cdot \frac{\sin\left(\frac{\Delta x}{2}\right)}{\left(\frac{\Delta x}{2}\right)}$$

$$= \lim_{\Delta x \to 0} \frac{-\cos\left(x+\frac{\Delta x}{2}\right)}{\sin(x+\Delta x)\sin x} \lim_{\frac{\Delta x}{2} \to 0} \frac{\sin\left(\frac{\Delta x}{2}\right)}{\left(\frac{\Delta x}{2}\right)}$$

$$= \frac{-\cos(x+0)}{\sin(x+0)\sin x} \times 1$$

$$= \frac{-\cos x}{\sin^2 x} = \frac{-1}{\sin x} \cdot \frac{\cos x}{\sin x} = -\text{cosec } x \cot x$$

$$\boxed{\frac{d}{dx}(\text{cosec } x) = -\text{cosec} x \cot x}$$

***Example* 3(d):** *Find the derivatives of the inverse trigonometric functions:*

i. $\sin^{-1} x$	ii. $\cos^{-1} x$
iii. $\tan^{-1} x$	iv. $\text{cosec}^{-1} x$
v. $\sec^{-1} x$	vi. $\cot^{-1} x$

Solution:

i. Let $y = \sin^{-1} x$

$\therefore\ x = \sin y$

Differentiating w.r.t. y,

$$\frac{dx}{dy} = \cos y = \sqrt{1-\sin^2 y} = \sqrt{1-x^2}$$

$$\therefore\ \frac{dy}{dx} = \frac{2}{x+2}$$

$$\boxed{\frac{d}{dx}(\sin^{-1} x) = \frac{1}{\sqrt{1-x^2}}}$$

ii. Let $y = \cos^{-1} x$

$\therefore\ x = \cos y$

Differentiating w.r.t. y,

$$\frac{dx}{dy} = -\sin y = -\sqrt{1-\cos^2 y}$$

$$= -\sqrt{1-x^2}$$

$$\therefore \quad \frac{dy}{dx} = \frac{-1}{\sqrt{1-x^2}}$$

$$\boxed{\frac{d}{dx}\left(\cos^{-1} x\right) = \frac{-1}{\sqrt{1-x^2}}}$$

iii. Let $y = \tan^{-1} x$

$\therefore x = \tan y$

Differentiating w.r.t. y,

$$\frac{dx}{dy} = \sec^2 y = 1+\tan^2 y$$

$$= 1+x^2$$

$$\therefore \quad \frac{dy}{dx} = \frac{1}{1+x^2}$$

$$\boxed{\frac{d}{dx}\left(\tan^{-1} x\right) = \frac{1}{1+x^2}}$$

iv. Let $y = \operatorname{cosec}^{-1} x$

$\therefore x = \operatorname{cosec} y$

Differentiating w.r.t. y,

$$\frac{dx}{dy} = -\operatorname{cosec} y \cot y$$

$$= -\operatorname{cosec} y \sqrt{\operatorname{cosec}^2 y - 1}$$

$$= -x\sqrt{x^2-1}$$

$$\therefore \quad \frac{dy}{dx} = \frac{-1}{x\sqrt{x^2-1}}$$

$$\boxed{\frac{d}{dx}\left(\operatorname{cosec}^{-1} x\right) = \frac{-1}{x\sqrt{x^2-1}}}$$

v. Let $y = \sec^{-1} x$

$\therefore x = \sec y$

Differentiating w.r.t. y,

$$\frac{dx}{dy} = \sec y \tan y$$

$$= \sec y \sqrt{\sec^2 y - 1}$$

$$= x \sqrt{x^2 - 1}$$

$$\therefore \quad \frac{dy}{dx} = \frac{1}{x\sqrt{x^2 - 1}}$$

$$\boxed{\frac{d}{dx}\left(\sec^{-1} x\right) = \frac{1}{x\sqrt{x^2 - 1}}}$$

vi. Let $y = \cot^{-1} x$

$\therefore x = \cot y$

Differentiating w.r.t. y,

$$\frac{dx}{dy} = -\operatorname{cosec}^2 y = -(1 + \cot^2 y)$$

$$= -(1 + x^2)$$

$$\therefore \quad \frac{dy}{dx} = \frac{-1}{1 + x^2}$$

Example 4: *Find the derivatives of the following functions with respect to x.*

i. 8 sin x *ii. 4 e^x* *iii. k log x*

Solution:

i. Let $y = 8 \sin x$

$$\frac{dy}{dx} = 8\frac{d(\sin x)}{dx} = 8\cos x$$

ii. Let $y = 4\, e^x$

$$\frac{dy}{dx} = 4\frac{d(e^x)}{dx} = 4\, e^x$$

iii. Let $y = k \log x$

$$\frac{dy}{dx} = k\frac{d(\log x)}{dx} = \frac{k}{x}$$

Rule–1

If u and v are differentiable functions of x,

$$\frac{d(u \pm v)}{dx} = \frac{du}{dx} \pm \frac{dv}{dx}$$

Proof:

Let $y = u \pm v$.

Consider a small increment Δx in x. Let Δy be the corresponding change in y, Δu be the corresponding change in u and Δv be the corresponding change in v.

Then,

$$y + \Delta y = (u + \Delta u) \pm (v + \Delta v) \text{ and}$$

$$\Delta y = (u + \Delta u) \pm (v + \Delta v) - (u \pm v)$$

$$= \Delta u \pm \Delta v$$

$$\text{Hence } \frac{dy}{dx} = \lim_{\Delta x \to 0} \frac{\Delta y}{\Delta x} = \lim_{\Delta x \to 0} \frac{\Delta u \pm \Delta v}{\Delta x}$$

$$= \lim_{\Delta x \to 0} \frac{\Delta u}{\Delta x} \pm \lim_{\Delta x \to 0} \frac{\Delta v}{\Delta x}$$

$$= \frac{du}{dx} \pm \frac{dv}{dx}$$

Remark:

If u_1, u_2,, u_n are differentiable functions of x, then

$$\frac{d(u_1 \pm u_2 \pm \ldots \pm u_n)}{dx} = \frac{du_1}{dx} \pm \frac{du_2}{dx} \pm \ldots \pm \frac{du_n}{dx}$$

Example 5: *Find the derivatives of the following functions with respect to x.*

i. $8\sin x + 3\sec x$

ii. $4\,e^x - 3\,x^3$

iii. $k \log x - c \tan x + 5\,a^x$

Solution:

i. Let $y = 8\sin x + 3\sec x$

$$\frac{dy}{dx} = 8\frac{d(\sin x)}{dx} + 3\frac{d(\sec x)}{dx}$$

$$= 8\cos x + 3\sec x \tan x$$

ii. Let $y = 4\,e^x - 2\,x^3$

$$\frac{dy}{dx} = 4\frac{d(e^x)}{dx} - 2\frac{d(x^3)}{dx}$$

$$= 4\,e^x - 6\,x^2$$

iii. Let $y = k \log x - c \tan x + 5\, a^x$

$$\frac{dy}{dx} = \frac{k}{x} = - c \sec^2 x + 5\, a^x \log a$$

Rule–2 (Product Rule)

If u and v are differentiable functions of x, then,

$$\frac{d(uv)}{dx} = u.\frac{dv}{dx} + v.\frac{du}{dx}$$

Proof:

Let $y = u\, v$

Consider a small increment Δx in x. Let Δy be the corresponding change in y, Δu be the corresponding change in u and Δv be the corresponding change in v.

Then,

$$y + \Delta y = (u+\Delta u)\,(v+\Delta v) \text{ and}$$

$$\Delta y = (u+\Delta u)\,(v+\Delta v) - (u\, v)$$

$$= u\Delta v + v\Delta u$$

Hence $$\frac{dy}{dx} = \lim_{\Delta x \to 0} \frac{\Delta y}{\Delta x} = \lim_{\Delta x \to 0} \frac{u\Delta v + v\Delta u}{\Delta x}$$

$$= u \lim_{\Delta x \to 0} \frac{\Delta v}{\Delta x} + v \lim_{\Delta x \to 0} \frac{\Delta u}{\Delta x}$$

$$= u\frac{dv}{dx} + v\frac{du}{dx}$$

Remark–1: Product Rule for differentiation can also be stated as:

Differential Coefficient of a product

= First(Derivative of the Second)+ Second(Derivative of the First)

Remark–2: If u, v and w are differentiable functions of x, then

$$\frac{d(uvw)}{dx} = vw,\frac{du}{dx} + uw.\frac{dv}{dx} + uv.\frac{dw}{dx}$$

Example 6: *Find the derivative of $x^3 \log x$ with respect to x.*

Solution: Let $y = x^3 \log x$

$$\frac{dy}{dx} = x^3 \frac{d(\log x)}{dx} + \log x\, \frac{d(x^3)}{dx}$$

$$= x^3 \frac{1}{x} + \log x.\ 3x^2$$

$$= x^2 + 3x^2 \log x$$

$$= x^2 (1 + 3 \log x)$$

Rule–3 (Quotient Rule)

If u and v are differentiable functions of x, and $v(x) \neq 0$, then

$$\frac{d}{dx}\left(\frac{u}{v}\right) = \frac{v\frac{du}{dx} - u\frac{dv}{dx}}{v^2}$$

(Note that $\frac{u}{v}$ is differentiable at all points x where $v(x) \neq 0$).

Proof:

Let $y = \frac{u}{v}$

Consider a small increment Δx in x. Let Δy be the corresponding change in y, Δu be the corresponding change in u and Δv be the corresponding change in v.

Then,

$$y + \Delta y = \frac{u + \Delta u}{v + \Delta v} \text{ and}$$

$$\Delta y = \frac{u + \Delta u}{v + \Delta v} - \frac{u}{v} = \frac{v(u + \Delta u) - u(v + \Delta v)}{v(v + \Delta v)}$$

$$= \frac{v\Delta u - u\Delta v}{v(v + \Delta v)}$$

$$\frac{\Delta y}{\Delta x} = \frac{1}{\Delta x}\frac{v\Delta u - u\Delta v}{v(v + \Delta v)} = \frac{v\frac{\Delta u}{\Delta x} - u\frac{\Delta v}{\Delta x}}{v(v + \Delta v)}$$

Hence $$\frac{dy}{dx} = \lim_{\Delta x \to 0} \frac{\Delta y}{\Delta x}$$

$$= \lim_{\Delta x \to 0} \frac{v\frac{\Delta u}{\Delta x} - u\frac{\Delta v}{\Delta x}}{v(v + \Delta v)}$$

$$= \frac{v\frac{du}{dx} - u\frac{dv}{dx}}{v(v + 0)} = \frac{v\frac{du}{dx} - u\frac{dv}{dx}}{v^2}$$

Remark: If Dr. denotes the denominator and Nr. Denotes the numerator of the quotient of two functions, then, Quotient Rule for differentiation can also be stated as:

Differential Coefficient of a Quotient =

$$\frac{\text{Dr.} \times \text{Derivative of the Nr.} - \text{Nr.} \times \text{Derivative of the Dr.}}{(\text{Dr.})^2}$$

Example 7: *Find the derivative of* $\frac{x}{\sin x}$ *with respect to x.*

Solution:

Let $y = \frac{x}{\sin x}$

$$\frac{dy}{dx} = \frac{\sin x.\frac{d}{dx}(x) - x.\frac{d}{dx}(\sin x)}{(\sin x)^2}$$

$$= \frac{\sin x.\, 1 - x.\cos x}{\sin^2 x} = \frac{\sin x - x.\cos x}{\sin^2 x}$$

Rule–4 (Chain Rule)

If y is a differentiable function of u and u is a differentiable function of x, then,

$$\frac{dy}{dx} = \frac{dy}{du}.\frac{du}{dx}$$

Proof:

Let $y = f(u)$ and $u = g(x)$. Consider a small increment Δx in x. Let Δy be the corresponding change in y and Δu be the corresponding change in u.

Then,

$$y + \Delta y = f(u + \Delta u)$$

$$\Delta y = f(u + \Delta u) - f(u)$$

$$\lim_{\Delta u \to 0} \Delta y = \lim_{\Delta u \to 0} [f(u + \Delta u) - f(u)] = 0 \quad [\because f(x) \text{ is continuous}]$$

Similarly, $\Delta u = g(x + \Delta x) - g(x)$

$$\lim_{\Delta x \to 0} \Delta u = \lim_{\Delta x \to 0} [g(x + \Delta x) - g(x)] = 0 \quad [\because g(x) \text{ is continuous}]$$

$$\frac{\Delta y}{\Delta x} = \frac{\Delta y}{\Delta u}.\frac{\Delta u}{\Delta x}$$

$$\frac{dy}{dx} = \lim_{\Delta x \to 0} \frac{\Delta y}{\Delta x}$$

$$= \lim_{\Delta x \to 0} \frac{\Delta y}{\Delta u} \frac{\Delta u}{\Delta x} = \lim_{\Delta x \to 0} \frac{\Delta y}{\Delta u} . \lim_{\Delta x \to 0} \frac{\Delta u}{\Delta x}$$

$$= \lim_{\Delta u \to 0} \frac{\Delta y}{\Delta u} . \lim_{\Delta x \to 0} \frac{\Delta u}{\Delta x} = \frac{dy}{du} . \frac{du}{dx}$$

Remark –1:

Chain Rule is also known as 'Function of a function Rule'.

Remark –2:

If $y = f(u)$ and $u = g(x)$, then chain rule states that:

Derivative of y w.r.t x = (Derivative of y w.r.t u) × (Derivative of u w.r.t x)

Remark –3:

Chain rule can be extended for finite number of functions.

For example,

if $y = f(u)$, $u = g(v)$ and $v = h(x)$, then,

$$\frac{dy}{dx} = \frac{dy}{du} . \frac{du}{dv} . \frac{dv}{dx}$$

Example 8(a): *Find the derivative of sin $(x^2 + 1)$ with respect to x.*

Solution: Let $y = \sin (x^2 + 1)$

$$\frac{dy}{dx} = \cos (x^2 + 1) \frac{d}{dx}(x^2 + 1) .$$

$$= \cos (x^2 + 1) . 2x = 2x \cos (x^2 + 1)$$

Example 8(b): *Find the derivatives of the following functions with respect to x from first principles.*

(i) $3x^2 + 5x - 8$ *(ii)* $\frac{1}{\sqrt{x}}$ *(iii)* $\sqrt{\sin x}$

(iv) $\tan^{-1} x$ *(v)* $\sin \sqrt{x}$

Solution: (i) Let $y = 3x^2 + 5x - 8$

Consider a small increment Δx in x and let Δy be the corresponding change in y. Then,

$$y + \Delta y = 3(x + \Delta x)^2 + 5(x + \Delta x) - 8$$

$$\Delta y = 3(x + \Delta x)^2 + 5(x + \Delta x) - 8 - (3x^2 + 5x - 8)$$

$$= 3x^2 + 6x\,\Delta x + (\Delta x)^2 + 5x + 5\,\Delta x - 8 - 3x^2 - 5x + 8$$

$$= 6x\,\Delta x + (\Delta x)^2 + 5\,\Delta x$$

$$\frac{\Delta y}{\Delta x} = 6x + \Delta x + 5$$

$$\frac{dy}{dx} = \lim_{\Delta x \to 0} \frac{\Delta y}{\Delta x} = 6x + 5$$

ii. Let $y = \dfrac{1}{\sqrt{x}}$

Consider a small increment Δx in x and let Δy be the corresponding change in y. Then,

$$y + \Delta y = \frac{1}{\sqrt{x + \Delta x}}$$

$$\Delta y = \frac{1}{\sqrt{x+\Delta x}} - \frac{1}{\sqrt{x}}$$

$$= \frac{\sqrt{x} - \sqrt{x+\Delta x}}{(\sqrt{(x+\Delta x)x}}$$

$$= \frac{(\sqrt{x} - \sqrt{x+\Delta x})(\sqrt{x} + \sqrt{x+\Delta x})}{(\sqrt{(x+\Delta x)x})(\sqrt{x} + \sqrt{x+\Delta x})}$$

$$= \frac{x - (x+\Delta x)}{(\sqrt{(x+\Delta x)x})(\sqrt{x} + \sqrt{x+\Delta x})}$$

$$= \frac{-\Delta x}{(\sqrt{(x+\Delta x)x})(\sqrt{x} + \sqrt{x+\Delta x})}$$

$$\frac{\Delta y}{\Delta x} = \frac{-1}{(\sqrt{(x+\Delta x)x})(\sqrt{x} + \sqrt{x+\Delta x})}$$

$$\frac{dy}{dx} = \lim_{\Delta x \to 0} \frac{\Delta y}{\Delta x}$$

$$= \lim_{\Delta x \to 0} \frac{-1}{(\sqrt{(x+\Delta x)x})(\sqrt{x} + \sqrt{x+\Delta x})}$$

$$= \frac{-1}{(\sqrt{(x+0)x})(\sqrt{x} + \sqrt{x+0})} = \frac{-1}{2x\sqrt{x}}$$

iii. Let $y = \sqrt{\sin x}$

$$y + \Delta y = \sqrt{\sin(x + \Delta x)}$$

$$\Delta y = 3(x + \Delta x)^2 + 5(x + \Delta x) - 8 - (3x^2 + 5x - 8)$$

$$= 3x^2 + 6x\,\Delta x + (\Delta x)^2 + 5x + 5\,\Delta x - 8 - 3x^2 - 5x + 8$$

$$= 6x\,\Delta x + (\Delta x)^2 + 5\,\Delta x$$

$$\frac{\Delta y}{\Delta x} = 6x + \Delta x + 5$$

$$\frac{dy}{dx} = \lim_{\Delta x \to 0} \frac{\Delta y}{\Delta x} = 6x + 5$$

iv. Let $y = \tan^{-1} x$

Then $x = \tan y$

$$x + \Delta x = \tan(y + \Delta y)$$

$$\Delta x = \tan(y + \Delta y) - \tan y$$

$$= \frac{\sin(y+\Delta y)}{\cos(y+\Delta y)} - \frac{\sin y}{\cos y}$$

$$= \frac{\sin(y+\Delta y)\cos y - \cos(y+\Delta y)\sin y}{\cos(y+\Delta y)\cos y}$$

$$= \frac{\sin(y+\Delta y - y)}{\cos(y+\Delta y)\cos y} = \frac{\sin(\Delta y)}{\cos(y+\Delta y)\cos y}$$

$$\frac{\Delta x}{\Delta y} = \frac{1}{\Delta y}\frac{\sin(\Delta y)}{\cos(y+\Delta y)\cos y}$$

$$= \frac{1}{\cos(y+\Delta y)\cos y}\frac{\sin(\Delta y)}{\Delta y}$$

$$\frac{dx}{dy} = \lim_{\Delta y \to 0}\frac{\Delta x}{\Delta y}$$

$$= \lim_{\Delta y \to 0}\frac{1}{\cos(y+\Delta y)\cos y}\frac{\sin(\Delta y)}{\Delta y}$$

$$= \lim_{\Delta y \to 0}\frac{1}{\cos(y+\Delta y)\cos y}\lim_{\Delta y \to 0}\frac{\sin(\Delta y)}{\Delta y}$$

$$= \frac{1}{\cos(y+0)\cos y} \times 1 = \frac{1}{\cos^2 y} = \sec^2 y$$

$$\therefore \quad \frac{dy}{dx} = \frac{1}{\sec^2 y} = \frac{1}{1+\tan^2 y} = \frac{1}{1+x^2}$$

v. Let $y = \sin\sqrt{x}$

Then $y + \Delta y = \sin\sqrt{x+\Delta x}$

$$\Delta y = \sin\sqrt{x+\Delta x} - \sin\sqrt{x}$$

$$= 2\cos\frac{\sqrt{x+\Delta x}+\sqrt{x}}{2}\sin\frac{\sqrt{x+\Delta x}-\sqrt{x}}{2}$$

$$= \cos\frac{\sqrt{x+\Delta x}+\sqrt{x}}{2}\frac{\sin\frac{\sqrt{x+\Delta x}+\sqrt{x}}{2}}{\frac{\sqrt{x+\Delta x}+\sqrt{x}}{2}}(\sqrt{x+\Delta x}+\sqrt{x})$$

$$= \cos\frac{\sqrt{x+\Delta x}+\sqrt{x}}{2}\frac{\sin\frac{\sqrt{x+\Delta x}+\sqrt{x}}{2}}{\frac{\sqrt{x+\Delta x}+\sqrt{x}}{2}}\frac{\left(\sqrt{x+\Delta x}+\sqrt{x}\right)\left(\sqrt{x+\Delta x}-\sqrt{x}\right)}{\sqrt{x+\Delta x}-\sqrt{x}}$$

$$= \cos\frac{\sqrt{x+\Delta x}+\sqrt{x}}{2}\frac{\sin\frac{\sqrt{x+\Delta x}+\sqrt{x}}{2}}{\frac{\sqrt{x+\Delta x}+\sqrt{x}}{2}}\frac{\Delta x}{\sqrt{x+\Delta x}-\sqrt{x}}\frac{\Delta y}{\Delta x}$$

$$= \cos\frac{\sqrt{x+\Delta x}+\sqrt{x}}{2}\frac{\sin\frac{\sqrt{x+\Delta x}+\sqrt{x}}{2}}{\frac{\sqrt{x+\Delta x}+\sqrt{x}}{2}}\frac{1}{\sqrt{x+\Delta x}-\sqrt{x}}$$

$$\frac{dy}{dx} = \lim_{\Delta x\to 0}\frac{\Delta y}{\Delta x}$$

$$= \lim_{\Delta x\to 0}\cos\frac{\sqrt{x+\Delta x}+\sqrt{x}}{2}\frac{\sin\frac{\sqrt{x+\Delta x}+\sqrt{x}}{2}}{\frac{\sqrt{x+\Delta x}+\sqrt{x}}{2}}\frac{1}{\sqrt{x+\Delta x}-\sqrt{x}}$$

$$= \lim_{\Delta x\to 0}\cos\frac{\sqrt{x+\Delta x}+\sqrt{x}}{2}\lim_{\Delta x\to 0}\frac{\sin\frac{\sqrt{x+\Delta x}+\sqrt{x}}{2}}{\frac{\sqrt{x+\Delta x}+\sqrt{x}}{2}}$$

$$\lim_{\Delta x\to 0}\frac{1}{\sqrt{x+\Delta x}-\sqrt{x}} = \cos\sqrt{x}.1.\frac{1}{2\sqrt{x}} = \frac{\cos\sqrt{x}}{2\sqrt{x}}$$

Example 8(c): *Differentiate with respect to x.*

i. $2x^3 + 3x^2 - 5x + 1$ *ii.* $Sin\ 2x + \log 3x$

iii. $\tan(x^2 + 2)$ *iv.* $cosec\ \sqrt{x}$

Solution:

i. Let $y = 2x^3 + 3x^2 - 5x + 1$

$$\frac{dy}{dx} = 6x^2 + 6x - 5$$

ii. Let $y = \text{Sin } 2x + \log 3x$

$$\frac{dy}{dx} = \cos 2x .2 + \frac{1}{3x}.3 = 2 \cos 2x + \frac{1}{x}$$

iii. Let $y = \tan (x^2 + 2)$

$$\frac{dy}{dx} = \sec^2 (x^2 + 2).\ 2x = 2x \sec^2 (x^2 + 2)$$

iv. Let $y = \text{cosec} \sqrt{x}$

$$\frac{dy}{dx} = -\ \text{cosec} \sqrt{x}\ \cot \sqrt{x}\ \cdot \frac{1}{2\sqrt{x}}$$

$$= \frac{-1}{2\sqrt{x}}\ \text{cosec} \sqrt{x}\ \cot \sqrt{x}\ .$$

Example 9: *Differentiate with respect to x.*

i. $(x + 1)\ e^x$ | *ii. $\sin 2x \log x^2$*

iii. $\cos(2 - x^2) \tan(x^2 + 2)$ | *iv. $(x^2 + 1) \sqrt{x^2 - 1}$*

v. $x \sin x$ | *vi. $k (x^2 + 1) \cot x$*

Solution:

i. Let $y = (x + 1)\ e^x$

$$\frac{dy}{dx} = (x + 1)\ e^x + e^x .1 = e^x (x + 2)$$

ii. Let $y = \text{Sin } 2x \log x^2$

$$\frac{dy}{dx} = \text{Sin } 2x\ \frac{1}{x^2}.\ 2x + \log x^2 \cos 2x.\ 2$$

$$= \frac{2}{x} \text{Sin } 2x + 2\log x^2 \cos 2x$$

iii. Let $y = \cos(2 - x^2) \tan (x^2 + 2)$

$$\frac{dy}{dx} = \cos(2 - x^2) \sec^2 (x^2 + 2)\ 2x + \tan (x^2 + 2) \times - \sin (2 - x^2) \times -2x$$

$$= 2x \cos(2 - x^2) \sec^2 (x^2 + 2) + 2x \tan (x^2 + 2) \sin (2 - x^2)$$

iv. Let $y = (x^2 + 1)$

$$\frac{dy}{dx} = (x^2+1)\frac{1}{2\sqrt{x^2-1}}\ 2x + \sqrt{x^2-1}\ 2x$$

$$= \frac{x(x^2+1)}{\sqrt{x^2-1}} + 2x\ \sqrt{x^2-1}$$

v. Let $y = x \sin x$

$$\frac{dy}{dx} = x\frac{d(\sin x)}{dx} + \sin x\ \frac{d(x)}{dx}$$

$$= x \cos x + \sin x\ .1$$

$$= x \cos x + \sin x$$

vi. Let $y = k\ (x^2+1) \cot x$

$$\frac{dy}{dx} = k\ [(x^2+1)\frac{d(\cot x)}{dx} + \cot x\ \frac{d(x^2+1)}{dx}]$$

$$= k\ [(x^2+1)(-\operatorname{cosec}^2 x) + \cot x\ .\ 2x]$$

$$= k\ [-(x^2+1)\operatorname{cosec}^2 x + 2x \cot x]$$

Example 10: *Differentiate with respect to x.*

i. $\frac{x^3}{\log x}$ *ii.* $\frac{x^3+2x-7}{x+5}$

iii. $\frac{x^2+1}{x^2-1}$ *iv.* $\frac{\log(x^2+1)}{3x^2+2x-1}$

v. $\frac{x+e^x}{1+\log x}$ *vi.* $\frac{10^x}{\sin x}$

Solution:

i. Let $y = \frac{x^3}{\log x}$

$$\frac{dy}{dx} = \frac{\log x.\frac{d}{dx}(x^3) - x^3.\frac{d}{dx}(\log x)}{(\log x)^2}$$

$$= \frac{\log x.(3x^2) - x^3.\frac{1}{x}}{(\log x)^2} = \frac{3x^2 \log x - x^2}{(\log x)^2}$$

$$= \frac{x^2(3\log x - 1)}{(\log x)^2}$$

ii. Let $y = \frac{x^3+2x-7}{x^2+5}$

$$\frac{dy}{dx} = \frac{(x^2+5)(3x^2+2)-(x^3+2x-7)\,2x}{(x^2+5)^2}$$

$$= \frac{(x^2+5)(3x^2+2)-\ (x^3+2x-7)\,2x}{(x^2+5)^2}$$

$$= \frac{x^4+13x^2+14x+10}{(x^2+5)^2}$$

iii. Let $y = \dfrac{x^2+1}{x^2-1}$

$$\frac{dy}{dx} = \frac{(x^2-1)2x-(x^2+1)2x}{(x^2-1)^2} = \frac{-4x}{(x^2-1)^2}$$

iv. Let $y = \dfrac{\log(x^2+1)}{3x^2+2x-1}$

$$\frac{dy}{dx} = \frac{(3x^2+2x-1)\dfrac{2x}{x^2+1}-\log(x^2+1)(6x+2)}{(3x^2+2x--1)^2}$$

$$= \frac{2x(3x^2+2x-1)-(6x+2)(x^2+1)\log(x^2+1)}{(x^2+1)(3x^2+2x-1)^2}$$

v. Let $y = \dfrac{x+e^x}{1+\log x}$

$$\frac{dy}{dx} = \frac{(1+\log x)(1+e^x)-(x+e^x)\dfrac{1}{x}}{(1+\log x)^2}$$

$$= \frac{x(1+\log x)(1+e^x)-(x+e^x)}{x(1+\log x)^2}$$

vi. Let $y = \dfrac{10^x}{\sin x}$

$$\frac{dy}{dx} = \frac{\sin x.\ 10^x \log 10 -\ 10^x \cos x}{\sin^2 x}$$

$$= \frac{10^x(\sin x.\log 10 - \cos x)}{\sin^2 x}$$

Example 11: *Differentiate the following functions with respect to x:*

i. log(sec x + tan x)   *ii.* $e^{\tan x}$

iii. $\sqrt{\log(x^3-2x+1)}$ *iv.* $\log\sqrt{\dfrac{1+\sin x}{1-\sin x}}$

v. $\left(x+\sqrt{x^2+a^2}\right)^n$ vi. $\sin(m \sin^{-1} x)$

Solution:

i. Let $y = \log(\sec x + \tan x)$

$$\frac{dy}{dx} = \frac{1}{\sec x + \tan x}\cdot(\sec x \tan x + \sec^2 x)$$

$$= \frac{\sec x(\sec x + \tan x)}{\sec x + \tan x} = \sec x$$

ii. Let $y = e^{\tan x}$

$$\frac{dy}{dx} = e^{\tan x} \cdot \frac{d}{dx}(\tan x) = e^{\tan x} .\sec^2 x$$

iii. Let $y = \sqrt{\log(x^3 - 2x + 1)}$

$$\frac{dy}{dx} = \frac{1}{2\sqrt{\log(x^3-2x+1)}}\frac{d}{dx}\log(x^3-2x+1)$$

$$= \frac{1}{2\sqrt{\log(x^3-2x+1)}}\frac{1}{(x^3-2x+1)}\frac{d}{dx}(x^3-2x+1)$$

$$= \frac{1}{2\sqrt{\log(x^3-2x+1)}}\frac{1}{(x^3-2x+1)}(3x^2-2)$$

$$= \frac{3x^2-2}{2\sqrt{\log(x^3-2x+1)}(x^3-2x+1)}$$

iv. Let $y = \log\sqrt{\frac{1+\sin x}{1-\sin x}}$

$$= \frac{1}{2}\left[\log(1+\sin x) - \log(1-\sin x)\right]$$

$$\frac{dy}{dx} = \frac{1}{2}\left[\frac{1}{1+\sin x}\cos x - \frac{1}{1-\sin x}(-\cos x)\right]$$

$$= \frac{1}{2}\left[\frac{\cos x}{1+\sin x} + \frac{\cos x}{1-\sin x}\right]$$

$$= \frac{1}{2}\left[\frac{\cos x - \cos x \sin x + \cos x + \cos x \sin x}{1-\sin^2 x}\right]$$

$$= \frac{1}{2}\left[\frac{2\cos x}{\cos^2 x}\right] = \frac{1}{\cos x} = \sec x$$

v. Let $y = \left(x + \sqrt{x^2 + a^2}\right)^n$

$$\frac{dy}{dx} = n\left(x + \sqrt{x^2 + a^2}\right)^{n-1}\left[1 + \frac{1}{2\sqrt{x^2 + a^2}}.2x\right]$$

$$\frac{dy}{dx} = n\left(x + \sqrt{x^2 + a^2}\right)^{n-1}\left[\frac{\sqrt{x^2 + a^2} + x}{\sqrt{x^2 + a^2}}\right]$$

$$= \frac{n\left(x + \sqrt{x^2 + a^2}\right)^n}{\sqrt{x^2 + a^2}}$$

vi. Let $y = \sin(m \sin^{-1}x)$

$$\frac{dy}{dx} = \cos(m \sin^{-1}x)\ m\frac{1}{\sqrt{1 - x^2}}$$

$$= \frac{m\cos(m \sin^{-1}x)}{\sqrt{1 - x^2}}$$

Example 12:

i. If $y = e^x + e^{-x}$, prove that $= \frac{dy}{dx} = \sqrt{y^2 - 4}$

ii. If $y = \sqrt{x^2 + a^2}$, prove that $y\frac{dy}{dx} - x = 0$

iii. If $y = \frac{x}{x+2}$, prove that $x\frac{dy}{dx} = (1 - y)\, y$

iv. If $y = \sqrt{\frac{1-x}{1+x}}$, prove that $(1 - x^2)\frac{dy}{dx} + y = 0$

Solution:

i. $y = e^x + e^{-x}$

$$\frac{dy}{dx} = e^x + e^{-x}.(-1) = e^x - e^{-x}$$

$$\sqrt{y^2 - 4} = \sqrt{(e^x + e^{-x})^2 - 4} = \sqrt{e^{2x} + e^{-2x} + 2 - 4}$$

$$= \sqrt{e^{2x} + e^{-2x} - 2} = \sqrt{(e^x - e^{-x})^2}$$

$$= e^x - e^{-x} = \frac{dy}{dx}$$

ii. $y = \sqrt{x^2 + a^2}$

$$\frac{dy}{dx} = \frac{1}{2\sqrt{x^2 + a^2}}.2x = \frac{x}{\sqrt{x^2 + a^2}}$$

$$y\frac{dy}{dx} - x = \sqrt{x^2 + a^2}\,\frac{x}{\sqrt{x^2 + a^2}} - x$$

$$= x - x = 0$$

iii. $y = \dfrac{x}{x+2}$

$$\frac{dy}{dx} = \frac{(x+2).1 - x.1}{(x+2)^2} = \frac{2}{(x+2)^2}$$

$$x\frac{dy}{dx} = \frac{2x}{(x+2)^2}$$

$$(1 - y)\,y = (1 - \frac{x}{x+2})\frac{x}{x+2} = \frac{2}{x+2}\,\frac{x}{x+2}$$

$$= \frac{2x}{(x+2)^2} = x\frac{dy}{dx}$$

iv. $y = \sqrt{\dfrac{1-x}{1+x}} = \dfrac{\sqrt{1-x}}{\sqrt{1+x}}$

$$\frac{dy}{dx} = \frac{\sqrt{1+x}\,\dfrac{-1}{2\sqrt{1-x}} - \sqrt{1-x}\,\dfrac{1}{2\sqrt{1+x}}}{1+x}$$

$$= \frac{-(1+x) - (1-x)}{2(1+x)\sqrt{1+x}\sqrt{1-x}}$$

$$= \frac{-1}{(1+x)\sqrt{1+x}\sqrt{1-x}}$$

$$(1 - x^2)\frac{dy}{dx} + y = (1 - x^2) + \frac{-1}{(1+x)\sqrt{1+x}\sqrt{1-x}} + \frac{\sqrt{1-x}}{\sqrt{1+x}}$$

$$= \frac{-(1+x)(1-x)}{(1+x)\sqrt{1+x}\sqrt{1-x}} + \frac{\sqrt{1-x}}{\sqrt{1+x}}$$

$$= \frac{-\sqrt{1-x}}{\sqrt{1+x}} + \frac{\sqrt{1-x}}{\sqrt{1+x}} = 0$$

Example 13: *Find* $\frac{dy}{dx}$ *if* $x^2 + y^2 = 2xy$

Solution: $x^2 + y^2 = 2xy$

Differentiating each term with respect to x,

$$2x + 2y\frac{dy}{dx} = 2(x\frac{dy}{dx} + y)$$

$$\frac{dy}{dx}(y - x) = y - x$$

$$\frac{dy}{dx} = \frac{y-x}{y-x} = 1$$

Example 14: *Differentiate* x^x *with respect to* x:

Solution: Let $y = x^x$

Taking logarithm on both sides, $\log y = \log (x^x)$

i.e., $\log y = x \log x$

Differentiating with respect to x,

$$\frac{1}{y}\frac{dy}{dx} = x\frac{1}{x} + \log x = 1 + \log x$$

$$\therefore \quad \frac{dy}{dx} = y(1 + \log x) = x^x (1 + \log x)$$

Example 15(a): *Differentiate* $\sin^{-1}(3x - 4x^3)$ *with respect to* x

Solution:

Let $\quad y = \sin^{-1}(3x - 4x^3)$

Put $\quad x = \sin\theta$;

Then,

$$3x - 4x^3 = 3\sin\theta - 4\sin^3\theta = \sin 3\theta$$

$$\Rightarrow \quad \sin^{-1}(3x - 4x^3) = \sin^{-1}(\sin 3\theta) = 3\theta$$

$$\therefore \quad y = \sin^{-1}(3x - 4x^3) = 3 = 3\sin^{-1}x$$

$$\frac{dy}{dx} = 3\frac{1}{\sqrt{1-x^2}} = \frac{3}{\sqrt{1-x^2}}$$

Example 15(b): *Find* $\frac{dy}{dx}$ *if* : $x = at^2$, $y = 2at$

Solution: $x = at^2 \quad \frac{dx}{dt} = 2at$

$$y = 2at \qquad \frac{dy}{dt} = 2a$$

$$\therefore \qquad \frac{dy}{dx} = \frac{dy/dt}{dx/dt} = \frac{2a}{2at} = \frac{1}{t}$$

Example 15(c): *Find* $\frac{dy}{dx}, \frac{d^2y}{dx^2}$ *and* $\frac{d^3y}{dx^3}$ *of* $x^2 + \sin 2x$

Solution: Let $y = x^2 + \sin 2x$

$$\frac{dy}{dx} = 2x + \cos 2x \;.\; 2 = 2x + 2\cos 2x$$

$$\frac{d^2y}{dx^2} = 2 + 2.\,(-\sin 2x.2) = 2 - 4\sin 2x$$

$$\frac{d^3y}{dx^3} = -4\cos 2x.2 = -8\cos 2x$$

Example 16: *Find* $\frac{dy}{dx}$ *and* $\frac{d^2y}{dx^2}$, *if* $x = ct,\ y = \frac{c}{t}$

Solution: $x = ct$

$$\frac{dx}{dt} = c$$

$$\therefore \qquad \frac{dt}{dx} = \frac{1}{c}$$

$$y = \frac{c}{t}$$

$$\frac{dy}{dt} = \frac{-c}{t^2}$$

$$\therefore \qquad \frac{dy}{dx} = \frac{dy/dt}{dx/dt} = \frac{-c}{t^2} \div c$$

$$= \frac{-c}{ct^2} = \frac{-1}{t^2}$$

$$\frac{d^2y}{dx^2} = \frac{d}{dx}\left(\frac{dy}{dx}\right) = \frac{d}{dt}\left(\frac{dy}{dx}\right)\frac{dt}{dx}$$

$$= \frac{d}{dt}\left(\frac{-1}{t^2}\right)\frac{1}{c} = \frac{2}{ct^3}$$

Example 17: *Find* $\frac{dy}{dx}$ *in each of the following:*

i. $\sqrt{x} + \sqrt{y} = \sqrt{a}$

ii. $x^2 + y^2 + 4x + 5y - 8 = 0$

iii. $ax^2 + 2hxy + by^2 + 2gx + 2fy + c = 0$

iv. $\sin x + \sin y = \sin p$

v. $2x^3 + y^3 + 4xy + y^2x + 5y = x^2 + y^2$

vi. $x\sqrt{1+x} + y\sqrt{1+y} = 0$

Solution:

i. $\sqrt{x} + \sqrt{y} = \sqrt{a}$

Differentiating each term with respect to x,

$$\frac{1}{2\sqrt{x}} + \frac{1}{2\sqrt{y}}\frac{dy}{dx} = 0$$

$$\frac{1}{\sqrt{y}}\frac{dy}{dx} = \frac{-1}{\sqrt{x}}$$

$$\frac{dy}{dx} = -\sqrt{\frac{y}{x}}$$

ii. $x^2 + y^2 - 4x + 5y - 8 = 0$

Differentiating each term with respect to x,

$$2x + 2y\frac{dy}{dx} - 4 + 5\frac{dy}{dx} = 0$$

$$\frac{dy}{dx}(2y + 5) = 4 - 2x$$

$$\frac{dy}{dx} = \frac{4-2x}{2y+5}$$

iii. $ax^2 + 2hxy + by^2 + 2gx + 2fy + c = 0$

Differentiating each term with respect to x,

$$2ax + 2h(x\frac{dy}{dx} + y) + 2by + 2g + 2f\frac{dy}{dx} = 0$$

$$\frac{dy}{dx}(hx + by + f) = -(ax + hy + g)$$

$$\frac{dy}{dx} = \frac{-(ax+hy+g)}{hx+by+f}$$

iv. $\sin x + \sin y = \sin \pi$

Differentiating each term with respect to x,

$$\cos x + \cos y\frac{dy}{dx} = 0$$

$$\frac{dy}{dx} = \frac{-\cos x}{\cos y}$$

v. $2x^3 + y^3 + 4xy + y^2x + 5y = x^2 + y^2$

Differentiating each term with respect to x,

$$6x^2 + 3y^2\frac{dy}{dx} + 4(x\frac{dy}{dx} + y) + y^2 + x.\frac{dy}{dx} + 5\frac{dy}{dx} = 2x + 2y\frac{dy}{dx}$$

$$\frac{dy}{dx}(3y^2 + 5x - 2y + 5) = -6x^2 + 2x - 4y$$

$$\frac{dy}{dx} = \frac{-6x^2 + 2x - 4y}{3y^2 + 5x - 2y + 5}$$

vi. $x\sqrt{1+y} + y\sqrt{1+x} = 0$

$$x\sqrt{1+y} = -y\sqrt{1+x}$$

Squaring both sides,

$$x^2(1+y) = y^2(1+x)$$

Differentiating each term with respect to x,

$$x^2(1+\frac{dy}{dx}) + (1+y)\,2x = y^2 + (1+x)\,2y\frac{dy}{dx}$$

$$\frac{dy}{dx}(x^2 - 2xy - 2y) = y^2 - x^2 - 2xy - 2x$$

$$\frac{dy}{dx} = \frac{y^2 - x^2 - 2xy - 2x}{x^2 - 2xy - 2y}$$

Example 18:

i. *If* $y = e^{x+e^{x+e^{x+\dots\text{to}\,\infty}}}$, *show that* $= \frac{dy}{dx} = \frac{y}{1-y}$

ii. *If* $\sin y = x\sin(a+y)$, *show that* $\frac{dy}{dx} = \frac{\sin^2(a+y)}{\sin a}$

iii. *If* $xy = 1$, *show that* $\frac{dy}{dx} + y^2 = 0$

Solution:

i. Given, $y = e^{x+e^{x+e^{x+\dots\text{to}\,\infty}}}$

i.e., $y = e^{x+y}$

Differentiating each term with respect to x,

$$\frac{dy}{dx} = e^{x+y}(1+\frac{dy}{dx})$$

$$\frac{dy}{dx} = (1-e^{x+y}) = e^{x+y}$$

$$\frac{dy}{dx} = \frac{e^{x+y}}{1-e^{x+y}} = \frac{y}{1-y}$$

ii. Given, $\sin y = x \sin(a+y)$

i.e., $x = \dfrac{\sin y}{\sin(a+y)}$

Differentiating each term with respect to y,

$$\frac{dx}{dy} = \frac{\sin(a+y)\cos y - \sin y \cos(a+y)}{\sin^2(a+y)}$$

$$= \frac{\sin(a+y-y)}{\sin^2(a+y)} = \frac{\sin a}{\sin^2(a+y)}$$

$$\therefore \quad \frac{dy}{dx} = \frac{\sin^2(a+y)}{\sin a}$$

iii. Given, $xy = 1$

i.e., $y = \dfrac{1}{x}$

Differentiating with respect to x,

$$\frac{dy}{dx} = \frac{-1}{x^2}$$

$$\therefore \quad \frac{dy}{dx} + y^2 = \frac{-1}{x^2} + \frac{1}{x^2} = 0$$

Example 19(a): *Differentiate with respect to x:*

i. $(\sin x)^{\log x}$

ii. $(\log x)^{x}$

iii. $(\sin^{-1} x)^{\cos^{-1} x}$

iv. $(x + \log x)^{\log x}$

v. $\left(x+\dfrac{1}{x}\right)^{\frac{1}{x}}$

Solution:

i. Let $y = (\sin x)^{\log x}$

Taking logarithm on both sides,

$$\log y = \log [(\sin x)^{\log x}]$$

$$= \log x (\log \sin x)$$

Differentiating with respect to x,

$$\frac{1}{y}\frac{dy}{dx} = \log x \frac{1}{\sin x} \cos x + \log \sin x \frac{1}{x}$$

$$= \log x \cot x + \frac{1}{x} \log \sin x$$

$$\therefore \quad \frac{dy}{dx} = y (\log x \cot x + \frac{1}{x} \log \sin x)$$

$$= (\sin x)^{\log x} (\log x \cot x + \frac{1}{x} \log \sin x)$$

ii. Let $y = (\log x)^x$

Taking logarithm on both sides,

$$\log y = \log (\log x)^x$$

$$= x \log (\log x)$$

Differentiating with respect to x,

$$\frac{1}{y}\frac{dy}{dx} = x \frac{1}{\log x}.\frac{1}{x} + \log (\log x).1$$

$$= \frac{1}{\log x} + \log (\log x)$$

$$\therefore \quad \frac{dy}{dx} = y [\frac{1}{\log x} + \log (\log x)]$$

$$= (\log x)^x [\frac{1}{\log x} + \log (\log x)]$$

iii. Let $y = (\sin^{-1} x)^{\cos^{-1} x}$

Taking logarithm on both sides,

$$\log y = \log (\sin^{-1} x)^{\cos^{-1} x}$$

$$= \cos^{-1} x \log (\sin^{-1} x)$$

Differentiating with respect to x,

$$\frac{1}{y}\frac{dy}{dx} = \cos^{-1} x \frac{1}{\sin^{-1} x}.\frac{1}{\sqrt{1-x^2}} + \log (\sin^{-1} x).\frac{-1}{\sqrt{1-x^2}}$$

$$= \frac{\cos^{-1} x}{\sqrt{1-x^2}\sin^{-1} x} - \frac{\sin x}{x}$$

$$\therefore \quad \frac{dy}{dx} = y\left(\frac{\cos^{-1} x}{\sqrt{1-x^2}\sin^{-1} x} - \frac{\log(\sin^{-1} x)}{\sqrt{1-x^2}}\right)$$

iv. Let $y = (x + \log x)^{\log x}$

Taking logarithm on both sides,

$$\log y = \log (x + \log x)^{\log x}$$

$$= \log x \log (x + \log x)$$

Differentiating with respect to x,

$$\frac{1}{y}\frac{dy}{dx} = \log x \frac{1}{x+\log x}\left(1+\frac{1}{x}\right) + \log (x + \log x)\ \frac{1}{x}$$

$$= \frac{(1+x)\log x}{x(x+\log x)} + \frac{\log(x+\log x)}{x}$$

$$\therefore \quad \frac{dy}{dx} = y\left(\frac{(1+x)\log x}{x(x+\log x)} + \frac{\log(x+\log x)}{x}\right)$$

$$= (x + \log x)^{\log x}\left(\frac{(1+x)\log x}{x(x+\log x)} + \frac{\log(x+\log x)}{x}\right)$$

v. Let $y = \left(x+\frac{1}{x}\right)^{\frac{1}{x}}$

Taking logarithm on both sides,

$$\log y = \log \left(x+\frac{1}{x}\right)^{\frac{1}{x}}$$

$$= \frac{1}{x}\log\left(x+\frac{1}{x}\right)$$

Differentiating with respect to x,

$$\frac{1}{y}\frac{dy}{dx} = \frac{1}{x}\left(\frac{1-\frac{1}{x^2}}{x+\frac{1}{x}}\right) + \log\left(x+\frac{1}{x}\right) \times \frac{-1}{x^2}$$

$$= \frac{x^2-1}{x^2(x^2+1)} - \frac{\log(x+1/x)}{x^2}$$

$$\therefore \quad \frac{dy}{dx} = y\left(\frac{x^2-1}{x^2(x^2+1)} - \frac{\log(x+1/x)}{x^2}\right)$$

$$= \left(x+\frac{1}{x}\right)^{\frac{1}{x}}\left(\frac{x^2-1}{x^2(x^2+1)}-\frac{\log(x+1/x)}{x^2}\right)$$

Example 19(b): *Differentiate the following functions with respect to x*

i. $x^x + (\log x)^x$

ii. $(\sin x)^{\log x} - (\sin^{-1} x)^x$

iii. $x^x + x^{\frac{1}{x}}$

iv. $x^{\sin x} + x^{\cos x}$

Solution:

i. Let $y = x^x + (\log x)^x = u + v$ (say)

$$\therefore \quad \frac{dy}{dx} = \frac{du}{dx}+\frac{dv}{dx} \qquad \ldots(1)$$

We have, $u = x^x$

Taking logarithm on both sides,

$$\log u = \log (x)^x$$
$$= x \cdot \log x$$

Differentiating with respect to x,

$$\frac{1}{u}\frac{du}{dx} = x.\frac{1}{x} + \log x = 1 + \log x$$

$$\therefore \quad \frac{du}{dx} = x^x(1 + \log x) \qquad \ldots(A)$$

Also, we have,

$$v = (\log x)^x$$

Taking logarithm on both sides,

$$\log v = \log (\log x)^x$$
$$= x \log (\log x)$$

Differentiating with respect to x,

$$\frac{1}{v}\frac{dv}{dx} = x.\frac{1}{\log x}.\frac{1}{x} + \log (\log x)$$

$$= \frac{1}{\log x} + \log (\log x)$$

$$\therefore \quad \frac{dv}{dx} = (\log x)^x \left(\frac{1}{\log x} + \log(\log x)\right) \quad \text{...(B)}$$

Substituting in (1),

$$\frac{dy}{dx} = \frac{du}{dx} + \frac{dv}{dx}$$

$$= x^x(1 + \log x) + (\log x)^x \left(\frac{1}{\log x} + \log(\log x)\right)$$

ii. Let $y = (\sin x)^{\log x} - (\sin^{-1} x)^x = u - v$ (say)

$$\therefore \quad \frac{dy}{dx} = \frac{du}{dx} - \frac{dv}{dx} \quad \text{...(1)}$$

We have, $u = (\sin x)^{\log x}$

Taking logarithm on both sides,

$$\log u = \log (\sin x)^{\log x}$$

$$= \log x [\log (\sin x)]$$

Differentiating with respect to x,

$$\frac{1}{u}\frac{du}{dx} = \log x . \frac{\cos x}{\sin x} + \log (\sin x) \ \frac{1}{x}$$

$$= \log x . + \frac{1}{x} \log (\sin x)$$

$$\therefore \quad \frac{du}{dx} = (\sin x)^{\log x} [\log x . + \frac{1}{x} \log (\sin x)] \quad \text{...(A)}$$

Also we have, $v = (\sin^{-1} x)^x$

Taking logarithm on both sides,

$$\log v = \log (\sin^{-1} x)^x$$

$$= x \log(\sin^{-1} x)$$

Differentiating with respect to x,

$$\frac{1}{v}\frac{dv}{dx} = x . \frac{1}{\sin^{-1} x} . \frac{1}{\sqrt{1-x^2}} + \log (\sin^{-1} x)$$

$$= \left(\frac{x}{\sqrt{1-x^2}\sin^{-1} x} + \log(\sin^{-1} x)\right)$$

$$\therefore \quad \frac{dv}{dx} = (\sin^{-1} x)^x \left(\frac{x}{\sqrt{1-x^2}\sin^{-1} x} + \log(\sin^{-1} x) \right) \qquad \ldots(B)$$

Substituting in (1),

$$\frac{dy}{dx} = \frac{du}{dx} - \frac{dv}{dx}$$

$$= (\sin x)^{\log x} [\log x . \cot x + \frac{1}{x} \log(\sin x)]$$

$$-(\sin^{-1} x)^x \left(\frac{x}{\sqrt{1-x^2}\sin^{-1} x} + \log(\sin^{-1} x) \right)$$

iii. Let $y = x^x + x^{\frac{1}{x}} = u + v$ (say)

$$\therefore \quad \frac{dy}{dx} = \frac{du}{dx} + \frac{dv}{dx} \qquad \ldots(1)$$

We have, $u = x^x$

Proceeding as in Qn.(i), we get

$$\frac{du}{dx} = x^x(1 + \log x) \qquad \ldots(A)$$

Also we have, $v = x^{\frac{1}{x}}$

Taking logarithm on both sides,

$$\log v = \log(x^{\frac{1}{x}})$$

$$= \frac{1}{x}\log x$$

Differentiating with respect to x,

$$\frac{1}{v}\frac{dv}{dx} = \frac{1}{x}.\frac{1}{x} + \log x . \frac{-1}{x^2} = \frac{1-\log x}{x^2}$$

$$\therefore \quad \frac{dv}{dx} = x^{\frac{1}{x}}\left(\frac{1-\log x}{x^2}\right) \qquad \ldots(B)$$

Substituting in (1),

$$\frac{dy}{dx} = \frac{du}{dx} + \frac{dv}{dx}$$

$$= x^x(1 + \log x) + x^{\frac{1}{x}}\left(\frac{1-\log x}{x^2}\right)$$

iv. Let $y = x^{\sin x} + x^{\cos x} = u + v$ (say)

$$\therefore \quad \frac{dy}{dx} = \frac{du}{dx} + \frac{dv}{dx} \qquad \ldots(1)$$

We have, $u = x^{\sin x}$

Taking logarithm on both sides,

$$\log u = \log (x^{\sin x})$$
$$= \sin x \log x$$

Differentiating with respect to x,

$$\frac{1}{u}\frac{du}{dx} = \sin x . \frac{1}{x} + \log x . \cos x$$
$$= \frac{\sin x}{x} + \log x \cos x$$
$$\therefore \quad \frac{du}{dx} = x^{\sin x}\left[\frac{\sin x}{x} + \log x \cos x\right] \qquad \ldots(A)$$

Also we have, $v = x^{\cos x}$

Taking logarithm on both sides,

$$\log v = \log (x^{\cos x})$$
$$= \cos x \log x$$

Differentiating with respect to x,

$$\frac{1}{v}\frac{dv}{dx} = \cos x . \frac{1}{x} + \log x (-\sin x) = \frac{\cos x}{x} - \log x \sin x$$
$$\therefore \quad \frac{dv}{dx} = \left[\frac{\cos x}{x} - \log x \sin x\right] \qquad \ldots(B)$$

Substituting in (1),

$$\frac{dy}{dx} = \frac{du}{dx} + \frac{dv}{dx}$$
$$= x^{\sin x}\left[\frac{\sin x}{x} + \log x \cos x\right] + \left[\frac{\cos x}{x} - \log x \sin x\right]$$

Example 19(c): *Differentiate the following with respect to x*

i. $\dfrac{(x+1)(x^2+2)(x^3+3)}{(x^4+4)}$

ii. $\sqrt{\dfrac{(x^2-1)^3}{(x+1)(x+2)(x+3)}}$

iii. $\sqrt[3]{\dfrac{(x^2-3)\sin(x+1)}{x\cos^2 x}}$

Solution:

i. Let $y = \dfrac{(x+1)(x^2+2)(x^3+3)}{(x^4+4)}$

Taking logarithm on both sides,

$\log y = \log(x+1) + \log(x^2+2) + \log(x^3+3) - \log(x^4+4)$

Differentiating with respect to x,

$$\frac{1}{y}\frac{dy}{dx} = \frac{1}{x+1} + \frac{2x}{x^2+2} + \frac{3x^2}{x^3+3} - \frac{4x^3}{x^4+4}$$

$$\therefore \quad \frac{dy}{dx} = y\left(\frac{1}{x+1} + \frac{2x}{x^2+2} + \frac{3x^2}{x^3+3} - \frac{4x^3}{x^4+4}\right)$$

$$= \frac{(x+1)(x^2+2)(x^3+3)}{(x^4+4)}$$

$$\left(\frac{1}{x+1} + \frac{2x}{x^2+2} + \frac{3x^2}{x^3+3} - \frac{4x^3}{x^4+4}\right)$$

ii. Let $y = \sqrt{\dfrac{(x^2-1)^3}{(x+1)(x+2)(x+3)}}$

Taking logarithm on both sides,

$$\log y = \frac{1}{2}\left(3\log(x^2-1) - \log(x+1) - \log(x+2) - \log(x+3)\right)$$

Differentiating with respect to x,

$$\frac{1}{y}\frac{dy}{dx} = \frac{1}{2}\left(\frac{3.2x}{x^2-1} - \frac{1}{x+1} - \frac{1}{x+2} - \frac{1}{x+3}\right)$$

$$\therefore \quad \frac{dy}{dx} = \frac{y}{2}\left(\frac{6x}{x^2-1} - \frac{1}{x+1} - \frac{1}{x+2} - \frac{1}{x+3}\right)$$

$$= \frac{1}{2}\sqrt{\frac{(x^2-1)^3}{(x+1)(x+2)(x+3)}}$$

$$\left(\frac{6x}{x^2-1} - \frac{1}{x+1} - \frac{1}{x+2} - \frac{1}{x+3}\right)$$

iii. Let $y = \sqrt[3]{\dfrac{(x^2-3)\sin(x+1)}{x\cos^2 x}}$

Taking logarithm on both sides,

$$\log y = \frac{1}{3}[\log(x^2-3) + \log\sin(x+1) - \log x - 2\log(\cos x)]$$

Differentiating with respect to x,

$$\frac{1}{y}\frac{dy}{dx} = \frac{1}{3}\left[\frac{2x}{x^2-3} + \frac{\cos(x+1)}{3\sin(x+1)} - \frac{1}{x} + \frac{2\sin x}{\cos x}\right]$$

$$\therefore \quad \frac{dy}{dx} = \frac{y}{3}\left[\frac{2x}{x^2-3} + \frac{\cos(x+1)}{3\sin(x+1)} - \frac{1}{x} + \frac{2\sin x}{\cos x}\right]$$

$$= \frac{1}{3}\sqrt[3]{\frac{(x^2-3)\sin(x+1)}{x\cos^2 x}}\left[\frac{2x}{x^2-3} + \frac{\cos(x+1)}{3\sin(x+1)} - \frac{1}{x} + \frac{2\sin x}{\cos x}\right]$$

Example 20:

i. If $e^{x-y} = x^y$, *prove that* $\dfrac{dy}{dx} = \dfrac{\log x}{(1+\log x)^2}$

ii. If $y = x^{x^{x\ldots\text{to}\,\infty}}$, *prove that* $\dfrac{dy}{dx} = \dfrac{y^2}{x(1-\log y)}$

iii. If $x^m y^n = (x+y)^{m+n}$, *prove that*

$\dfrac{dy}{dx} = \dfrac{y}{x}$, *given* $\dfrac{m}{n} \neq \dfrac{x}{y}$

Solution:

i. Given $e^{x-y} = x^y$

Taking logarithm on both sides,

$$\log e^{x-y} = \log x^y$$

i.e., $x - y = y\log x \; (\because \log e = 1)$

i.e., $x = y + y\log x = y(1+\log x)$

i.e., $y = \dfrac{x}{1+\log x}$

$$\therefore \quad \frac{dy}{dx} = \frac{(1+\log x).1 - x.\frac{1}{x}}{(1+\log x)^2}$$

$$= \frac{\log x}{(1+\log x)^2}$$

ii. Given $y = x^{x^{x \dots \text{to} \infty}}$

i.e., $y = x^y$

Taking logarithm on both sides,

$$\log y = \log x^y$$

i.e., $\log y = y \log x$

Differentiating with respect to x,

$$\frac{1}{y}\frac{dy}{dx} = y.\frac{1}{x} + \log x.\frac{dy}{dx}$$

$$\frac{dy}{dx}\left(\frac{1}{y} - \log x\right) = \frac{y}{x}$$

$$\therefore \quad \frac{dy}{dx} = \frac{y}{x}\left(\frac{y}{1-y\log x}\right)$$

$$= \frac{y^2}{x(1-\log y)} \; (\because y \log x = \log y)$$

iii. Given $x^m y^n = (x+y)^{m+n}$

Taking logarithm on both sides,

$\log (x^m y^n) = \log (x+y)^{m+n}$

i.e., $m \log x + n \log y = (m+n) \log(x+y)$

Differentiating with respect to x,

$$m\frac{1}{x} + n\frac{1}{y}\frac{dy}{dx} = (m+n)\frac{1}{x+y}\left(1+\frac{dy}{dx}\right)$$

$$\therefore \quad \frac{dy}{dx}\left(\frac{n}{y} - \frac{m+n}{x+y}\right) = \left(\frac{m+n}{x+y} - \frac{m}{x}\right)$$

i.e., $$\frac{dy}{dx}\left(\frac{n(x+y)-(m+n)y}{y(x+y)}\right) = \frac{(m+n)x - m(x+y)}{y(x+y)}$$

i.e., $$\frac{dy}{dx}\left(\frac{nx-my}{y(x+y)}\right) = \frac{nx-my}{x(x+y)}$$

i.e., $$\frac{dy}{dx} = \frac{y}{x} \quad (\because \frac{m}{n} \neq \frac{x}{y} \Rightarrow nx - my \neq 0)$$

Example 21: *Differentiate the given functions accordingly:*

i. *Differentiate x log x with respect to (x + log x)*

ii. *Differentiate x^x with respect to x log x*

iii. *Differentiate* $\sin^{-1}\left(\dfrac{2x}{1+x^2}\right)$ *with respect to x*

Solution:

i. Let $u = x \log x$ and $v = x + \log x$

$$\frac{du}{dx} = x\frac{1}{x} + \log x = 1 + \log x$$

$$\frac{dv}{dx} = 1 + \frac{1}{x} = \frac{x+1}{x}$$

$$\therefore \frac{du}{dv} = \frac{du/dx}{dv/dx} = \frac{x(1+\log x)}{x+1}$$

ii. Let $u = x^x$ and $v = x \log x$

$$\log u = \log x^x = x \log x$$

$$\frac{1}{u}\frac{du}{dx} = x\frac{1}{x} + \log x = 1 + \log x$$

$$\therefore \frac{du}{dx} = u(1 + \log x) = x^x (1 + \log x)$$

$$v = x \log x$$

$$\frac{dv}{dx} = x\frac{1}{x} + \log x = 1 + \log x$$

$$\therefore \frac{du}{dv} = \frac{du/dx}{dv/dx} = \frac{x^x(1+\log x)}{(1+\log x)} = x^2$$

Aliter:

Let $u = x^x$ and $v = x \log x$

$\log u = \log x^x = x \log x = v$

i.e., $u = e^v$

Differentiating with respect to v,

$$\frac{du}{dv} = e^v = u = x^x$$

iii. Put $x = \tan\theta$

Then,

$$\frac{2x}{1+x^2} = \frac{2\tan\theta}{1-\tan^2\theta} = \sin 2\theta$$

$$\Rightarrow \sin^{-1}\left(\frac{2x}{1+x^2}\right) = \sin^{-1}(\sin 2\theta) = 2\theta$$

$$\therefore\ y = \sin^{-1}\left(\frac{2x}{1+x^2}\right) = 2\theta = 2\tan^{-1}x$$

$$\frac{dy}{dx} = 2\frac{1}{1+x^2} = \frac{2}{1+x^2}.$$

Example 22: *Differentiate the following functions with respect to x*

i. $\cos^{-1}(4x^3 - 3x)$

ii. $\tan^{-1}\left(\frac{3x-x^3}{1-3x^2}\right)$

Solution:

i. Let $y = \cos^{-1}(4x^3 - 3x)$

Put $x = \cos\theta$;

Then, $4x^3 - 3x = 4\cos^3\theta - 3\cos\theta = \cos 3\theta$

$\Rightarrow$ $\cos^{-1}(4x^3 - 3x) = \cos^{-1}(\cos 3\theta) = 3\theta$

$\therefore$ $y = \cos^{-1}(4x^3 - 3x) = 3\theta = 3\cos^{-1}x$

$$\therefore\ \frac{dy}{dx} = 3\frac{-1}{\sqrt{1-x^2}} = \frac{-3}{\sqrt{1-x^2}}$$

ii. Let $y = \tan^{-1}\left(\frac{3x-x^3}{1-3x^2}\right)$

Put $x = \tan\theta$;

Then, $\left(\frac{3x-x^3}{1-3x^2}\right) = \frac{3\tan\theta - \tan^3\theta}{1-3\tan^2\theta} = \tan 3\theta$

$$\Rightarrow\ \tan^{-1}\left(\frac{3x-x^3}{1-3x^2}\right) = \tan^{-1}(\tan 3\theta) = 3\theta$$

$$\therefore\ y = \tan^{-1}\left(\frac{3x-x^3}{1-3x^2}\right) = 3\theta = 3\tan^{-1}x$$

$$\frac{dy}{dx} = 3\frac{1}{1+x^2} = \frac{3}{1+x^2}$$

Example 23: *Differentiate the following functions with respect to x*

i. $\cos^{-1}\left(\frac{x-1/x}{x+1/x}\right)$

ii. $\tan^{-1}\left(\frac{1-\cos x}{\sin x}\right)$

Solution:

i. Let $$y = \cos^{-1}\left(\frac{x-1/x}{x+1/x}\right) = \cos^{-1}\left(\frac{x^2-1}{x^2+1}\right)$$

$$= \cos^{-1}\left(-\frac{1-x^2}{1+x^2}\right) = \pi - \cos^{-1}\left(\frac{1-x^2}{1+x^2}\right)$$

Put $x = \tan\theta$;

Then $$\frac{1-x^2}{1+x^2} = \frac{1\text{-}\tan^2\theta}{1+\tan^2\theta} = \cos 2q$$

$$\Rightarrow \quad \pi - \cos^{-1}\left(\frac{1-x^2}{1+x^2}\right) = \pi - \cos^{-1}(\cos 2\theta)$$

$$= \pi - 2\theta$$

$$\therefore \quad y = \cos^{-1}\left(\frac{x-1/x}{x+1/x}\right) = \pi \,\square\, 2\theta = \pi \,\square\, 2\tan^{-1}x$$

$$\frac{dy}{dx} = 0 - 2\frac{1}{1+x^2} = \frac{-2}{1+x^2}$$

ii. Let $$y = \tan^{-1}\left(\frac{1-\cos x}{\sin x}\right)$$

$$= \tan^{-1}\left(\frac{2\sin^2(x/2)}{2\sin(x/2)\cos(x/2)}\right)$$

$$= \tan^{-1}(\tan(x/2)) = \frac{x}{2}$$

$$\therefore \quad \frac{dy}{dx} = \frac{1}{2}$$

Example 24: *Find* $\frac{dy}{dx}$:

i. $x = a(\theta - \sin\theta)$, $y = a(1 - \cos\theta)$

ii. $x = a[\cos\theta + \log\tan(\theta/2)]$, $y = a\sin\theta$

iii. $x = \frac{2t}{1+t^2}, y = \frac{1-t^2}{1+t^2}$

iv. $x = a \cos^3\theta$, $y = b \sin^3\theta$

v. $x = a(\cos t + t \sin t)$, $y = a(\sin t - t \cos t)$

Solution:

i. $x = a(\theta - \sin\theta) \frac{dx}{d\theta} = a(1 - \cos\theta)$

$y = a(1 - \cos\theta) \frac{dy}{d\theta} = a \sin\theta$

$$\therefore \quad \frac{dy}{dx} = \frac{dy/d\theta}{dx/d\theta} = \frac{a\sin\theta}{a(1\text{-}\cos\theta)}$$

$$= \frac{2\sin(\theta/2)\cos(\theta/2)}{2\sin^2(\theta/2)} = \cot(\theta/2)$$

ii. $x = a[\cos q + \log \tan()]$

$$\frac{dx}{d\theta} = a\left(-\sin\theta + \frac{1}{\tan(\theta/2)}\sec^2(\theta/2).\frac{1}{2}\right)$$

$$= a\left(-\sin\theta + \frac{1}{2\sin(\theta/2)\cos(\theta/2)}\right)$$

$$= a\left(-\sin\theta + \frac{1}{\sin\theta}\right) = a\left(\frac{1-\sin^2\theta}{\sin\theta}\right) = \frac{a\cos^2\theta}{\sin\theta}$$

$$y = a\sin\theta \; ; \; \frac{dy}{d\theta} = a\cos\theta$$

$$\therefore \quad \frac{dy}{dx} = \frac{dy/d\theta}{dx/d\theta} = \frac{a\cos\theta}{\left(\frac{a\cos^2\theta}{\sin\theta}\right)} = \tan\theta$$

iii. $x = \frac{2t}{1+t^2}$

$$\frac{dx}{dt} = \frac{(1+t^2)2 - 2t.2t}{(1+t^2)^2} = \frac{2(1-t^2)}{(1+t^2)^2}$$

$$y = \frac{1-t^2}{1+t^2}$$

$$\frac{dy}{dt} = \frac{(1+t^2)(-2t)-(1-t^2)2t}{(1+t^2)^2} = \frac{-4t}{(1+t^2)^2}$$

$$\therefore \quad \frac{dy}{dx} = \frac{dy/dt}{dx/dt} = \frac{-4t}{(1+t^2)^2} \div \frac{2(1-t^2)}{(1+t^2)^2}$$

$$= \frac{-4t}{(1+t^2)^2} \times \frac{(1+t^2)^2}{2(1-t^2)} = \frac{2t}{t^2-1}$$

iv. $x = a \cos^3\theta$

$$\frac{dx}{d\theta} = a.3 \cos^2\theta .(-\sin \theta) = -3a \cos^2 \theta \sin \theta$$

$$y = b \sin^3 \theta$$

$$\frac{dy}{d\theta} = b.3 \sin^2 \theta .\cos \theta = 3b \sin^2 \theta \cos \theta$$

$$\therefore \quad \frac{dy}{dx} = \frac{dy/d\theta}{dx/d\theta} = \frac{3b\sin^2\theta \cos\theta}{-3a\cos^2\theta\sin\theta} = -\frac{b}{a}\tan \theta$$

v. $x = a(\cos t + t \sin t)$

$$\frac{dx}{dt} = a[-\sin t + (t \cos t + \sin t.\ 1)]$$

$$= a[-\sin t + t \cos t + \sin t]$$

$$= at \cos t$$

$$y = a(\sin t - t \cos t)$$

$$\frac{dy}{dt} = a[\cos t - (t \times - \sin t + \cos t.\ 1)]$$

$$= a[\cos t + t \sin t - \cos t]$$

$$= at \sin t$$

$$\therefore \quad \frac{dy}{dx} = \frac{dy/dt}{dx/dt} = \frac{at \sin t}{at \cos t} = \tan t$$

Example 25(a): *Find the second and third derivatives of x log x*

Let y = x log x

$$\frac{dy}{dx} = x.\frac{1}{x} + \log x = 1 + \log x$$

$$\frac{d^2y}{dx^2} = \frac{1}{x}$$

$$\frac{d^3y}{dx^3} = \frac{-1}{x^2}$$

Example 25(b): *Find second derivatives of the following functions:*

i. $x^2 - 2x + 1$

ii. $\cos x - \sin 2x$

iii. $x^3 + \tan^2 x$

iv. $e^{\sqrt{x+1}}$

v. $\sqrt{1-x}$

vi. $\log(1 - 2x)$

vii. $\frac{\log x}{x}$

Solution:

i. Let $y = x^2 - 2x + 1$

$$\frac{dy}{dx} = 2x - 2$$

$$\therefore \quad \frac{d^2y}{dx^2} = 2$$

ii. Let $y = \cos x - \sin 2x$

$$\frac{dy}{dx} = -\sin x - \cos 2x \,.\, 2 = -\sin x - 2\cos 2x$$

$$\therefore \quad \frac{d^2y}{dx^2} = -\cos x - 2(-\sin 2x).2 = -\cos x + 4\sin 2x$$

iii. Let $y = x^3 + \tan^2 x$

$$\frac{dy}{dx} = 3x^2 + 2\tan x \,.\, \sec^2 x$$

$$\therefore \quad \frac{d^2y}{dx^2} = 6x + 2(\tan x.\ 2\sec x.\ \sec x \tan x + \sec^2 x.\sec^2 x)$$

$$= 6x + 4\tan^2 x \sec^2 x + 2\sec^4 x$$

iv. Let $y = e^{\sqrt{x+1}}$

$$\frac{dy}{dx} = e^{\sqrt{x+1}} \frac{1}{2\sqrt{x+1}} = \frac{e^{\sqrt{x+1}}}{2\sqrt{x+1}}$$

$$\therefore \quad \frac{d^2y}{dx^2} = \frac{1}{2}\left(\frac{\sqrt{x+1}e^{\sqrt{x+1}}\frac{1}{2\sqrt{x+1}} - e^{\sqrt{x+1}}\frac{1}{2\sqrt{x+1}}}{x+1}\right)$$

$$= \frac{e^{\sqrt{x+1}}(\sqrt{x+1}-1)}{4(x+1)^{3/2}}$$

v. Let $y = \sqrt{1-x}$

$$\frac{dy}{dx} = \frac{-1}{2\sqrt{1-x}} = \frac{-1}{2}(1-x)^{-\frac{1}{2}}$$

$$\therefore \quad \frac{d^2y}{dx^2} = \frac{-1}{2}\left(\frac{-1}{2}(1-x)^{\frac{-3}{2}}\right) = \frac{-(1-x)^{\frac{-3}{2}}}{4}$$

vi. Let $y = \log(1 - 2x)$

$$\frac{dy}{dx} = \frac{1}{1-2x}.(-2) = \frac{-2}{1-2x}$$

$$\therefore \quad \frac{d^2y}{dx^2} = \frac{2}{(1-2x)^2}.(-2) = \frac{-4}{(1-2x)^2}$$

vii. Let $y = \dfrac{\log x}{x}$

$$\frac{dy}{dx} = \frac{x.\frac{1}{x} - \log x.1}{x^2} = \frac{1-\log x}{x^2}$$

$$\therefore \quad \frac{d^2y}{dx^2} = \frac{x^2(-\frac{1}{x}) - (1-\log x)2x}{x^4}$$

$$= \frac{-3+2\log x}{x^3}$$

Example 25(c): *Find* $\dfrac{dy}{dx}$ *and* $\dfrac{d^2y}{dx^2}$:

i. $x = a(\theta + \sin\theta)$, $y = a(1 - \cos\theta)$

ii. $x = \dfrac{1}{2}\left(t + \dfrac{1}{t}\right)$, $y = \dfrac{1}{2}\left(t + \dfrac{1}{t}\right)$

Solution:

i. $x = a(\theta + \sin\theta)\ \dfrac{dx}{d\theta} = a(1 + \cos\theta)$

$y = a(1 - \cos\theta)\ \dfrac{dy}{d\theta} = a \sin\theta$

$$\therefore \quad \frac{dy}{dx} = \frac{dy/d\theta}{dx/d\theta} = \frac{a\sin\theta}{a(1+\cos\theta)}$$

$$= \frac{2\sin(\theta/2)\cos(\theta/2)}{2\cos^2(\theta/2)} = \tan(\theta/2)$$

$$\frac{d^2y}{dx^2} = \frac{d}{dx}\left(\frac{dy}{dx}\right) = \frac{d}{d\theta}\left(\frac{dy}{dx}\right)\frac{d\theta}{dx}$$

$$= \frac{d}{d\theta}[\tan(\theta/2)].\frac{1}{a(1+\cos\theta)}$$

$$= \sec^2(\theta/2)\frac{1}{2}\frac{1}{2a\ \cos^2(\theta/2)}$$

$$= \frac{1}{4a}\sec^4(\theta/2)$$

ii. $x = \frac{1}{2}\left(t+\frac{1}{t}\right)$

$$\frac{dx}{dt} = \frac{1}{2}\left(1-\frac{1}{t^2}\right) \quad \therefore \quad \frac{dt}{dx} = \frac{2t^2}{t^2-1}$$

$$y = \frac{1}{2}\left(t-\frac{1}{t}\right)$$

$$\frac{dy}{dt} = \frac{1}{2}\left(1+\frac{1}{t^2}\right)$$

$$\therefore \quad \frac{dy}{dx} = \frac{dy/dt}{dx/dt} = \frac{1}{2}\left(1+\frac{1}{t^2}\right) \div \frac{1}{2}\left(1-\frac{1}{t^2}\right)$$

$$= \frac{t^2+1}{t^2-1}$$

$$\frac{d^2y}{dx^2} = \frac{d}{dx}\left(\frac{dy}{dx}\right) = \frac{d}{dt}\left(\frac{dy}{dx}\right)\frac{dt}{dx}$$

$$= \frac{d}{dt}\left(\frac{t^2+1}{t^2-1}\right)\frac{2t^2}{t^2-1}$$

$$= \left(\frac{(t^2-1)2t-(t^2-1)2t}{(t^2-1)^2}\right)\frac{2t^2}{t^2-1}$$

$$= \left(\frac{-4t}{(t^2-1)^2}\right)\frac{2t^2}{t^2-1} = \frac{-8t^3}{(t^2-1)^3}$$

Example 26:

i. *If* $y = x^4 - 2x^3 + 3x + 1$, *prove that* $\frac{d^4y}{dx^4} - 24 = 0$

ii. *If* $y = e^{-x\cos x}$, *prove that* $y_4 + 4y = 0$

iii. *If* $y = a\cos(\log x) + b\sin(\log x)$,

prove that $x^2\frac{d^2y}{dx^2} + x\frac{dy}{dx} + y = 0$

iv. *If* $y = (\sin^{-1} x)^2$, *prove that* $(1 - x^2)y_2 - x\,y_1 = 2$

Solution:

i. $y = x^4 - 2x^3 + 3x + 1$

$$\frac{dy}{dx} = 4x^3 - 6x^2 + 3$$

$$\frac{d^2y}{dx^2} = 12x^2 - 12x$$

$$\frac{d^3y}{dx^3} = 24x - 12$$

$$\frac{d^4y}{dx^4} = 24$$

$$\therefore \frac{d^4y}{dx^4} - 24 = 0$$

ii. $y = e^{-x}\cos x$

$$y_1 = \frac{dy}{dx} = e^{-x}(-\sin x) - \cos x e^{-x}$$

$$= -e^{-x}(\sin x + \cos x)$$

$$y_2 = -[e^{-x}(\cos x - \sin x) - (\sin x + \cos x)e^{-x}]$$

$$= e^{-x}(-\cos x + \sin x + \sin x + \cos x)$$

$$= e^{-x}(2\sin x)$$

$$y_3 = 2[e^{-x}\cos x - \sin x\, e^{-x}]$$

$$= 2e^{-x}[\cos x - \sin x\,]$$

$$y_4 = 2[e^{-x}(-\sin x - \cos x\,) - (\cos x - \sin x)e^{-x}]$$

$$= 2e^{-x}(-\sin x - \cos x - \cos x + \sin x\ )$$

$$= 2e^{-x}(-2\cos x) = -4e^{-x}\cos x$$

$$\therefore\ y_4 + 4y = -4e^{-x}\cos x + 4e^{-x}\cos x = 0$$

iii. $y = a\cos(\log x) + b\sin(\log x)$

$$\frac{dy}{dx} = -a\sin(\log x)\frac{1}{x} + b\cos(\log x)\frac{1}{x}$$

$$\therefore\quad x\frac{dy}{dx} = -a\sin(\log x) + b\cos(\log x)$$

Differentiating both sides with respect to x,

$$x\frac{d^2y}{dx^2} + \frac{dy}{dx}.1 = -a\cos(\log x)\frac{1}{x} - b\sin(\log x)\frac{1}{x}$$

$$\therefore\quad x^2\frac{d^2y}{dx^2} + x\frac{dy}{dx} = -a\cos(\log x) - b\sin(\log x) = -y$$

$$\text{i.e., } x^2\frac{d^2y}{dx^2} + x\frac{dy}{dx} + y = 0$$

iv. $y = (\sin^{-1} x)^2$

$$y_1 = 2(\sin^{-1} x)\ \frac{1}{\sqrt{1-x^2}}$$

Squaring and Cross multiplying,

$$(1-x^2)(y_1)^2 = 4(\sin^{-1} x)^2 = 4y$$

Differentiating both sides with respect to x,

$$(1-x^2)2y_1y_2 + y_1^2(-2x) = 4y_1$$

Dividing throughout by $2y_1$,

$$(1-x^2)y_2 - xy_1 = 2$$

Example 27:

i. If $y = \log(x + \sqrt{x^2+a^2})$,

prove that $(x^2 + a^2)\dfrac{d^2y}{dx^2} + x\dfrac{dy}{dx} = 0$

ii. If $y = \sin(m\sin^{-1} x)$, *prove that,*

$(1 - x^2)y_2 - xy_1 + m^2y = 0$

iii. *If* $y = \left(x+\sqrt{1+x^2}\right)^m$,

prove that $(1 + x^2)\frac{d^2y}{dx^2} + x\frac{dy}{dx} - m^2 y = 0$

iv. *If* $y = e^x \sin x$,

prove that $\frac{d^2y}{dx^2} - 2\frac{dy}{dx} + 2y = 0$

Solution:

i. $y = \log(x + \sqrt{x^2+a^2})$

$$\frac{dy}{dx} = \frac{1}{x+\sqrt{x^2+a^2}}\left(1+\frac{2x}{2\sqrt{x^2+a^2}}\right)$$

$$= \frac{1}{x+\sqrt{x^2+a^2}}\left(\frac{x+\sqrt{x^2+a^2}}{\sqrt{x^2+a^2}}\right)$$

$$= \left(\frac{1}{\sqrt{x^2+a^2}}\right)$$

i.e., $y_1 = \left(\frac{1}{\sqrt{x^2+a^2}}\right)$

Squaring and Cross multiplying,

$$(x^2 + a^2)\, y_1^2 = 1$$

Differentiating with respect to x,

$$(x^2 + a^2)2\, y_1\, y_2 + y_1^2 .2x = 0$$

Dividing throughout by $2y_1$,

$$(x^2 + a^2)\, y_2 + x\, y_1 = 0$$

or, $$(x^2 + a^2)\frac{d^2y}{dx^2} + x\frac{dy}{dx} = 0$$

ii. $y = \sin(m \sin^{-1} x)$

$$y_1 = \cos(m \sin^{-1} x)\, m\, \frac{1}{\sqrt{1-x^2}}$$

Cross multiplying,

$$y_1\sqrt{1-x^2} = m \cos(m \sin^{-1} x)$$

Differentiating,

$$y_1 \frac{-2x}{2\sqrt{1-x^2}} + \sqrt{1-x^2}.y_2 = -m \sin(m \sin^{-1} x)\, m \frac{1}{\sqrt{1-x^2}}$$

$$\text{i.e., } y_1 \frac{-x}{\sqrt{1-x^2}} + \sqrt{1-x^2}.y_2 = -m^2 \frac{y}{\sqrt{1-x^2}}$$

Multiplying throughout by $\sqrt{1-x^2}$,

$$(1 - x^2)y_2 - x\,y_1 + m^2 y = 0$$

iii. $y = \left(x+\sqrt{1+x^2}\right)^m$

$$y_1 = m\left(x+\sqrt{1+x^2}\right)^{m-1}\left(1+\frac{2x}{2\sqrt{1+x^2}}\right)$$

$$= m\left(x+\sqrt{1+x^2}\right)^{m-1}\left(\frac{x+\sqrt{1+x^2}}{\sqrt{1+x^2}}\right)$$

$$= m\left(x+\sqrt{1+x^2}\right)^{m}\left(\frac{1}{\sqrt{1+x^2}}\right) = \frac{my}{\sqrt{1+x^2}}$$

$$\text{i.e., } y_1 = \frac{my}{\sqrt{1+x^2}}$$

Squaring and Cross multiplying,

$$(1 + x^2)\,y_1^2 = m^2 y^2$$

Differentiating with respect to x,

$$(1 + x^2)2\,y_1\,y_2 + y_1^2\,.2x = m^2\,.2\,y\,y_1$$

Dividing throughout by $2y_1$,

$$(1 + x^2)\,y_2 + y_1\,x = m^2\,.\,y$$

i.e., $$(1 + x^2)\,y_2 + x\,y_1 - m^2 y = 0$$

or, $$(1 + x^2)\frac{d^2y}{dx^2} + x\frac{dy}{dx} - m^2 y = 0$$

iv. $y = e^x \sin x$

$$y_1 = e^x \cos x + e^x \sin x = e^x(\cos x + \sin x)$$

$$y_2 = e^x(-\sin x + \cos x) + e^x(\cos x + \sin x)$$

$= e^x(-\sin x + \cos x + \cos x + \sin x) = 2e^x \cos x$

$\therefore y_2 - 2y_1 + 2y = 2e^x \cos x - 2e^x(\cos x + \sin x) + 2e^x \sin x$

$= 2e^x(\cos x - \cos x - \sin x + \sin x) = 0$

i.e., $\frac{d^2y}{dx^2} - 2\frac{dy}{dx} + 2y = 0$

EXERCISES

1. If $y = x + \sqrt{x^2 - a^2}$, prove that $(x^2 - a^2)\frac{d^2y}{dx^2} + x\frac{dy}{dx} = y$
2. Differentiate $e^{\sin^{-1}x}$ with respect to $\sin^{-1} x$
3. If $y = x^2 \cos x$, prove that $x^2 y_2 - 4x y_1 + (x^2 + 6) y = 0$.
4. Differentiate $\sin^{-1}(2x\sqrt{1-x^2})$ with respect to $\sec^{-1}\frac{1}{\sqrt{1-x^2}}$

   (vi) Differentiate $\tan^{-1}\frac{2x}{1-x^2}$ with respect to $\sin^{-1}\frac{2x}{1+x^2}$
5. Find $\frac{dy}{dx}$, if:

   (i) $x = a\sin^2\theta,\ y = b\cos^2\theta$

   (ii) $x = \frac{3at}{1+t^2},\ y = \frac{3at^2}{1+t^2}$

   (iii) $x = a[\cos\theta + (1/2)\log\tan^2(\theta/2)],\ y = a\sin\theta$

   (iv) $x = a\left(t+\frac{1}{t}\right),\ y = a\left(t-\frac{1}{t}\right)$

   (v) $x = \frac{e^t + e^{-t}}{2},\ y = \frac{e^t + e^{-t}}{2}$
6. (i) If $y = x\cos x$, prove that $x^2 y_2 - 2x y_1 + (x^2 + 2) y = 0$
7. If $y = \cos^{-1} x$, prove that $(1 - x^2)y_2 - x y_1 = 0$
8. If $y = e^{\sin^{-1}x}$, prove that $(1 - x^2)y_2 - x y_1 - y = 0$
9. If $y = \sin(\log x)$, prove that $x^2\frac{d^2y}{dx^2} + x\frac{dy}{dx} + y = 0$
10. (i) $y = x^{xy}$, prove that $\frac{dy}{dx} = \frac{y^2(1+\log x)}{1-xy\log x}$

(ii) If $y = \dfrac{x^3}{(x-a)(x-b)(x-c)}$,

prove that $\dfrac{dy}{dx} = \dfrac{y}{x}\left(\dfrac{a}{a-x} + \dfrac{b}{b-x} + \dfrac{c}{c-x}\right)$

(iii) If $y^x = e^{y-x}$, prove that $\dfrac{dy}{dx} = \dfrac{(1+\log y)^2}{\log y}$

(vii) If $y = \sqrt{\tan x + \sqrt{\tan x + \sqrt{\tan x + \text{....to } \infty}}}$,

prove that = $\dfrac{dy}{dx} = \dfrac{\sec^2 x}{2y-1}$

(iv) If $y = \left(x^x\right)\sqrt{x}$, prove that $\dfrac{dy}{dx} = x^{(x+1/2)}\left(\dfrac{2x+1}{2x} + \log x\right)$

11. (i) If $x\,y^2 = 1$, prove that $2\dfrac{dy}{dx} + y^3 = 0$

(ii) If $\log\sqrt{x^2+y^2} = \tan^{-1}\left(\dfrac{y}{x}\right)$, prove that $\dfrac{dy}{dx} = \dfrac{x+y}{x-y}$

(iii) If $y\sqrt{x^2+1} = \log\,(\sqrt{x^2+1} - x)$,

prove that $(x^2+1)\dfrac{dy}{dx} + xy + 1 = 0$

(iv) If $y = \sqrt{x + \sqrt{x + \sqrt{x + \text{....to } \infty}}}$, prove that $\dfrac{dy}{dx} = \dfrac{1}{2y-1}$

12. Find $\dfrac{dy}{dx}$, if:

(i) $x^{2/3} + y^{2/3} = a^{2/3}$ (ii) $\dfrac{x^2}{a^2} + \dfrac{y^2}{b^2} = 1$

(iii) $\sin xy + \cos(x+y) = 1$ (iv) $\tan^{-1}(x^2+y^2) = a$

(v) $\sec\left(\dfrac{x+y}{x-y}\right) = a$

13. Differentiate with respect to x:

(i) $x^{\sqrt{x}}$ (ii) x^{x^x}

(iii) $\left(x^x\right)^x$ (iv) $(\log x)^x + x^{\log x}$

(v) $x^x + x^{1/x}$ (vi) $\dfrac{\sqrt{1-x^2}(2x-3)^{1/2}}{(x^2+2)^{2/3}}$

(ii) If $\sin(x+y) = -5$, show that $\frac{dy}{dx} = -1$

14. (i) Differentiate $\log(1+x^2)$ with respect to x^2

(ii) Differentiate $\sqrt{1+x}$ with respect to x^2

15. Differentiate $\sin^{-1}\sqrt{1-x^2}$ with respect to $\cos^{-1} x$

16. If $y = x \sin(a+y)$,

prove that $\frac{dy}{dx} = \frac{\sin^2(a+y)}{\sin(a+y) - y\cos(a+y)}$

6

Mapping or Functions

INTRODUCTION OF MAPPING

Let a be any set. Let the function $f: A \to A$ be defined by the formula $f(x) = x \; \forall x \in A$, that is, let each element of A be mapped on itself. Then f is called the identity function, or the identity transformation on A. We denote this function by I_A. Thus, if I_A denotes the identity mapping of a set A, we have $I_A(x) = x \; \forall \; x \in A$.

Suppose, by some rule or other, we assign to each element of A a unique element of B. Here two non-empty sets. Let A = {a, b, c, d} and B = {{x, y, z}. Suppose a is associated to x, b is associated to y, c is associated to x and d is associated to x. The set

{(a, x), (b, y), (c, x), (d, x)}

of such assignments is called a function or a mapping from A to B. If we denote this set by f, then we write

$f: A \to B$

Which is read as "f is a function of A to B" or "f is a mapping from A to B". It must be noted that there may be some elements of the set B which are not associated to any element of the set A. Each element of the set A must be associated to one and only one element of the set B. Two or more elements of the set A may be associated to the same element of the set B.

Example:

Let A = {1, 2, 3, 4, 5}. Then f = {(1, 1), (2, 2), (3, 3), (4, 4), (5, 5)} is an identity mapping of the set A. Identity mapping is always one-one and onto.

CONSTANT FUNCTION

A function $f: A \to B$ is called a constant function if the some element $b \in B$ is assigned to every element in A. In other words, $f: A \to B$ is a constant function if the range of f consists of only one element.

Example: *Let $f : R \to R$ be defined by the formula $f(x) = 5$. Then f is a constant function since 5 is assigned to every element.*

ONLY ONE-ONE AND ONTO MAPPINGS POSSESS INVERSE MAPPINGS

If the mapping $f : A \to B$ is not onto, the correspondence f^{-1} will not be a mapping from B to A. The reason is that in this case there will be some elements in B which will have no f^{-1} image in A. Again if $f : A \to B$ is not one-one, even then the correspondence f^{-1} will not be a mapping from B to A. The reason is that in this case some elements in B will be associated to more than one element in A. Thus, in this case some elements in B will not have a unique f^{-1} image in A.

Theorem 1:

If $f : A \to B$ be one-one and onto, then the inverse mapping of f is unique.

Proof:

Let $f : A \to B$ be one-one and onto. Let $g : B \to A$ and $h : B \to A$ be two inverse mappings of A. To prove that $g = h$.

Let b be any arbitrary element $\in$ B. Let $g(b) = a_1$ and $h(b) = a_2$. Since g is an inverse mapping of f, therefore, $g(b) = a_1 \Rightarrow f(a_1) = b$. Also since h is an inverse mapping of f, therefore $h(b) = a_2 \Rightarrow f(a_2) = b$. But f is a one-one mapping.

$\therefore f(a_1) = b$ and $f(a_2) = b \Rightarrow a_1 = a_2 \Rightarrow g(b) = h(b)$. Hence $g = h$.

Theorem 2:

Let A and B be two sets. If $f : A \to B$ is one-one onto, then $f^{-1} : B \to A$ is also one-one onto.

Proof:

First, let us prove that the mapping f^{-1} is one-one. Let y_1 and y_2 be any two element of B.

Suppose $f^{-1}(y_1) = x_1$, and $f^{-1}(y_2) = x_2$ where $x_1, x_2 \in A$. Then by the definition of the mapping f^{-1}, $f(x_1) = y_1$, and $f(x_2) = y_2$.

Now $f^{-1}(y_1) = f^{-1}(y_2) \Rightarrow x_1 = x_2$

$\Rightarrow f(x_1) = f(x_2)$ [$\because$ f is a mapping from A to B]

$\Rightarrow y_1 = y_2$.

$\therefore$ The mapping f^{-1} is one-one.

Now to prove that the mapping f^{-1} is onto A. Let x be any arbitrary element of A. Since f is a mapping from A to B, therefore there exists an element $y \in B$ such that $y = f(x)$ or $x = f^{-1}(y)$. Thus x is the f^{-1} image of the element $y \in B$. Hence the mapping f^{-1} is also onto.

SOME SETS OF NUMBERS

We shall here given a list of important sets of numbers along with the symbols we shall often use to denote them.

1. The set N = {1, 2, 3, 4, 5,...} of all natural numbers. The natural numbers are also called counting numbers. The number 0 is not an element of the set of natural numbers.
2. The set I = {..., –3, –2, –1, 0, 1, 2, 3,...} of all integers. The integers are also called whole numbers. Then set of integers is also denoted by Z.
3. The set I_+ = {1, 2, 3, 4,...} of all positive integers. The number 0 is neither positive nor negative.
4. The set Q = {x : x = p/q, where p and q are integers and $q \neq 0$} of all rational numbers.
5. The set Q_0 of all non-zero rational numbers.
6. The set R of all real numbers *i.e.,*
$$R = \{x : x = ab_1b_2b_2...b_n...\},$$
where $a \in I$ and b's are any of 0, 1, 2, 3,..., 9; n can be as large as we please and can go upto infinity. Here every real number has been expressed in the form of decimals with ten as base.
7. The set c of all complex numbers, *i.e.,* C = {x : x = a + ib, where a and b are real numbers and $i = \sqrt{(-1)}$}.

INTERVALS DEFINED AS SETS OF REAL NUMBERS

Consider the following sets of real numbers:

$$A = \{x : x \in R \text{ and } 2 \leq x \leq 7\}$$
$$B = \{x : x \in R \text{ and } 2 < x < 7\}$$
$$C = \{x : x \in R \text{ and } 2 < x \leq 7\}$$
$$D = \{x : x \in R \text{ and } 2 \leq x < 7\}.$$

Such sets are called intervals.

The set A is a closed interval and is denoted by A = [2, 7]. The set B is an open interval and is denoted by B = (2, 7) or by B = [2, 7]. The set C is an open closed interval and is denoted by C = (2, 7) or by C = [2, 7].

The set D is a closed open interval and is denoted by D = [2, 7] or by D = [2, 7].

TRANSFORMATIONS OR OPERATORS

If the domain and co-domain of a function f are both the same set, say $f : A \to A$, then f is often called an operator or a transformation on A.

FUNCTIONS AS SETS OF ORDERED PAIRS

If A and Bare any two non-empty sets, then a mapping f of A to B is a subset f of $A \times B$ satisfying the following conditions:

(i) for each $a \in A$, $(a, b) \in f$ for some $b \in B$;

(ii) if $(a, b) \in f$ and $(a, b') \in f$, then $b = b'$.

The first condition ensures that we have a rule that assigns to each element $a \in A$ some element $b \in B$. Thus, each element in A will have image. The second condition guarantees that the image is unique. Accordingly f is a function from A to B.

Note: If $f : A \to B$, it is important to distinguish between a function f and the value f (x) of f for any element x. While f is a subset of $A \times B$, f (x) is an element of the set B.

Example 1: *Let A = {a, b, c, d} and B = {a, b, c}. Then f = {(a, b), (b, c) (c, c), (d, b)} is a function from A to B, since to each element ∈ A we have assigned a unique element of B.*

Here f (a) = b = c, f (c) = c and f (d) = b. Then domain of f is the set A and the range of f i.e., f (A) = {b, c}.

Example 2: *Let A = {a = {a, b, c} and B = {x, y, z}. Then*

$$f = \{(a, x), (b, y), (a, z), (c, z)\}$$

is not a function from A to B. The reason is that two elements, x and z∈B are assigned to the same element a∈A.

Example 3: *Let A = {1, 2, 3} and B ={1, 2, 3, 4, 5, 6}. Let f assign to each member in A its square. Then f is not a function from A to B since no member of B is assigned to the element 3 ∈ A.*

EQUALITY OF TWO FUNCTIONS

Definition: Two functions f and g of $A \to B$ are said to be equal iff $f(x) = g(x)\ \forall\ x \in A$ and we write f = g. For two unequal mappings from A to B, here must exist at least on element $x \in A$ such that $f(x) \neq g(x)$.

TYPES OF FUNCTIONS

(a) 'Into' and 'Onto' Mapping

If the mapping $f : A \to B$ is such that there is at least one element in B which is not the f-image of any element in A, then we say that f is a mapping of A 'into' B. In this case the range of f is a proper subset of the co-domain of f *i.e.*, $f(A) \subseteq B$. Thus in 'into' mappings at least one element of the co-domain B is left uncovered by the f-images of the domain A.

If the mapping $f : A \to B$ is such that each element in B is the f-image of at least one element in A, then we say that f is a mapping of A 'onto' B. In this case the range of f is equal to the co-domain of f i.e., $f(A) = B$. Thus in 'onto' mappings the co-domain B is completely covered by the f-images of the domain A. If f is a mapping of A onto B, we write

$$f : A \xrightarrow{\text{onto}} B.$$

An onto mapping is also called as surjection.

Note: It should be noted that in common use whenever we say that f is a mapping of A into B, we usually include in it the possibility that the mapping f may be onto B also.

(b) 'One-one' and 'Many-one' Mappings

A mapping $f : A \to B$ is said to be one-one or one-to-one (abbreviated 1-1) if different elements in A have different f-images in B i.e., if

$$f(x) = f(x') \Rightarrow x = x' \ (x \text{ and } x' \in A).$$

One-to-one mapping is also sometimes known as injection.

In one-one mappings an element in B has only one pre-image in A. If f is a one-one mapping of A to B, we write

$$f : A \xrightarrow[1-1]{} B.$$

A mapping $f : A \to B$ is said to be many-one if two (or more than two) distinct elements in A have the same f-image in B i.e.,

$$f(x) = f(x'),\ x \neq x'.$$

In many-one mappings some elements in B have more than one pre-image in A.

One-one Onto Mappings: *If $f : A \to B$ is 1-1 and onto B, then f is called a one-to-one correspondence between A and B.*

One-one onto mapping is sometimes also known as *bijection*. If $f : A \to B$ 1-1 and onto B, we write

$$f : A \xrightarrow[1-1]{\text{onto}} B.$$

FUNCTION

Definition: *Let A and B be two given sets. Suppose there exists a correspondence denoted by f, which associates to each member of A a unique member of b. Then f is called a function or a mapping of A to B. The mapping f of A to B is denoted by*

$$f : A \to B \Rightarrow \text{by } A \xrightarrow{f} B$$

The set A is called the domain of the function f, and B is called the *co-domain* of f. Further if $a \in A$, then the element in B which is assigned to a is called the f image of a or the value of the function f for a and is denoted by f (a). The element a may be referred to as the **pre-image** of f (a).

DENUMERABLE SETS

We are familiar with the set of natural numbers, $N = \{1, 2, 3, 4, ...\}$. If a set A is equivalent to N. *i.e.,* if $\exists$ a one-one mapping from A onto N, then A is called *denumerable.*

A set is called *countable* if it is finite or denumerable and a set is called *non-denumerable* or *uncountable* if it is infinite and if it is not equivalent to N, *i.e.,* if it is not *countable.*

PRODUCT/COMPOSITE OF MAPPING

Here f be a function of X to Y and let g be a function of Y to Z. Here the domain of the function g is the co-domain of the function f. Let x be any arbitrary element in X *i.e.,* let x Î X. Then the image of x under f *i.e.,* f (x) is in Y. Since g : Y ® Z and f (x) Î Y, therefore we can find the image of f (x) under g *i.e.,* we can find g [f (x)] which will be in Z also unique. Thus, we have a rule which assigns to each element x Î X a unique element g[f (x)] Î Z. In this way we have a function of X to Z. This new function is called the product function or composite function of f and g and it is denoted by (g o f) Þ (gf).

Definition:

Let $f : X \to Y$ and $g : Y \to Z$.

Then the composite of the functions f and g denoted by (g o f), is a mapping of $X \to Z$ given by

$$(g \circ f) : X \to Z \text{ such that}$$

$$(g \circ f)(x) = g[f(x)], \forall x \in X.$$

Example 1: *Let $f : R \to R$ be given by the formula*

$$f(x) = \sin x, (x \in R)$$

and the map $g : R \rightarrow R$ *be given by the formula*

$$g(x) = x^2, \ (x \in R)$$

The composite map $(g \circ f) : R \rightarrow R$ *is given by the formula* $(g \circ f)(x) = g[f(x)] = g(\sin x) = (\sin x)^2 = \sin^2 x, \ (x \in R)$.

Note: If $f : A \rightarrow A$ and $g : A \rightarrow A$, then we can find both the composite mappings g o f and f o g. But in general $(g \circ f) \neq (f \circ g)$. The composite map $(f \circ g) : R \rightarrow R$ is given by the formula $(f \circ g)(x) = f[g(x)] = f(x^2) = \sin x^2$, $(x \in R)$

We see that $(f \circ g) : R \rightarrow R$ and $(g \circ f) : R \rightarrow R$ are defined by different formulas.

Example 2: *Let* $X = \{1, 2, 3, 4\}$. $Y = \{a, b, c, d\}$ *and* $Z = \{l, m, n\}$.

Let $f : X \rightarrow Y$ *and* $g = Y \rightarrow Z$ *be defined by*

$f = \{(1, a), (2, c) (3, b), (4, a)\}$ *and* $g = (a, l), (b, l), (c, m), (d, m)\}$. *Then the composite mapping* $(g \circ f) : X \rightarrow Z$ *may be computed in the following manner:*

$(g \circ f)(1) = g[f(1)] = g(a) = l$, $(g \circ f)(2) = g[f(2)] = g(c) = m$,

$(g \circ f)(3) = g[f(3)] = g(b) = l$, $(g \circ f)(4) = g[f(4)] = g(a) = l$,

Thus $(g \circ f) : X \rightarrow Z$ is given by

$$g \circ f = \{(1, l), (2, m), (3, l), (4, l)\}.$$

Theorem 1:

If $f : X \rightarrow Y$ *be a one-one and onto mapping, then*

$$f \circ f^{-1} = I_Y \text{ and } f^{-1} \circ f = I_X,$$

where I_X *and* I_Y *are the identity mappings of the sets X and Y respectively.*

Proof:

We shall first prove that $f^{-1} f = I_X$.

Since $f : X \rightarrow y$ and $f^{-1} : Y \rightarrow X$, therefore $(f^{-1} \circ f) : X \rightarrow X$.

Let x be any arbitrary element $\in X$ and let $f(x) = y$, where $y \in Y$. Then by the definition of the mapping f^{-1}, $f^{-1}(y) = x$.

We have $(f^{-1} \circ f)(x) = f^{-1}[f(x)] = f^{-1}(y) = x$. Thus the function $f^{-1} \circ f$ maps very element $x \in X$ onto itself. Therefore $f^{-1} \circ f = I_X$.

Now we shall prove that $f \circ f^{-1} = I_Y$. Since $f^{-1} : Y \rightarrow X$ and $f : X \rightarrow Y$, therefore $((f \circ f^{-1}), Y \rightarrow Y$.

Let y be any arbitrary element $\in Y$ and let $f^{-1}(y) = x$, where $x \in X$ Then $f(x) = y$.

We have $(f \circ f^{-1})(y) = f[f^{-1}(y)] = f(x) = y$. Thus, the function fof^{-1} maps every element $y \in Y$ onto itself. Therefore $f \circ f^{-1} = I_Y$.

Theorem 2:

Let $f : X \to Y$ and $g : Y \to X$. Then g is the inverse function of f i.e., $g f^{1}$, if the product function $(g \circ f) : Y \to X$ is the identity function on Y.

Proof:

In order to show that f is invertible, we are to show that f is one-one and onto.

1. **f is one-one:** Let $x_1, x_2 \in X$. Then

$$f(x_1) = f(x_2) \Rightarrow g[f(x_1)] = g[f(x_2)]$$
$$\Rightarrow (g \circ f)(x_1) = (g \circ f)(x_2)$$
$$\Rightarrow I(x_1) = 1(x_2) \qquad [\because g \circ f = I]$$
$$\Rightarrow x_1 = x_2$$

$\therefore$ f is one-one

2. **f is onto:** Let y be any element of Y. Since g is a function from Y to X, therefore $g(y) \in X$

Let $g(y) = x$. Then

$$g(y) = x \Rightarrow f[g(y)] = f(x)$$
$$\Rightarrow (f \circ g)(y) = f(x)$$
$$\Rightarrow I(y) = f(x) \qquad [\because g \circ f = I]$$
$$\Rightarrow y = f(x).$$

Thus $y \in Y \Rightarrow \# x \in X$ such that $f(x) = y$. Therefore f is onto. Since f is 1-1 onto, therefore f is invertible *i.e.,* f^{-1} exists.

3. Now we shall prove that $f^{-1} = g$. We have

$$f \circ g = I \Rightarrow f^{-1} \circ (f \circ g) = f^{-1} \circ I$$
$$\Rightarrow (f^{-1} \circ f) \circ g = f^{-1} \Rightarrow I \circ g = f^{-1} \Rightarrow g = f^{-1}.$$

Theorem 3:

Associativity of Composites of Functions

Let $f : A \to B$, $g : B \to C$ and $h : C \to D$, then

$$(h \circ g) \circ f = h \circ (g \circ f).$$

Proof:

It can be easily seen that both $(h \circ g) \circ f$ and $h \circ (g \circ f)$ are mappings of A to C. These two mappings will be equal if they assign the same image to each element x in the domain A, *i.e.,* if

$$[(h \circ g) \circ f]\,(x) = [h \circ (g \circ f)]\,(x).$$

We have, be definition of composites of mappings,

$$[(h \circ g)\, f]\,(x) = (h \circ g)\,[f\,(x)] = h\,[g\,\{f\,(x)\}] = h\,[(g \circ f)\,(x)]$$
$$= [h \circ (g \circ f)]\,(x).$$

Hence $(h \circ g) \circ f = h \circ (g \circ f)$.

Theorem 4:

If $f : X \to Y$, and $g : Y \to Z$ be two one-one onto maps, then $g \circ f : X \to Z$ is also one-one onto and

$$(g \circ f)^{-1} : Z \to X = (f^{-1} \circ g^{-1}) : Z \to X.$$

Proof:

We shall prove that

(i) $g \circ f$ is one-one, (ii) $g \circ f$ is onto, (iii) $(g \circ f)^{-1} \circ g^{-1}$.

(ii) Let x_1 and x_2 be any two elements in X. The mapping $g \circ f$ will be one-one if we show that the images of x_1 and x_2 under this mapping are equal only if $x_1 = x_2$. We have

$(g \circ f)\,(x_1) = (g \circ f)\,(x_2) \Rightarrow g\,[f\,(x_1)] = g\,[(x_2)]$

[by the definition of the composite mapping $g \circ f$]

$\Rightarrow f\,(x_1) = f\,(x_2)$ [$\because$ $f\,(x_1), f\,(x_2), \in Y$ and the mapping $g : Y \to Z$ is one-one]

$\Rightarrow x_1 = x_2$ [$\because$ f is one-one]

$\therefore$ the mapping $g \circ f$ is one-one.

(ii) Let z be any arbitrary element of Z. Since g is an onto mapping of Y to Z, therefore $\exists\ y \in Y$, such that $\exists\ x \in X$, such the $f\,(x) = y$. Thus, for every $z \in Z\ \exists\ x \in X$, such that $z = g\,(y) = g\,[f\,(x) = (g \circ f)\,(x)$. Therefore $g \circ f$ is a mapping of X onto Z.

Thus $g \circ f$ is one-one onto and hence is inversible.

(iii) Let z be any arbitrary element $\in Z$, such that $z = g\,(y)$, $(y \in Y, z \in Z)$. Then $y = g^{-1}\,(z)$.

Also let $y = f\,(x)$, $(y \in Y, x \in X)$. Then $x = f^{-1}\,(y)$.

We have $(g \circ f)\,(x) = g\,[f\,(x)] = g\,(y) = z$.

Since $g \circ f$ if one-one onto.

$\therefore$ $(g \circ f)\,(x) = z \Rightarrow (g \circ f)^{-1}\,(z) = x$.

Also $(f^{-1} \circ g^{-1})\,(z) = f^{-1}\,[g^{-1}\,(z)] = f^{-1}\,(y) = x$.

Thus " $z \in Z, (g \circ f)^{-1} (z) = (f_{-1} \circ g^{-1}) (z)$.

Hence the mappings $(g \circ f)^{-1}$ and $f^{-1} \circ g^{-1}$ are equal.

Theorem 5:

If $f : X \rightarrow Y$, then $I_Y \circ f = f$ and $f \circ I_X = f$ i.e., then product of any function with the identity function is the function itself.

Proof:

Let x be any arbitrary element $\in$ X and let

$$f (x) = y, (x \in X, y \in Y).$$

Since $f : X \rightarrow Y$ and $I_Y : Y \rightarrow X$, therefore $(I_Y \circ f) : X \rightarrow Y$.

We have $(I_Y \circ f) (x) = I_Y [f (x)] = I_Y (y) = y$.

[$\because$ I_Y is the identity mapping of Y]

Also $f (x) = y$.

Thus $\forall x \in X, (I_Y \circ f) (x) = f (x)$. Therefore $I_Y \circ f = f$.

Again since $I_x : X \rightarrow X$ and $f : X \rightarrow Y$, therefore $f \circ I_X : X \rightarrow Y$.

We have $(f \circ I_X) (x) = f [I_X (x)] = f (x)$

[$\because$ I_X is the identity mapping of X]

Thus $\forall x \in X, (f \circ I_X) (x) = f (x)$. Therefore $f \circ I_X = f$.

HOW TO PROVE THAT A GIVEN FUNCTION F IS ONE-ONE

If we are given a function f and we are to prove that it is one-one, we can do so by showing that if $f (x_1) = f (x_2)$ then $x_1 = x_2$, where x_1 and x_2 are arbitrary elements of the domain of f. We can also prove it by showing that if $x_1 \neq x_2$, then $f (x_1) \neq f (x_2)$.

HOW TO PROVE THAT A GIVEN FUNCTION F : A → B IS ONTO

To prove that f is onto, we show that if $y \in B$, $\exists x \in A$ such that $y = f(x)$. Then $y \in B \Rightarrow y \in f(A)$. Having chosen y arbitrarily, every element of B is an element of f (A) and hence $B \subseteq f (A)$. But $f (A) \subseteq B$. Therefore $B = f (A)$ and the function f is onto B.

INVERSE IMAGE OF AN ELEMENT

Let f be a function of A to B and let $b \in B$. Then the inverse image of the element be under f denoted by $f^{-1} (b)$ consists of those elements in

A which have b as their f-image. Symbolically, if $f : A \rightarrow B$, then $f^{-1}(b) = \{x : x \in A \text{ and } f(x) = b\}$.

f^{-1} is read as "f inverse". $f^{-1}(b)$ is always a subset of A.

COUNTING PRINCIPLE

If A, B and C are finite sets, and U be the finite universal set, then

(i) $n(A \cup B) = n(A) + n(B) - n(A \cap B)$

(ii) $n(A \cup B) = n(A) + n(B) \Leftrightarrow$ A, B are disjoint non-void sets

(iii) $n(A - B) = n(A) - n(A \cap B)$, *i.e.*,
$n(A - B) + n(A \cap B) = n(A)$.

(iv) $n(A \Delta B)$ = No. of elements which belong to exactly one of A or B
$= n((A - B) È (B - A)$
$= n(A - B) + n(B - A)$ ((A – B) and (B – A) and disjoint)
$= n(A) - n(A \cap B) + n(B) - n(A \cap B)$
$= n(A) + n(B) - 2n(A \cap B)$

(v) $n(A \cup B \cup C) = n(A) + n(B) + n(C) - n(A \cap B) - n(B \cap C) - n(A \cap C) + n(A \cap B \cap C)$

(vi) No. of elements in exactly two of the sets A, B, C
$= n(A \cap B) + n(B \cap C) + n(C \cap A) - 3n(A \cap B \cap C)$

(vii) No. of elements in exactly one of the sets A, B, C.
$= n(A) + n(B) + n(C) - 2n(A \cap B) - 2n(B \cap C) - 2n(A \cap C) + 3n(A \cap B \cap C)$

(viii) $n(A' \cup B') = n((A \cap B)')$
$= n(U) - n(A \cap B)$

(ix) $n(A' \cap B') = n((A \cup B)')$
$= n(U) - n(A \cup B)$.

DUALITY PRINCIPLE

It may be noticed that if any law of the algebra of sets, universal set U is replaced by ϕ, ϕ by U, $\cup$ by $\cap$ and $\cap$ by $\cup$ wherever these occur, the new statement thus, obtained is also a law of the algebra of sets. This fact is known as the duality principle, and any law obtained as a result of its application is called the dual of the original law. It is sometimes called primal law. For example, the dual of the law.

$$A \cup (B \cap C) = (A \cup B) \cap (A \cup C)$$

is $$A \cap (B \cup C) = (A \cap B) \cup (A \cap C)$$

It will be observed that this principle is always true.

DEDUCTION AND INDUCTION

Deduction: The process of deducing particular results from a general result is called deduction. For example, we know that the sum of the first n natural numbers is . From this result, we deduce that the sum of the first 50 natural numbers is 1275.

Induction: The process of establishing a valid general result from particular results is called induction. The principle of mathematical indiction is used to establish the validity of a general result involving natural numbers.

Consider the following statement :

$$P(n) = \text{“}n(n + 1) \text{ is even”}.$$

When $n = 2$, we have $P(2) = 2(3) = 6$ is even.

We cannot continue like this by testing for all natural numbers $n \in N$. We can prove that the result is true by using the following argument. Let n be even. Then n (n + 1) is even. Let n be odd. Then (n + 1) is even. Therefore n(n +1) is even. The result is proved. However, all the statements as given above may not be as simple to prove.

PRINCIPLE OF MATHEMATICAL INDUCTION

Let P(n), $n \in N$ be a statement such that

(i) P(1) is true, and (ii) truth of P(r) implie the truth of P(r + 1). Then, by the principle of mathematical induction, the statement P(n) is true for all $n \in N$. Obviously, the principle of mathematical induction involves the following steps :

(i) First, we prove that the result is true for $n = 1$,

(ii) Then, we assume that the result is true for $n = r$.

(iii) Finally, we prove that the result is true for $n = r + 1$.

Then, we conclude by the principle of mathematical induction that the statement is true for all $n \in N$.

Example 1: *Let R be the set of real numbers and let $f: R \to R$ be defined by the formula $f(x) = x^2$. Then $f^{1}(9) = \{3, -3\}$, since 9 is the f-image of both 3 and –3 and there is no other real number whose square is 9. Also we observe that $f^{1}(-5) = \emptyset$, since there is no real number whose square is –5.*

Example 2: *Let* $A = \{a, b, c, d, e\}$, *and* $B = \{x, y, z\}$.

Let $f : A \rightarrow B$ *be defined by*

$f = \{(a, x), (b, x), (c, y), (d, x), (e, y)\}$.

Then $f^{-1}(z)$ *is the null set* $\varnothing$, *since no element of A has z as its f-image.*

$f^{-1}(y) = \{c, e\}$ *since both c and e have y as their f-image.*

Also $f^{-1}(x) = \{a, b, d\}$.

ALGEBRAIC STRUCTURES

A set equipped with one or more binary compositions is known as an algebraic structure. Thus, if S is any set and $*$ is a binary composition on the set S, then the pair S, $*$) is an *algebraic structure.*

TYPES OF BINARY OPERATIONS

Commutative Operations: *A binary operation* o *on a set A is called commutative if* $a \circ b = b \circ a$, *for every* $a, b, \in A$.

Example 1: *Subtraction is not a commutative binary composition on the set R.* $5 \in R$, $12 \in R$ *but* $5 - 12 \neq 12 - 5$. *Thus* $a - b \neq b - a$, $(\forall a\ b \in R)$.

Example 2: *Addition and multiplication are commutative binary operations on the set of real numbers since*

$$a + b = b + a \text{ and } ab = ba, \text{ (for all } a, b, \in R).$$

Example 3: *The multiplication composition on the set M of all* 2×2 *real matrices is not commutative because there exist matrices A, B belonging to M such that* $AB \neq BA$.

Example 4: *Union and intersection of sets are commutative binary compositions on the set of all subsets of a set S.*

Associative Operations: *A binary operation* o *on a set A is called associative if* $(a \circ b) \circ c = a \circ (b \circ c)$, *(for every* $a, b, c, \in A$*).*

Notice that o will not be associative if there exists even one triad a, b, c, of the elements of A such that

$$(a \circ b) \circ c \neq a \circ (b \circ c).$$

Example 5: *Let* o *be a binary operation on R defined by*

$$a \circ b = a + 2b \text{ for all } a, b, \in R.$$

Then $(a \circ b) \circ c = (a + 2b) \circ c = a + 2b + 2c$,

and $a \circ (b \circ c) = a \circ (b + 2c) = a + 2(b + 2c) = a + 2b + 4c$.

Thus $(a \circ b) \circ c \neq a \circ (b \circ c)$, *for every* $a, b, c, \in R$.

Hence o is not an associative binary operation on R.

Distributive Operations: *Let A be a set on which two binary operations denoted by o and ∗ are defined. The operation ∗ is said to be left distributive with respect to o if*

$$a * (b \circ c) = (a * b) \circ (c * a), \text{ for all } a, b, c, \in A. \quad ...(1)$$

and is said to be right distributive with respect to o if

$$(b \circ c) * a = (b * a) \circ (c * a), \text{ for all } a, b, c, \in A. \quad ...(2)$$

When both (1) and (2) hold, we say simply that ∗ is distributive with respect to o.

If o is commutative, then (2) will hold if (1) holds and vice-versa.

Example 6: *The operations of union and intersection of sets distribute over each other, since*

$$A \cup (B \cap C) = (A \cup B) \cap (A \cup C)$$

and $$A \cap (B \cup C) = (A \cap B) \cup (A \cap C).$$

Example 7: *For the binary operation of addition on N, there is no identity element, but 1 is an identity element with respect to multiplication. We have* $1 \cdot a = a \cdot 1 = a$, *for every* $a \in N$.

Example 8: *Let Q be the set of all rational numbers. Then 0 is an identity element for the binary operation of addition on Q since* $0 + a = a + 0 = a$ *for very* $a \in Q$.

Example 9: *For the set R of all real numbers, multiplication* $(* = \cdot)$ *is distributive with respect to addition* $(\circ = +)$, *since*

$$a \cdot (b + c) = a \cdot b + a \cdot c \text{ and } (b + c) \cdot a = b \cdot a + c \cdot a \ \forall a, b, c, \in R.$$

But addition is not distributive with respect to multiplication, since $a + (b \cdot c) \neq (a + b) \cdot (a + c)$, *for every* $a, b, c \in R$.

Compositions with identity element: Let o denote a composition on a set S. Then o is said to be a composition with identity element if ∃ an element e ∈ S such that

$$x \circ e = x = e \circ x, \ \forall x \in S.$$

The element e, if it exists, is called the identity element and (S, o) is then an algebraic structure with identity element.

Inversible Elements of Algebraic Structures with Identity

An element a of a set A is said to be inversible for a binary operation o with identity e if $\exists\, b \in A$ *such that*

$$a \circ b = e = b \circ a.$$

Also then b is said to be an inverse of a and is denoted by a $^{-1}$.

The inversible elements in A re also called the units in A.

The identity element is always inversible and is its own in verse, since e o e = e o e = e. Thus $e^{-1} = e$.

Example 1: *For the operation of multiplication on N 1 is the identity and on element except 1 is inversible.*

Example 2: *The inverse with respect to addition, or additive inverse of* $x \in R$ *is* $-x$*, since* $x + (-x) = (-x) + x = 0$*. Where 0 is the additive identity element of R.*

Example 3: *For the operation of multiplication on Q 1 is the identity and every element other then 0 is inversible.*

Theorem:

For an associative binary operation o *on A with identity, an inversible element possesses a unique inverse.*

Proof:

Let e be the identity of A with respect to the binary operation o . Let b and b' be the inverses of the same element a; then to prove that b = b'

We have (b o a) o b' = e o b' = b'

and b o (a o b') = b o e = b.

But the operation o is associative, therefore

(b o a) o b' = b o (a o b').

Hence b = b',

IMAGE OF A SUBSET

Definition:

Let be a function of A to B. Let x be any subset of A. Then the set Y = {y : $\in$ *B and y = f (x) for some x* $\in$ *X} is called the image of X under f and we write f (x) = Y.*

INVERSE IMAGE OF A SUBSET

Let f : A → B and let C be a subset of B *i.e.,* $C \subseteq B$. Then the inverse image of C under f, denoted by f^{-1} (C), consists of those elements in A which are mapped into some element in C. Symbolically,

$$f^{-1}(C) = \{x : x \in A \text{ and } f(x) \in C\}.$$

f^{-1} (C) is always a subset of A. In particular f^{-1} [f (A)] = A, where f (A) is the range of the function f : A → B. Also f^{-1} (B) = A.

INVERSE FUNCTION

Let f : A → B be a one-one onto mapping. Let b be any arbitrary element of B. Since the mapping f is onto B, therefore there will be at least one element in A, say a, such that

$$b = f(a),\ b \in B,\ a \in A.$$

Since the mapping f is also one-one, therefore there will be only one element a in A such that b = f (a). Let us denote a by f^{-1} (b). Thus we see that if f : A → B is one-one onto we can define a new correspondence, denoted by f^{-1}, which associates to each element in B a unique element in A. Accordingly f^{-1} is a function of B to A.

Hence, if f : A → B is one-one onto, the f^{-1} B → A. The mapping f^{-r} is called the inverse mapping of the mapping f.

Definition: Let f : A → B be a one-one onto mapping. Then the mapping, $f^{-1} : B \to A$, which associates to each element $b \in B$ the element $a \in A$, such that f (a) = b, is called the inverse mapping of the mapping f : A → B.

RANGE OF A FUNCTION

Let f be a mapping of A to B *i.e.,* let f : A → B. Each element of A has unique image and each element in B need not appear as the image of an element in A. There can be more than one elements of A which have the same image in B. We define the *range* of f to consist of those elements in B which appear as the image of at least one element in A. Hence, we often speak of the range of a function as the *image of its domain.*

We denote the range of f : A → B by f (A).

Thus $f(A) = \{f(x) : x \in A\}$. Obviously $f(A) \subseteq B$.

Also we can say that

range of f = f (A) = {y : y ∈ B

and y = f (x) for some x ∈ A}.

CARDINALLY EQUIVALENT SETS

Two sets A and B are said to have the same number of elements iff a one-to-one mapping of A onto B exists. Such sets are said to be cardinally equivalent and we write A~B. Cardinally equivalent sets are to have the same cardinal number or the same cardinality.

BINARY OPERATION OR BINARY COMPOSITION

Definition: *A binary operation "o" on a non-empty set A is a mapping which associates with each ordered pair (a, b) of elements of A, a uniquely defined element $c \in A$. Thus "o" is a mapping of the product set $A \times A$ to A. Symbolically, a map $o : A \times A \to A$, is called a binary operation on the set A.*

The o-image of the element $(a, b) \in A \times A$ is denoted by $a \text{ o } b$.

If a set A is closed with respect to the composition o, then we say that o is a binary operation on the set A.

Example 1: *The product of any two rational numbers is a unique rational number. Therefore multiplication is a binary operation on the set Q of all rational numbers.*

Example 2: *Let S be any set and P (S) be the set of all subsets of S i.e., P (S) be the power set of S. If S_1 and $S_2 \in P(S)$ then both $S_1 \cup S_2$ and $S_1 \cap S_2$ are members of P (S). Therefore both union and intersection of sets are binary operations on the set P (S).*

Example 3: *We know that the multiplication of any two $n \times n$ real matrices is also an $n \times n$ real matrix. Therefore if M be the set of all $n \times n$ real matrices, then multiplication of matrices is a binary composition on M.*

Example 4: *The composition of addition is a binary operation on the set of even natural numbers. Then sum of two even natural numbers is also an even natural number. Thus the set of even natural numbers is closed under the composition of addition. On the other hand, addition is not a binary operation on the set of odd natural numbers. The sum of two odd natural numbers is an even natural number. Thus the sum of two odd natural numbers does not belong to the set of odd natural numbers. In this way the set of odd natural numbers is not closed under addition.*

Example 5: *The set of natural numbers N is not closed under subtraction. $4 \in N$ and $3 \in N$, but $3 - 4 \notin N$. Thus, subtraction is not a binary operation on N. On the other hand, the students may verify that subtraction is a binary operation on I.*

SOLVED EXAMPLES

Example 1: *If $f : R \to R$ be defined by $f(x) = |x|$, then prove that $f \circ f = f$.*

Solution: Since $f : R \to R$, therefore the composite mapping f o f is also a mapping from R to R *i.e.*, $f \circ f : R \to R$.

Now if x be any element of R, we have

$(f \circ f)(x) = f[f(x)] = f(|x|)$

$= ||x|| \; |x| = f(x).$

Thus we have $(f \circ f)(x) = f(x), \forall x \in R$.

Hence be the definition of the equality of two mappings, we have $f \circ f = f$.

Example 2: *Give an example to show that composition of two functions is not in general commutative.*

Solution: Consider the two mappings f and g from R to R given. These mappings f and g are such that $f \circ g \neq g \circ f$.

Hence, the composite of mappings is not a commutative operation.

Example 3(a): *Let S be a non-empty set and ∗ be a binary operation on S defined by $x * y = x : x, y \in S$.*

Determine whether ∗ is commutative and associative.

Solution: Let $x, y \in S$. Then

$$x * y = x \text{ and } y * x = y.$$

Thus if x and y are distinct elements of S, we have

$$x * y \neq y * x.$$

∴ the operation on S defined by ∗ is not commutative.

Again let x, y, z be any three elements of S. Then

$$(x * y) * z = x * z = x$$

and $\quad x * (y * z) = x * y = x.$

Thus $(x * y) * z = x * (y * z) \; \forall x, y, z \in S$.

Hence the operation on S defined by ∗ is associative.

Example 3(b): *Find whether the following algebraic structures have identity element or not:*

1. (N, +) 2. (N, ·) 3. (I, +); 4. (I, ·); 5. (R, ·); 6. (P (S), ∪);
7. (P (S), ∩).

Solution:

1. The algebraic structure (N, +) does not possess identity element. For ordinary addition of numbers, the number 0 is identity and $0 \notin N$.
2. The algebraic structure (N, ·) possesses identity element and it is 1. We have $1 \in N$ and

$$1a = a = a1 \; \forall a \in N.$$

3. The algebraic structure (I, +) possesses identity element and it is 0. We have $0 \in I$ and

$$0 + a = a = a + 0 \quad \forall\ a \in I.$$

4. The algebric structure (I, ·) possesses identity element and it is 1. We have $1 \in I$ and $1a = a = a1\ \forall\ a \in I$.
5. The algebraic structure (R, ·) possesses identity element and it is 1..
6. The algebraic structure (P (S), ∪) possesses identity element and it is the null set ∅. We have

$$\varnothing \in P(S) \text{ because } \varnothing \subseteq S.$$

Also $$\varnothing \cup A = A = A \cup \varnothing,\ \forall\ A \in P(S).$$

7. The algebraic structure (P (S), ∩) possesses identity element and it is the set S itself. We have

$$S \in P(S) \text{ because } S \subseteq S.$$

Also $S \cap A = A = A \cap S,\ \forall\ A \in P(S)$ *i.e.*, $\forall\ A \subseteq S$

Example 3(c): *Define binary operation. Show that the relation* o *given by a* o *b* = a^2 *is a binary operation on the set of natural numbers. Is this binary operation associative.*

Solution: Let a, b, ∈ N where N is the set of natural numbers. Then a^b is a unique natural numbers. Thus,

$$a \circ b = a^b \in N,\ \forall\ a, b \in N.$$

∴ 'o' is a binary operation on the set N.

Now let a, b, c, ∈ N. Then

$$(a \circ b) \circ c = (a^b) \circ c = (a^b) = a^{bc} \quad ...(1)$$

and $$a \circ (b \circ c) = a \circ (b^c) = (a)^{bc} \quad ...(2)$$

From (1) and (2), we observe that it is not necessary that a o (b o c) = (a o b) o c ∀ a, b, c ∈ N. Hence, the operation on N defined by 'o' is not associative.

For a concrete example, take 2, 2, 3 ∈ N.

We have $(2 \circ 2) \circ 3 = (2) \circ 3 = 4 \circ 3 = 4^3 = 64$

and $2 \circ (2 \circ 3) = 2 \circ (2^3) = 2 \circ 8 = 2^8 = 256.$

Thus $(2 \circ 2) \circ 3 \neq 2 \circ (2 \circ 3).$

Example 3(d): *Consider the binary operation* $* : Q \times Q \to Q$ *which 1 defined by* $a * b = a + b + ab$, *for every a, b, ∈ Q.*

1. *Is ∗ commutative?*
2. *Is ∗ associative?*
3. *Find the identity element for ∗.*
4. *Do any of the elements in Q have an inverse and what is it?*

Solution:

1. We have $a * b = a + b + ab$

and $b * a = b + a + ba$.

But $a + b = b + a$ and $ab = ba$.

$\therefore$ $a * b = b * a$ and thus $*$ is commutative.

2. We have $a * (b * c) = a * (b + c + bc)$

$= a + (b + c + bc) + a(b + c + bc)$

$= a + b + c + bc + ab + ac + abc.$

Also $(a * b) * c = (a + b + ab) * c$

$= (a + b + ab) + c + (a + b + ab)c$

$= a + b + c + ab + ac + bc + abc.$

$\therefore$ $a * (b * c) = (a * b) * c$ and so $*$ is associative.

3. Let e be the identity element for $*$. Then for every $a \in Q$, we must have $a * e = e * a = a$.

But $*$ is commutative, therefore $a * e = e * a$.

Now $a * e = a$, $(\forall a \in Q)$

$\Rightarrow a + e + ae = a \Rightarrow e + ae = 0 \Rightarrow e(1 + a) = 0$

$\Rightarrow e = 0$, since $1 + a \neq 0$, for every $a \in Q$.

Now $0 \in Q$ and we have

$0 * a = 0 + a + 0.a = a = a * 0 \; \forall a \in Q.$

Hence 0 is the identity element.

4. Since 0 is the identity element, therefore for a to have an inverse b, we must have $a * b = 0$.

Now $a * b = 0 \Rightarrow a + b + ab = 0 \Rightarrow a + b(a + 1) = 0$

$\Rightarrow a = -b(a + 1) \Rightarrow b = -\dfrac{a}{a+1}$, if $a \neq -1$.

Thus if $a \neq -1$, then a has an inverse and it is $-a/(a + 1)$.

Example 4: *Let N_0 and N_e denote the set of all odd natural numbers and the set of all even natural numbers respectively:*

1. *Is addition a binary operation on N_0?*
2. *Is multiplication a binary operation on N_0?*
3. *Is addition a binary operation on N_e? If yes, find whether this operation has identity element or not.*

Solution:

1. The sum of two odd natural numbers is an even natural number. Thus if x, y $\in N_0$, then $x + y \notin N_0$. Therefore addition is not a binary operation on the set N_0.
2. The product of two odd natural numbers is also an odd natural number and is unique. Thus,
$$x, y \in N_0, \forall x, y \in N_0.$$
Therefore multiplication is a binary operation on the set N_0.
3. The sum of two even natural numbers is an even natural number and is unique. Thus
$$x + y \in N_e, \forall x, y \in N_e.$$
Therefore addition is a binary operation on the set N_e.

However, the set N_e does not possess identity element for addition because the number 0 which is identity for ordinary addition of numbers is not an element of the set N_e.

Example 5: *Let the functions $f : R \to R$ and $g : R \to R$ be defined by $f(x) = 2x$, $g(x) = x^2 + 2$, $\forall x \in R$. Check the functions f and g for being*

(i) one-to-one

(ii) onto.

Find the formulae defining the functions f o g and g o g and obtain the values of (f o g) (2) and (g o g) (1).

Solution: The mapping f is both one-to-one and onto as shown below.

f is one-to-one: Let $x_1, x_2 \in R$. Then

$$f(x_1) = f(x_2) \Rightarrow 2x_1 = 2x_2 \Rightarrow x_1 = x_2.$$

$\therefore$ The mapping f is one-to-one.

f is onto: Let y be any element of R. Then y/2 is also an element of R and we have

$$f(y/2) = 2. (y/2) = y.$$

Thus, each element in R is the f-image of some element in R. Hence the mapping f is onto.

The mapping g is neither one-to-one nor onto as shown below.

g is not one-to-one: We have $1 \in R$ and $-1 \in R$. By the definition of the mapping g, we have

$$g(1) = 1^2 + 2 = 3, \ g(-1) = (-1)^2 + 2 = 3.$$

Thus, 1 and –1 are two distinct elements in R and we have g (1) = g (–1). Therefore the mapping g is not one-to-one.

The mapping g is not onto: We have $-5 \in R$ and there exists no $x \in R$ such that $g(x) = x^2 + 2 = -5$. Therefore the mapping g is not onto.

To find the formula for the mapping $f \circ g : R \to R$.

If $x \in R$, then we have

$$(f \circ g)(x) = f[g(x)] = f(x^2 + 2) = 2(x^2 + 2)$$
$$= 2x^2 + 4.$$

$\therefore f \circ g : R \to R$ is defined by the formula

$$(f \circ g)(x) = 2x^2 + 4, \forall x \in R.$$

Using this formula, we have

$$(f \circ g)(2) = 2.(2)^2 + 4 = 12.$$

Now to find the formula defining the mapping $g \circ g : R \to R$.

If $x \in R$, then we have

$$(g \circ g)(x) = g[g(x)] = g(x^2 + 2)$$
$$= (x^2 + 2)^2 + 2 = x^4 + 4x^2 + 6.$$

$\therefore g \circ g : R \to R$ is defined as

$$(g \circ g)(x) = x^4 + 4x^2 + 6, \forall x \in R.$$
$$(g \circ g)(1) = 1^4 + 4.1^2 + 6 = 11.$$

Example 6: *If $f : X \to Y$ and $g : Y \to Z$ are any two mappings, show that if $g \circ f : Y \to Z$ is one-to-one, then f is one-to-one.*

Solution: Let x_1, x_2 be any two elements of X.

Then $f(x_1) = f(x_2)$

$\Rightarrow g[f(x_1)] = g[f(x_2)]$

$\Rightarrow (g \circ f)(x_1) = (g \circ f)(x_2)$

$\Rightarrow x_1 = x_2$, because the mapping $g \circ f$ is given to be one-to-one.

Hence the mapping f is one-to-one.

Example 7(a): *Let $f : R \to R$ be given by the formula $f(x) = x^2$, $(x \in R)$ and $g : R \to R$ be give by the formula $g(x) = x + 3$, $(x \in R)$. Calculate $(f \circ g)(2)$ and $(g \circ f)(2)$. Hence show that*

$$f \circ g \neq g \circ f.$$

Solution: We have $(f \circ g)(2) = f[g(2)] = f(2 + 3)$

$-f(5) = 5^2 = 25.$

Also $(g \circ f)(2) = g[f(2)] = g(2^2) = g(4) = 4 + 3 = 7.$

Now both f o g and g o f are mappings from R to R. Since $2 \in R$ is such that $(f \circ g)(2) \neq (g \circ f)(2)$, therefore the mappings f o g and g o f are not equal. Hence $f \circ g \neq g \circ f$.

Example 7(b): *If $f : X \to Y$ and $A, B \subseteq X$. then prove that*

$$f(A \cup B) = f(A) \cup f(B).$$

Solution: Let y be any arbitrary element of $f(A \cup B)$

Then $y \in f(A \cup B)$

$\Rightarrow \exists\, x \in A \cup B$ such that $f(x) = y$

$\Rightarrow \exists\, x \in A$ or $x \in B$ such that $f(x) = y$

$\Rightarrow y \in f(A)$ or $y \in f(B)$

$\Rightarrow y \in f(A) \cup f(B)$.

$\therefore\ f(A \cup B) \subseteq f(A) \cup f(B)$. ...(1)

Again let y be any arbitrary element of $f(A) \cup f(B)$.

Then $y \in f(A) \cup f(B)$

$\Rightarrow y \in f(A)$ or $y \in f(B)$

$\Rightarrow \exists\, x \in A$ or $x \in B$ such that $f(x) = y$

$\Rightarrow \exists\, x \in A \cup B$ such that $f(x) = y$

$\Rightarrow y \in f(A \cup B)$

$\therefore\ f(A) \cup f(B) \subseteq f(A \cup B)$. ...(2)

From (1) and (2), we conclude that

$$f(A \cup B) = f(A) \cup f(B).$$

Example 7(c): *Let $X = \{x : x \in R$*

and $-\pi/2 \leq x \leq \pi/2$ i.e., let

$$X = [-\pi/2, \pi/2] \text{ and}$$

$$Y = \{y : y \in R \text{ and } -1 \leq y \leq 1\}$$

i.e., let $Y = [-1, 1]$.

Show that the funtion $f : X \to Y$ defined by

$$f(x) = \sin x,\ (x \in X', \text{ is one-one onto.}$$

Also give the inverse mapp $f^{-1} : Y \to X$.

Solution: Let m and n be any two different real numbers lying in the closed interval $[-\pi/2, \pi/2]$. We know that any two different real numbers lying in the closed interval $[-\pi/2, \pi/2]$ have not the same sine.

$\therefore\ m \neq n \Rightarrow \sin m \neq \sin n \Rightarrow f(m) \neq f(n)$. Hence f is one-one.

Again if y is any arbitrary real number lying in the closed interval $[-1, 1]$, $\exists$ a real number x lying in the closed interval $[-\pi/2, \pi/2]$ such that isn $x = y$.

Thus, every element y in Y is the f-image of some element x in X. Hence f is onto.

Thus $f : X \to Y$ is one-one onto, therefore f has an inverse function $f^{-1} : X \to Y$.

Let y be the image of x under the function f. Then

$$y = f(x) = \sin$$

Consequently, x will be the image of y under the inverse function f^{-1} *i.e.,*

$$x = f^{-1}(y).$$

Solving for x in terms of y in the equation $y = \sin x$, we get

$$x = \sin^{-1} y.$$

Thus $f^{-1}(y) = \sin^{-1} y$, $(y \in Y)$ is the formula defining the inverse function $f^{-1} : Y \to X$.

Example 8: *Decide whether or not the following are functions from A to B where A = {1, 2, 3, 4, 5} and B = {a, b, c, d, e}.*

If they are functions, give the range of each. If they are not, tell, why?

1. f = ({1, a), (2, b), (3, b), (5, e)}.
2. g = {(1, e), (5, d), (3, a), (2, b), (1, d), (4, a)}.
3. h = {(5, a), (1, e), (4, b), (3, c), (2, d)}.

Solution:

1. Since the element $4 \in A$ is not associated to any element $\in B$, therefore f is not a function from A to B.
2. The element $1 \in A$ is asociated to two different elements e and $d \in B$. Therefore g is not a function from A to B.
3. Each element of A is associated to a unique element of B. Therefore h is a function from $A \to B$. The range of h is the set of the h-images of all elements of A. So range of h = h (A) = {a, e, b, c, d} = B.

Example 9: *Let A = (–2, –1, 0, 1, 2}. Let the function $f : A \to R$ be defined by the promuls $f(x) = x^2 + 1$. Find the range of f.*

Solution: The range of f consists of those elements of R which appear as f-image of different elements of A. So we calculate the f-image of each element of A.

$$f(-2) = (-2)^2 + 1 = 5, \; f(-1) = (-1)^2 + 1 = 2,$$

$f(0) = (0)^2 + 1 = 1$, $f(1) = (1)^2 + 1 + 1 = 2$.

$f(2) = (2)^2 + 1 = 4 + 1 = 5$.

Thus, the range of f is the set {5, 2, 1, 2, 5}, *i.e.*, the set {5, 2, 1}.

Example 10: *Let N be the set of natural numbers and A be the set of even natural numbers.*

Let $f : N \to A$ be defined by the formula $f(x) = 2x$, $x \in N$. Show that the mapping f is one-one onto. Find the formula that defines the inverse function f^{-1}.

Solution: First to prove that f is a one-one mapping.

Suppose m and n are any two different elements $\in$ N.

Then $m \neq n \Rightarrow 2m \neq 2n \Rightarrow f(m) \neq f(n)$.

Thus different elements belonging to N have different f-images in A. Hence f is one-one.

Now to prove the f is an onto mapping. Let y be any arbitrary element in A *i.e.*, let y be any even natural number. Then f (y/2) = y and y/2 is a natural number *i.e.*, $y/2 \in N$. Thus each element in A is the f-image of some element in N. Hence f is onto.

Since $f : N \to A$ is one-one onto, therefore f has an inverse function $f^{-1} : A \to N$.

Let y be the image of x under the function f. Then

$$y = f(x) = 2x.$$

Consequently, x will be the image of y under the inverse function f^{-1} *i.e.*,

$$x = f^{-1}(y).$$

Solving for x in terms of y in the equation y = 2x, we get

$$x = y/2.$$

Then $$f^{-1}(y) = y/2.$$

Hence $f^{-1}(y) = y/2$, $(y \in A)$ is a formula defining the inverse function $f^{-1} : A \to N$.

Example 11: *Let S be the set of all triangles and R_+ be the set of positive real numbers. Prove that the map $f : S \to R_+$ given by*

$f(\Delta)$ = area of the Δ, $(\Delta \in S)$, is many-one and onto.

Solution: Let y be any arbitrary element in R_+ *i.e.*, let y be any positive real number. Then $\exists$ a triangle whose area is equal to y. Thus, every element in R_+ is the f-image of some element in S. Therefore f is onto.

Again there can be two or more triangles which have the same area. Therefore f is many-one.

Hence $f : S \to R_+$ is many-one onto.

Example 12: *Let Q be the set of rational numbers. Let $f : Q \to Q$ be defined by*

$$f(x) = 2x + 3, \ (x \in Q).$$

Show that f is one-one and onto. Also find a formula that defines the inverse function f^{-1}.

Solution: Let m and n be any two different elements in Q.

Then $\quad m \neq n \Rightarrow 2m \neq 2n$

$$\Rightarrow 2m + 3 \neq 2n + 3$$

$$\Rightarrow f(m) \neq f(n).$$

Thus different elements in Q have different f-images in Q. Hence f is one-one.

Let y be any arbitrary element in Q.

If $y = 2x + 3$, we have

$x = (y - 3)/2$ which is also a rational number.

Thus $\left(\frac{y-3}{2}\right) = y$ *i.e.*, any arbitrary element y in Q is the f-image of the element $(y - 2)/2 \in Q$. Hence f is onto

Since $f : Q \to Q$ is one-one onto, therefore f has an inverse function $f^{-1} : Q \to Q$.

Let y be the image of x under the function f. Then

$$y = f(x) = 2x + 3.$$

Consequently, x will be the image of y under the inverse function f^{-1} *i.e.*, $x = f^{-1}(y)$.

Solving for x in terms of y in the equation $y = 2x + 3$, we get

$$x = (y - 3)/2.$$

Thus, $f^{-1}(y) = (y - 3)/2, (y \in Q)$ is the formula defining the inverse function $f^{-1} : Q \to Q$.

Note: In order to prove that the mapping f is one-one, we can also argue like this:

Let m and n be any two elements in Q. Then

$$f(m) = f(n) \Rightarrow 2m + 3 = 2n + 3$$

$\Rightarrow 2m = 2n \Rightarrow m = n.$

Therefore f is one-one.

Example 13: *Let R_0 be the set of all non-zero real numbers. Let $f : R_0 \to R_0$ be defined by the formula $f(x) = 1/x$, $x \in R_0$. Show that f is one-one and onto mapping.*

Solution: Let m and n be any two different elements in R_0.

Then $\quad m \neq n \Rightarrow 1/m \neq 1/n$

$\Rightarrow f(m) \neq f(n).$

Thus different elements in R_0 have different f-images in R_0. Hence f is one-one.

Now let y be an arbitrary element in R_0. If y 1/x, we have x = 1/y which is a non-zero real number if y is a non-zero real number. Thus f (1/y) = y *i.e.*, any arbitrary element y in R_0 is the f-image of the element 1/y in R_0. Hence f is onto.

Thus $\quad f : R_0 \to R_0$ is one-one onto.

It should be noted that f (x) = 1/x cannot be a mapping from the set of real numbers to the set of real numbers. Then reason is that the image of 0 will not exist because f (0) = 1/0 and 1/0 is not a real number.

Example 14: *Let C be the set of complex numbers. Prove that the map $f : C \to R$ given by $f(z) = |z|$, $z \in C$ is neither one-one nor onto.*

Solution: Let z = x + iy be any complex number, where $x, y \in R$ and i $\sqrt{(-1)}$.

Then $|z| = \sqrt{(x^2 + y^2)}$ Also $|z|$ is always a non-negative real number *i.e.*, $|z| \geq 0$.

If m be any negative real number $\in$ R, then there exists no complex number $z \in C$ such that $|z| = m$.

Thus m is not the f-image of any complex number $\in$ C. Hence f is not onto but is into.

Also $\quad z_1 = 2 + i3 \in C$

and $\quad z_2 = 2 - i3 \in C.$

Then $\quad |2 + i3| = \sqrt{13}$

and $\quad |2 - i3| = \sqrt{13}.$

Thus $\quad z_1 \neq z_2$, although $|z_1| = |z_2|$ *i.e.*, $f(z_1) = f(z_2)$.

Thus we see that two different complex numbers z_1 and $z_2 \in C$ have the same f-image in R. Hence f is not one-one.

Example 15: *Show that the mapping $f : I_+ \to I_+$ defined by $f(x) = x^2$, $x \in I_+$ where I_+ is the set of positive integers, is one-one into.*

Solution: It is given that f is a mapping from I_+ to I_+ defined by

$$f(x) = x^2 \ \forall \ x \in I_+.$$

We have $I_+ = \{1, 2, 3, 4,...\}$.

f is one-one. Let x_1 $x_2 \in I_+$. We have

$$f(x_1) = f(x_2) \Rightarrow x_1^2 = x_2^2 \text{ [by def. of f]}$$

$$\Rightarrow x_1 = x_2 \Rightarrow \text{f is one-one.}$$

f is into. Now we shall prove that f is into and not onto. We have $2 \in I_+$. Now there exists no $x \in I_+$ such that $x^2 = 2$. Thus $2 \in I_+$ is not the f-image of any $x \in I_+$. Therefore the mapping f is into and not onto.

Example 16: *Classify the following as mappings of $R \to R$:*

(a) $f(x) = \log x$,

(b) $f(x) = +\sqrt{x}$.

(c) $f(x) = \tan x$, $(x \in R)$.

Solution:

(a) Let $x \in R$ and let x be a negative real number. Then log x is not a real number, *i.e.*, log x is not an element of R. Thus the negative real numbers in R will have no f-image in R. Also $\log 0 = -\infty$ which is not an element of R. Thus, the real number 0 in R has no f-image in R. Hence f is not a function from R to R. However f is a function from R_+ to R.

(b) Let x be a negative real number. Then $\sqrt{x}$ is not a real number *i.e.*, $\sqrt{x} \notin R$. Thus the negative real numbers in R will have no f-images in R. Hence f is not a function from R to R. If S denotes the set of all non-negative real numbers, then f defined by this formula is a function from S to R.

(c) Let $x = \pi/2$. Then $\tan \pi/2 =$ which is not an element of R. Thus $\pi/2 \in R$ has no f-image in R. In general if $x \in R$ is such that $x = n.\frac{\pi}{2}$ Where n is an odd integer, then x has no f-image in R. Hence f is not a function from R to R.

However, if $X = \{x : x \in R \text{ and } -\pi/2 < x < \pi/2\}$, then f is a function from X to R, Obviously this function will be one-one and onto.

Example 17: *If R is the set of real numbers discuss the mapping*

$f: R \to R$ *where* $f(x) = x^2, x \in R$.

What is the domain and the range of the mapping f ?

Solution: If x is any real number, the x^2 is also a real number and it will be unique. Thus each element $x \in R$ has a unique f-image in R. Therefore $f: R \to R$.

Since there is no real number whose square is negative, therefore any negative number in R is not the f-image of any element in R. Consequently f is a mapping of R into R and not from R onto R.

Also we see that $(2)^2 = 4$ and $(-2)^2 = 4$. Thus the elements 2 and –2 in R have the same f-image 4 in R. Hence f is a many-one into mapping of R to R.

Domain of f: Since $f: R \to R$, therefore domain of f is R.

Range of f: If $x \in R$, then $f(x) = x^2$ is a non-negative real number. Further if y is any non-negative real number, then $\exists$ a real number $\sqrt{y}$ such that $f(\sqrt{y}) = (\sqrt{y})^2 = y$. Thus, every non-negative real number is the f-image of some real number. Therefore the range of f is the set of non-negative real numbers.

Example 18: *Let A = { a, b, c,} and B = {e, f}. Write down all possible functions from A to B.*

Solution: While forming a mapping from A to B, each element of A can be associated to an element of B in two ways. Thus we can have $2 \times 2 \times 2$ *i.e.,* 8 distinct mappings from A to B.

These are as given below:

{(a, e), (b, e), (c, e)}, {(a, e), (b, e), (c, f)},

{(a, e), (b, f), (c, e)}, {(a, e), (b, f), (c, f)},

{(a, f), (b, e), (c, e)}, {(a, f), (b, e), (c, f)},

{(a, f), (b, f), (c, e)}, {(a, f), (b, f), (c, f)}.

Example 19: *Each of the following formulas defines a function from R to R. Find the range of each function.*

1. $f(x) = x^3$;
2. $g(x) = \sin x$;
3. $h(x) = x^2 + 1$.

Solution:

1. We know that every real number a has a real cube root $\left(\sqrt[3]{a}\right)$. Therefore if a be any arbitrary element $\in$ R, then f $\left(\sqrt[3]{a}\right) = \left(\sqrt[3]{a}\right)^3 = a$ *i.e.,* $\sqrt[3]{a} \in$ R is the pre-image of a $\in$ R. Since a is any arbitrary element $\in$ R, therefore the range of f is R. This function will be an onto function.
2. The sine of any real number lies in the closed interval [–1, 1]. Also all the numbers in this interval will be the sine of some real numbers. Hence the range of g is the closed interval [–1, 1]. This function will be an into function.
3. If we add 1 to the square of each real number, we get the set of numbers which are greater then or equal to 1. Let y be any real number greater than or equal to 1 and let y = h (x) = $x^2 + 1$. Then x = $\pm\sqrt{(y-1)}$ are also real numbers.

Thus, every real number which is greater than or equal to 1 is the h-image of some or other real number. Hence the range of h is the infinite interval (1, ∞).

Example 20: *Show that the mapping* $f : R \to R$ *defined by*

$$f(x) = \cos x \ \forall x \in R$$

is neither one-one nor onto. Modify the domain and co-domain of this mapping so that it may be both one-to-one and onto.

Solution: We have f : R → R such that

$$f(x) = \cos x \ \forall \ x \in R.$$

f is not one-one. We have 0, $2\pi \in$ R

and $\quad f(0) = \cos 0 = 1,$

$\quad f(2\pi) = \cos 2\pi = 1.$

Thus, 0 and 2π are two distinct elements in the domain R of f and they have the same f-image 1 in the co-domain R of f. Therefore f is may-one and not one-one.

f is not onto: We know that if x $\in$ R, then $-1 \le \cos x \le 1$. Therefore if y $\in$ the co-domain R of f and y > 1 or y < –1, then there exists no x $\in$ the domain R of f such that f (x) = cos x = y. Therefore the mapping f is into and not onto.

Now let X = $\{x : x \in R$ and $0 \le x \le \pi\}$

and $\quad$ Y = $\{y : y \in R$ and $-1 \le y \le 1\}$.

If we take X as the domain of f and Y as the co-domain of f, then the mapping f : X → Y such that

$$f(x) = \cos x \ \forall \ x \in X$$

is one-one and onto.

Example 21: *Let R be the set of all real numbers. Using the fact that every cubic equation with real coefficients has a real root, show that $x \to x^2 - x$ defines a mapping of R onto R. Is this a one-one mapping ?*

Solution: If $x \in R$, then $x^3 - x \in R$ and is unique. Therefore $x \to x^2 - x$ defines a mapping of R to R.

Let y be any arbitrary element $\in$ R *i.e.,* let y be any real number. Then $x^3 - x = y$ is a cubic equation with real coefficients. It will have at least one real root. Thus for any $y \in R$ there exists $x \in R$ such that $x^3 - x = y$. Therefore $x \to x^2 - x$ defines a mapping of R onto R.

Again $1 \to 1^3 - 1$ *i.e.,* $1 \to 0$

and $-1 \to (-1)^3 - (-1)$ *i.e.,* $-1 \to 0$.

Thus, the two elements 1 and $-1 \in R$ map onto the same element $0 \in R$. Hence the mapping is not one-one.

Example 22: *If S and T are non-empty sets, prove that there exists a one-to-one correspondence between S × T and T × S.*

Solution: If $a \in S$ and $b \in T$, then the ordered pair $(a, b) \in S \times T$ and the ordered pair $(b, a) \in T \times S$. Let f be a mapping from S × T to T × S defined by the formula

$$f(a, b) = (b, a) \ \forall \ (a, b) \in S \times T.$$

Here f (a, b) denotes the image under the mapping f of the element $(a, b) \in S \times T$.

Obviously the mapping f is well-defined.

f is one-one. Let $(a, b), (c, d) \in S \times T$. Then $f(a, b) = f(c, d)$

$\Rightarrow \quad (b, a) = (d, c)$ [by def. of f]

$\Rightarrow \quad b = d, a = c$

$\Rightarrow \quad (a, b) = (c, d)$

$\Rightarrow$ f is one-one.

f is onto: Let (y, z) be any element of T × S. Then $y \in T, z \in S$ Therefore $(z, y) \in S \times T$. We have $f(z, y) = (y, z)$.

Thus, (y, z) T × S ⇒ ∃ (z, y) ∈ S × T such that f (z, y) = (y, z). Therefore f is onto.

Thus, f is a one-one function from S × T onto T × S. Therefore f gives a one-to-one correspondence between S × T and T × S.

Example 23: *Let A = {1, 2, 3, 4, 5} and let the functions*

$$f : A \to A,\ g : A \to A$$

be defined as

$$f(1) = 3,\ f(2) = 5,\ f(3) = 3,\ f(4) = 1,\ f(5) = 2,$$

$$g(1) = 4,\ g(2) = 1,\ g(3) = 1,\ g(4) = 2,\ g(5) = 3.$$

Find the composite functions g o *f and f* o g.

Solution: We have

$$(g \circ f)(1) = g[f(1)] = g(3) = 1$$
$$(g \circ f)(2) = g[f(2)] = g(5) = 3$$
$$(g \circ f)(3) = g[f(3)] = g(3) = 1$$
$$(g \circ f)(4) = g[f(4)] = g(1) = 4$$
$$(g \circ f)(5) = g[f(5)] = g(2) = 1.$$

Also

$$(f \circ g)(1) = f[g(1)] = f(4) = 1$$
$$(f \circ g)(2) = f[g(2)] = f(1) = 3$$
$$(f \circ g)(3) = f[g(3)] = f(1) = 3$$
$$(f \circ g)(4) = f[g(4)] = f(2) = 5$$
$$(f \circ g)(5) = f[g(5)] = f(3) = 3.$$

Example 24: *Let the functions f : R → R, g : R → R be defined by*

$$f(x) = 2x + 1,\ g(x) = x^2 - 2.$$

Find the formulae for g o *f and f* o g.

Solution: Let us first find the formula for (g o f) : R → R.

We have $(g \circ f)(x) = g[f(x)] = g(2x + 1)$

$$= (2x + 1)^2 - 2 = 4x^2 + 4x - 1.$$

Now let us find the formula for f o g. We have

$(f \circ g)(x) = f[g(x)] = f(x^2 - 2) = 2(x^2 - 2) + 1 = 2x^2 - 3.$